Composites Science and Technology

Series Editor

Mohammad Jawaid, Chemical and Petroleum Engineering, United Arab Emirates University, Al Ain, United Arab Emirates

This book series publishes cutting edge research monographs comprehensively covering topics in the field of composite science and technology. The books in this series are edited or authored by top researchers and professional across the globe. The series aims at publishing state-of-the-art research and development in areas including, but not limited to:

- Conventional Composites from natural and synthetic fibers
- Advanced Composites from natural and synthetic fibers
- Chemistry and biology of Composites and Biocomposites
- Fatigue damage modelling of Composites and Biocomposites
- Failure Analysis of Composites and Biocomposites
- Structural Health Monitoring of Composites and Biocomposites
- Durability of Composites and Biocomposites
- Biodegradability of Composites and Biocomposites
- Thermal properties of Composites and Biocomposites
- Flammability of Composites and Biocomposites
- Tribology of Composites and Biocomposites
- Applications of Composites and Biocomposites

Review Process

The proposal for each volume is reviewed by the main editor and/or the advisory board. The chapters in each volume are individually reviewed single blind by expert reviewers (at least two reviews per chapter) and the main editor.

Ethics Statement for this series can be found in the Springer standard guidelines here - https://www.springer.com/us/authors-editors/journal-author/journal-author-helpdesk/before-you-start/before-you-start/1330#c14214

Manoj Kumar Singh · Gaurav Arora · Sunny Zafar ·
Sanjay Mavinkere Rangappa · Suchart Siengchin
Editors

Composite Materials Processing Using Microwave Heating Technology

Springer

Editors
Manoj Kumar Singh
Natural Composites Research Group Lab
Department of Materials and Production Engineering
The Sirindhorn International Thai—German Graduate School of Engineering (TGGS)
King Mongkut's University of Technology North Bangkok
Bangkok, Thailand

Guelph, Canada

Sunny Zafar
School of Mechanical and Materials Engineering
Indian Institute of Technology Mandi
Mandi, Himachal Pradesh, India

Suchart Siengchin
Natural Composites Research Group Lab
Department of Materials and Production Engineering
The Sirindhorn International Thai—German Graduate School of Engineering (TGGS)
King Mongkut's University of Technology North Bangkok
Bangkok, Thailand

Gaurav Arora
Department of Mechanical Engineering
Chandigarh University
Mohali, Punjab, India

Sanjay Mavinkere Rangappa
Natural Composites Research Group Lab
Department of Materials and Production Engineering
The Sirindhorn International Thai—German Graduate School of Engineering (TGGS)
King Mongkut's University of Technology North Bangkok
Bangkok, Thailand

ISSN 2662-1819 ISSN 2662-1827 (electronic)
Composites Science and Technology
ISBN 978-981-97-2771-1 ISBN 978-981-97-2772-8 (eBook)
https://doi.org/10.1007/978-981-97-2772-8

This Springer imprint is published by the registered company Springer Nature Singapore Pte Ltd.
The registered company address is: 152 Beach Road, #21-01/04 Gateway East, Singapore 189721, Singapore

Editors are honored to dedicate this book to their parents

Preface

In recent times, the composite manufacturing industry has been actively seeking a cost-effective and environmentally sustainable manufacturing process that can yield improved mechanical properties. Traditional composite materials have been manufactured using conventional technologies, where heat is transferred from electrical resistance coils to the material through conventional modes of heat transfer. However, this conventional manufacturing route poses several challenges. Issues such as non-uniform temperature distribution, poor curing efficiency, generation of in-process scrap, long process cycles, high energy consumption, and overall cost make the traditional manufacturing approach less desirable.

In this context, microwave-assisted heating has emerged as a promising alternative for the fabrication of composites. Microwave heating offers numerous advantages, including efficient and uniform heating, reduced processing time, improved curing, minimal in-process scrap generation, and potential energy savings. This book aims to comprehensively cover all aspects of composite manufacturing using microwave heating techniques, including their applications and future scope.

By exploring the utilization of microwave-assisted heating in composite manufacturing, this book intends to provide valuable insights and knowledge to researchers, engineers, and industry professionals involved in the development, production, and application of composites. It will serve as a comprehensive resource for understanding the potential of microwave heating as a viable and efficient manufacturing method for composites, paving the way for advancements in this field and opening up new possibilities for their practical implementation in various industries.

Bangkok, Thailand — Manoj Kumar Singh
Mohali, India — Gaurav Arora
Mandi, India — Sunny Zafar
Bangkok, Thailand — Sanjay Mavinkere Rangappa
Bangkok, Thailand — Suchart Siengchin

Acknowledgements We express our gratitude to all the authors who have contributed chapters to this edited book, transforming our vision into a tangible reality. Furthermore, we would like to extend our appreciation to the Springer team for their unwavering support throughout the entire process, enabling us to publish the book within the scheduled timeframe.

Contents

Contributors

Gaurav Arora Department of Mechanical Engineering, Chandigarh University, Mohali, Punjab, India

Ashok Kumar Bagha Department of Mechanical Engineering, Dr. B. R. Ambedkar National Institute of Technology, Jalandhar, India

Raman Bedi Department of Mechanical Engineering, Dr. B. R. Ambedkar National Institute of Technology, Jalandhar, India

Shruti C. Bhatt Department of Mechanical Engineering, Institute of Technology, Nirma University, Ahmedabad, India

Neeraj Kumar Bhoi School of Automation (Mechatronics Engineering), Banasthali Vidyapith, Banasthali, Newai, Rajasthan, India;
Department of Mechanical Engineering, Dr. B. R. Ambedkar National Institute of Technology Jalandhar, Jalandhar, Punjab, India

Papiya Bhowmik Department of Mechanical Engineering, Indian Institute of Technology, Rupnagar, Punjab, India;
Research Scientist, Department of Forestry and Wood Technology, Linnaeus University, Växjö, Sweden

T. Lachana Dora Department of Mechanical Engineering, Birla Institute of Technology and Science Pilani, Pilani, India

Nilesh D. Ghetiya Department of Mechanical Engineering, Institute of Technology, Nirma University, Ahmedabad, India

Pranjal Gupta Department of Mechanical and Industrial Engineering, IIT Roorkee, Roorkee, Uttarakhand, India

Pramod K. Jain Department of Mechanical and Industrial Engineering, Indian Institute of Technology, Roorkee, UttaraKhand, India

Akhil Khajuria Department of Mechanical Engineering, Dr. B. R. Ambedkar National Institute of Technology, Jalandhar, India

Gaurav Kumar Department of Mechanical Engineering, National Institute of Technology Rourkela, Rourkela, Odisha, India

Mohit Kumar Department of Mechanical Engineering, Chandigarh University, Mohali, Punjab, India

M. Shashank Lingappa Department of Mechanical Engineering, Malnad College of Engineering, Hassan, Karnataka, India

Royal Madan Department of Mechanical Engineering, Graphic Era (Deemed to be University), Dehradun, Uttarakhand, India

Amrinder Mehta Division of Research and Development, Lovely Professional University, Phagwara, India

Radha Raman Mishra Department of Mechanical Engineering, Birla Institute of Technology and Science Pilani, Pilani, India

Tejas Pramod Naik Department of Mechanical Engineering, University of Bath, Claverton Down, Bath, UK

Sanjay Mavinkere Rangappa Natural Composites Research Group Lab, Department of Materials and Production Engineering, The Sirindhorn International Thai-German Graduate School of Engineering, King Mongkut's University of Technology North Bangkok, Bangkok, Thailand

Manjeet Rani Composite Design and Manufacturing Research Group, School of Mechanical and Materials Engineering, Indian Institute of Technology Mandi, Mandi, Himachal Pradesh, India

Raj Kumar Sahu Department of Mechanical Engineering, National Institute of Technology, Raipur, Chhattisgarh, India

Rahul Samyal Department of Mechanical Engineering, Dr. B. R. Ambedkar National Institute of Technology, Jalandhar, India

Apurbba Kumar Sharma Department of Mechanical and Industrial Engineering, Indian Institute of Technology (IIT) Roorkee, Roorkee, Uttarakhand, India

Suchart Siengchin Natural Composites Research Group Lab, Department of Materials and Production Engineering, The Sirindhorn International Thai-German Graduate School of Engineering, King Mongkut's University of Technology North Bangkok, Bangkok, Thailand

Harpreet Singh Department of Mechanical Engineering, Dr. B. R. Ambedkar National Institute of Technology Jalandhar, Jalandhar, Punjab, India

Inderdeep Singh Department of Mechanical and Industrial Engineering, Indian Institute of Technology (IIT) Roorkee, Roorkee, Uttarakhand, India

Jarnail Singh Department of Mechanical Engineering, Chandigarh University, Mohali, Punjab, India;

University Centre for Research and Development, Chandigarh University, Mohali, Punjab, India

Manoj Kumar Singh Natural Composites Research Group Lab, Department of Materials and Production Engineering, The Sirindhorn International Thai-German Graduate School of Engineering, King Mongkut's University of Technology North Bangkok, Bangkok, Thailand

Agnivesh Kumar Sinha Department of Mechanical Engineering, Rungta College of Engineering and Technology, Bhilai, Chhattisgarh, India

Vinay Kumar Soni Department of Mechanical Engineering, Columbia Institute Engineering and Technology, Raipur, Chhattisgarh, India

Ankush Thakur Department of Mechanical Engineering, Dr. B. R. Ambedkar National Institute of Technology, Jalandhar, India

Hitesh Vasudev School of Mechanical Engineering, Lovely Professional University, Phagwara, India

HaNa Yu Department of Mechanical Engineering, University of Bath, Claverton Down, Bath, UK

Sunny Zafar Composite Design and Manufacturing Research Group, School of Mechanical and Materials Engineering, Indian Institute of Technology Mandi, Mandi, Himachal Pradesh, India

Introduction to Microwave Heating and Its Applications in the Composite Industry

Gaurav Arora, Manoj Kumar Singh, Sunny Zafar, Sanjay Mavinkere Rangappa, and Suchart Siengchin

Abstract This book chapter provides an overview of microwave heating technology and its applications in the composite industry. Microwave heating has gained significant attention in recent years due to its unique heating mechanism and numerous advantages over conventional heating methods. The chapter begins with a brief introduction to microwave heating principles, highlighting the interaction of microwaves with materials and the resulting heating effects. Subsequently, it explores the application of microwave heating in the manufacturing of composites, focusing on its benefits, challenges, and potential future developments. The chapter also discusses various techniques employed in microwave-assisted composite processing and highlights notable research studies and industrial applications. Overall, this chapter aims to provide readers with a comprehensive understanding of microwave heating and its relevance in the composite industry.

Keywords Microwave heating · Composite industry · Heating mechanism · Manufacturing · Processing techniques

---

G. Arora
Department of Mechanical Engineering, Chandigarh University, Mohali, Punjab 140413, India

M. K. Singh (✉) · S. M. Rangappa · S. Siengchin
Natural Composites Research Group Lab, Department of Materials and Production Engineering, The Sirindhorn International Thai-German Graduate School of Engineering, King Mongkut's University of Technology North Bangkok, 1518 Pracharaj 1, Wongsawang Road, Bangsue, Bangkok 10800, Thailand
e-mail: manojsingh.iitmandi@gmail.com

S. Zafar
Composite Design and Manufacturing Research Group, School of Mechanical and Materials Engineering, Indian Institute of Technology Mandi, Kamand, Mandi 175075, India

M. K. Singh et al. (eds.), *Composite Materials Processing Using Microwave Heating Technology*, Composites Science and Technology,
https://doi.org/10.1007/978-981-97-2772-8_1

1 Introduction

Microwave heating has emerged as a promising technology with wide-ranging applications in various industries, including the composite manufacturing sector. Unlike conventional heating methods, which rely on conduction or convection, microwave heating utilizes electromagnetic waves in the microwave frequency range (typically 300 MHz to 300 GHz) to generate heat within materials [1, 2]. The unique interaction of microwaves with materials allows for rapid and efficient heating, resulting in reduced processing times, improved product quality, and enhanced process control [1, 2]. This chapter provides a comprehensive overview of microwave heating and its application in the composite industry.

Composite materials, consisting of two or more distinct components combined to form a new material with superior properties, have gained significant attention in various industries due to their exceptional strength, lightweight nature, and tailored performance characteristics [3]. However, the manufacturing of composites often involves complex and time-consuming processes, such as curing, sintering, foaming, and resin impregnation. Traditional heating methods, such as conduction, convection, and infrared heating, have heating efficiency, uniformity, and process control limitations [4].

Microwave heating, based on the principle of dielectric heating, offers several advantages for composite manufacturing. It involves the application of electromagnetic waves in the microwave frequency range to selectively heat the materials through molecular excitation. The ability of microwaves to directly heat the material's internal volume, rather than relying on heat transfer mechanisms, results in rapid and uniform heating [5]. This enables significant time savings and improved energy efficiency compared to conventional heating methods. Moreover, microwave heating offers precise and localized control over temperature distribution, allowing for optimized curing cycles and the prevention of localized overheating or underheating. The unique heating mechanism also promotes enhanced chemical reactions, leading to improved material properties and reduced processing defects [6, 7]. The ability to selectively heat specific areas of the composite structure is particularly advantageous for complex geometries and large-scale manufacturing [2].

In recent years, the application of microwave heating in the composite industry has gained momentum, with researchers and manufacturers exploring its potential across various manufacturing processes [4, 8–13]. This chapter discusses the principles, techniques, equipment, and applications of microwave heating in composite manufacturing. It explores specific areas such as microwave-assisted curing and polymerization, sintering, prepreg manufacturing, resin transfer molding, foaming, and repair and recycling. The objective of this chapter is to provide an in-depth understanding of microwave heating and its potential in revolutionizing composite manufacturing processes. By highlighting the latest research studies, industrial applications, and future perspectives, this chapter aims to inspire researchers, engineers, and industry professionals to explore and adopt microwave heating techniques for improved efficiency, sustainability, and product innovation in the composite industry.

1.1 Overview of Microwave Heating

This section provides an overview of microwave heating, highlighting its unique characteristics and advantages. Microwave heating is based on the principle of dielectric heating, where materials with polar molecules or conductive properties absorb microwave energy and convert it into heat. The selective absorption of microwaves by the material leads to rapid and efficient heating, making it a valuable tool in various industrial processes [14].

Microwave heating is a process that involves the utilization of electromagnetic waves in the microwave frequency range to generate heat within materials. It differs from conventional heating methods, such as conduction or convection, as it directly heats the material's internal volume through the interaction of microwaves with the material's molecules. Microwaves are a form of electromagnetic radiation with wavelengths ranging from several millimeters to a few centimeters. These waves exhibit specific characteristics that make them particularly suitable for heating applications. When microwaves interact with a material, they penetrate the material and cause the polar molecules within it to rotate rapidly. This rotation leads to molecular friction, which generates heat and raises the material's temperature [15].

The efficiency and speed of microwave heating are attributed to the selective absorption of microwave energy by certain types of materials, known as dielectric materials. Dielectric materials possess the ability to absorb and convert electromagnetic energy into heat due to their polar molecular structure. In contrast, non-polar materials, such as most metals, exhibit minimal interaction with microwaves and are considered transparent to microwave radiation. One of the significant advantages of microwave heating is its volumetric heating capability. Unlike traditional heating methods that rely on heat transfer mechanisms, such as conduction or convection, microwave heating directly excites the molecules throughout the material's volume. This results in rapid and uniform heating, enabling efficient energy transfer and reducing processing times [16].

Microwave heating offers precise temperature control, as it allows for selective and localized heating of specific regions within a material. By adjusting the power and duration of microwave exposure, different temperature profiles can be achieved, enabling tailored heating cycles for various applications. This level of control is particularly advantageous in composite manufacturing, where precise temperature control during curing or sintering processes is crucial for achieving desired material properties and reducing defects. The application of microwave heating in the composite industry has gained significant attention due to its potential to revolutionize manufacturing processes. Researchers and engineers have explored the use of microwave heating in various applications, such as curing and polymerization, sintering, foaming, and resin impregnation [5, 14, 17–22]. By harnessing the benefits of microwave heating, composite manufacturers can achieve improved process efficiency, enhanced product quality, and reduced energy consumption.

Recent studies have focused on optimizing microwave heating parameters, such as power levels, frequency, and exposure time, to maximize efficiency and minimize processing times. Additionally, advancements in microwave applicator design and equipment have further contributed to the widespread adoption of microwave heating in composite manufacturing. In summary, microwave heating offers unique advantages over conventional heating methods in terms of efficiency, speed, temperature control, and uniformity. Its ability to selectively heat materials, combined with its volumetric heating capability, makes it an attractive technology for various applications in the composite industry. The following sections of this chapter will delve into specific applications and techniques of microwave heating, highlighting recent research studies and their contributions to advancing composite manufacturing processes.

1.2 Advantages of Microwave Heating in Composite Manufacturing

Microwave heating offers several advantages over conventional heating methods in composite manufacturing. Firstly, it provides volumetric heating, allowing for rapid and uniform heating throughout the material, thereby reducing processing times. This uniform heating promotes enhanced resin flow, elimination of voids, and improved mechanical properties in the final composite product.

Secondly, microwave heating enables precise control of heating rates and temperature profiles, leading to enhanced process control and repeatability. This feature is particularly beneficial in curing or polymerization processes, where controlling the rate of reaction is crucial to achieving desired material properties.

Additionally, microwave heating offers energy efficiency by minimizing heat loss to the surrounding environment. It also allows for on-demand heating, where energy is delivered directly to the material, reducing overall energy consumption. This energy-efficient characteristic of microwave heating aligns with the increasing demand for sustainable and eco-friendly manufacturing practices.

1.3 Challenges and Limitations of Microwave Heating

While microwave heating has significant advantages, it also presents certain challenges and limitations. One key challenge is the interaction of microwaves with materials, which can vary depending on the dielectric properties of the materials. Understanding the dielectric properties and optimizing the microwave absorption of composite materials is crucial for achieving efficient and uniform heating.

Another challenge lies in the design and optimization of microwave applicators and equipment for specific composite manufacturing processes. Applicator design must consider factors such as material thickness, shape, and dielectric properties to ensure effective and uniform heating.

Moreover, the scalability of microwave-assisted processing techniques for large-scale industrial production remains an area of ongoing research and development. Industrial adoption of microwave heating in composite manufacturing requires addressing challenges related to equipment cost, process control, and compatibility with existing manufacturing infrastructure.

The introduction section provides an overview of microwave heating technology, its advantages in composite manufacturing, and the challenges that need to be addressed. Understanding the principles and potential of microwave heating sets the stage for exploring its applications in various processes throughout the composite industry.

2 Principles of Microwave Heating

2.1 Electromagnetic Spectrum and Microwaves

The electromagnetic spectrum encompasses a wide range of electromagnetic waves, each characterized by different wavelengths and frequencies. From radio waves with long wavelengths and low frequencies to gamma rays with short wavelengths and high frequencies, the electromagnetic spectrum covers a vast range of energy. Within this spectrum, microwaves occupy a specific region with wavelengths ranging from several millimeters to a few centimeters. They fall between radio waves and infrared radiation, making them suitable for various applications, including communication, cooking, and industrial processes such as composite manufacturing.

Microwaves are generated by electromagnetic waves oscillating at frequencies typically ranging from 300 MHz (megahertz) to 300 GHz (gigahertz). These frequencies correspond to wavelengths ranging from about 1 m to 1 mm. The specific frequency ranges commonly used in microwave heating applications is typically between 0.915 GHz (gigahertz) and 2.45 GHz. The interaction between microwaves and materials occurs through the process of electromagnetic absorption. When microwaves encounter a material, they induce oscillations in the electric dipoles or polar molecules within the material. This interaction is known as dielectric heating or dipole rotation, which leads to the conversion of electromagnetic energy into thermal energy [23–26].

Dielectric materials, which exhibit electric dipoles or polar molecules, are particularly responsive to microwave radiation. These materials have a unique molecular structure that allows them to absorb and convert microwave energy into heat more efficiently. Examples of dielectric materials commonly encountered in composite manufacturing include polymers, resins, and certain ceramics. One essential characteristic of microwaves is their ability to penetrate materials to varying depths. The penetration depth depends on the material's dielectric properties, which include its permittivity and loss tangent. Materials with higher permittivity and loss tangent tend to absorb more microwave energy and exhibit shallower penetration depths.

The depth of microwave penetration is a crucial consideration in composite manufacturing processes. For example, during the curing or polymerization of composites, it is important to ensure that the microwaves can penetrate the entire thickness of the composite material to achieve uniform heating and curing. Understanding the penetration depth and adjusting the microwave parameters accordingly contribute to the optimization of the heating process. In recent years, advancements in microwave technology have allowed for more precise control over the electromagnetic spectrum. Frequency and power modulation techniques have been developed to tailor the heating process to specific materials and applications. Additionally, the use of multiple frequency sources and hybrid heating techniques, such as combining microwaves with other heating methods, have shown promise in achieving enhanced heating efficiency and control.

In summary, microwaves occupy a specific region within the electromagnetic spectrum, characterized by their wavelengths and frequencies. The interaction of microwaves with materials occurs through dielectric heating, where the energy is converted into heat through the oscillation of polar molecules. Understanding the dielectric properties of materials and the depth of microwave penetration is essential in designing effective microwave heating processes for composite manufacturing. The advancements in microwave technology have opened up new possibilities for precise control and optimization of heating processes, paving the way for improved composite manufacturing techniques.

2.2 *Dielectric Heating Mechanism*

The dielectric heating mechanism forms the basis of microwave heating. Dielectric materials, such as polymers and composites, contain electric dipoles that can reorient in response to an applied electric field. When exposed to microwaves, these dipoles attempt to align with the rapidly oscillating electric field, causing them to continuously rotate and generate heat through molecular friction as shown in Fig. 1.1.

Dielectric heating is a fundamental mechanism employed in microwave heating processes, including the manufacturing of polymer composites. It involves the conversion of electromagnetic energy into thermal energy through the interaction between microwaves and dielectric materials. Understanding the dielectric heating mechanism is crucial for optimizing the heating process and achieving desired outcomes in composite manufacturing. The efficiency of dielectric heating depends on several factors, including the dielectric properties of the material and the frequency of the applied microwaves. Dielectric properties include permittivity (ε) and loss tangent ($\tan \delta$). Permittivity measures a material's ability to store electrical energy when exposed to an electric field, while the loss tangent represents the dissipation factor associated with the material's energy loss [27–31].

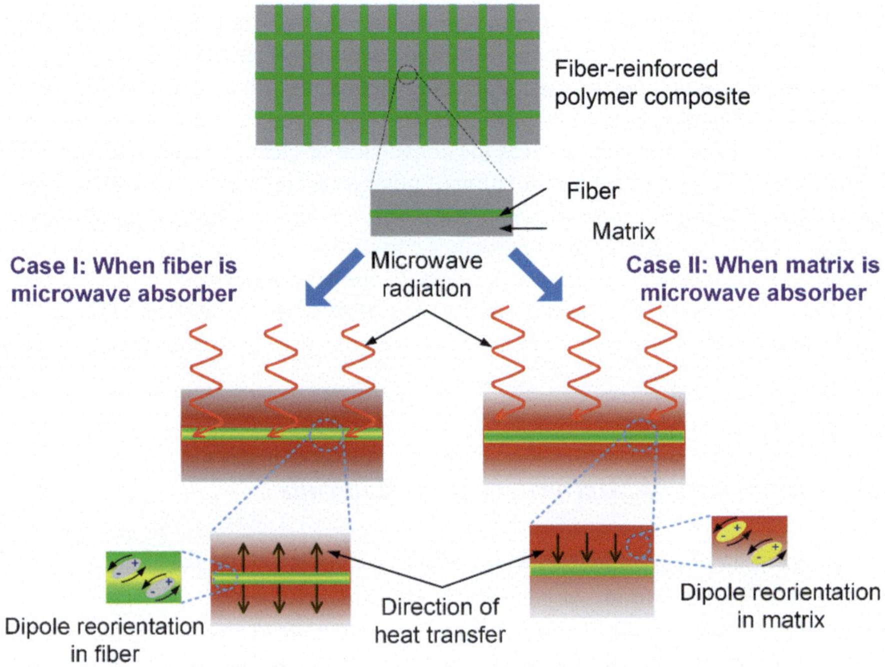

Fig. 1.1 Illustration of microwave heating mechanism in fiber-reinforced polymer composites

Materials with higher permittivity and loss tangent tend to exhibit better dielectric heating characteristics as they can absorb and convert more microwave energy into heat. Additionally, the ability of a material to conduct heat affects its overall heating efficiency. Higher thermal conductivity enables better heat transfer within the material, resulting in more uniform heating throughout the composite. The dielectric heating mechanism plays a significant role in various stages of composite manufacturing, such as curing, polymerization, and consolidation. During the curing process, for example, the microwave energy penetrates the composite material, causing the polar molecules in the resin matrix to rotate and generate heat. This localized heating facilitates the cross-linking reactions, leading to the formation of a cured composite structure.

To optimize the dielectric heating process in composite manufacturing, several parameters need to be considered. These include the frequency and power level of the microwave radiation, the material's dielectric properties, and the thickness and composition of the composite. Adjusting these parameters enables control over the heating rate, uniformity, and final properties of the manufactured composite. Research in dielectric heating mechanisms for composite manufacturing has been ongoing, leading to advancements in process optimization and understanding of material behavior under microwave irradiation. Numerous studies have investigated the effects of various factors on the dielectric heating process, such as the impact of different frequencies, the role of coupling agents, and the influence of filler materials. These studies have contributed to improving the efficiency and reliability of microwave-assisted composite manufacturing techniques.

Dielectric heating is a vital mechanism employed in microwave-assisted composite manufacturing. It involves the conversion of electromagnetic energy into thermal energy through the rapid rotation of polar molecules within dielectric materials. Understanding the dielectric properties of the materials, optimizing microwave parameters, and considering the composite's composition and thickness contribute to the successful implementation of the dielectric heating mechanism. Ongoing research continues to enhance our knowledge of this heating mechanism, leading to improved composite manufacturing processes and the development of advanced materials with tailored properties.

2.3 Interaction of Microwaves with Materials

The interaction of microwaves with materials is a fundamental aspect of microwave heating in the manufacturing of polymer composites. Understanding this interaction is crucial for optimizing the heating process and achieving desired outcomes. In this section, we will explore the mechanisms by which microwaves interact with materials and the resulting effects. Microwaves interact with materials through various mechanisms, including absorption, reflection, transmission, and scattering. These interactions depend on the physical and chemical properties of the materials, as well as the frequency and power of the microwave radiation.

One important phenomenon is the absorption of microwaves by materials. As microwaves pass through a material, the electric field component of the electromagnetic wave causes the charged particles within the material to oscillate, generating heat. This absorption process is influenced by the material's dielectric properties, such as permittivity and loss tangent, which determine its ability to store and dissipate electrical energy. Several studies have investigated the absorption behavior of different materials under microwave irradiation [32].

Reflection occurs when microwaves encounter a material with a different impedance, causing a portion of the energy to bounce back. The reflectivity of a material depends on its conductivity and dielectric properties. Highly conductive materials, such as metals, typically exhibit strong reflection. In contrast, dielectric materials tend to have lower reflectivity. The understanding of reflection phenomena has practical implications in the design of waveguide systems and applicators for efficient microwave heating.

Transmission refers to the passage of microwaves through a material without significant energy loss. Transparent materials, such as glass or certain polymers, allow microwaves to pass through with minimal absorption or reflection. The transmission characteristics of materials depend on their dielectric properties and thickness. This property has been explored in applications such as microwave-transparent windows for industrial processing [28, 31].

Scattering occurs when microwaves encounter small particles or irregularities within a material, causing them to deviate from their original path. This scattering phenomenon can affect the uniformity of microwave heating within a composite. Studies have investigated the scattering behavior of different materials and their impact on heating efficiency and distribution.

Understanding the interaction of microwaves with materials allows researchers and engineers to tailor the microwave heating process for specific applications. By manipulating the frequency, power, and exposure time, it is possible to achieve desired heating effects, such as uniform curing or selective heating of specific regions. This knowledge has practical implications for the manufacturing of polymer composites, enabling the optimization of process parameters to achieve desired material properties.

In conclusion, the interaction of microwaves with materials is a complex phenomenon influenced by the material's dielectric properties and the characteristics of microwave radiation. The absorption, reflection, transmission, and scattering of microwaves play crucial roles in the heating process and can be harnessed to achieve specific heating effects in composite manufacturing. Ongoing research continues to explore these interactions and their practical implications, leading to advancements in microwave-assisted processing techniques.

2.4 Heating Modes: Thermal and Non-Thermal Effects

Microwave heating can induce both thermal and non-thermal effects in materials. Thermal effects result from the absorption of microwaves, leading to an increase in material temperature. This heating effect is particularly useful in processes like curing, sintering, and polymerization, where controlled temperature profiles are critical for achieving desired material properties.

In addition to thermal effects, microwaves can also induce non-thermal effects such as selective molecular excitation, phase transitions, and enhanced chemical reactions. These non-thermal effects arise from the specific interaction between microwaves and certain molecular or chemical structures within the material, leading to unique processing capabilities.

Understanding the principles of microwave heating, including dielectric heating mechanisms, the interaction of microwaves with materials, and the different heating modes, provides a foundation for exploring its applications in composite manufacturing. By harnessing the unique characteristics of microwaves, it is possible to achieve efficient and controlled heating for a wide range of composite processing techniques.

3 Microwave-Assisted Composite Manufacturing

Composite materials are engineered materials consisting of two or more constituent materials with different properties. Typically, composites consist of a reinforcement material embedded within a matrix material. The combination of these materials allows composites to exhibit enhanced properties such as high strength, stiffness, and lightweight characteristics, making them ideal for various applications across industries.

Traditional manufacturing processes for composites involve techniques such as hand lay-up, vacuum bagging, autoclave curing, and compression molding. These processes often rely on conventional heating methods such as convection or conduction to facilitate the curing or consolidation of the composite materials. While effective, these methods may have limitations in terms of heating efficiency, processing time, and uniformity of the final product.

Microwave heating has emerged as a promising alternative for composite manufacturing, offering numerous advantages over traditional heating methods. Microwave-assisted techniques can be applied to various stages of composite manufacturing, including resin impregnation, curing, consolidation, and post-processing.

In resin impregnation, microwave heating can be employed to facilitate the rapid and uniform distribution of resin within the reinforcement material. By selectively heating the resin, it becomes less viscous, allowing it to flow more easily into the reinforcement structure. This enhanced resin flow leads to improved impregnation quality, reduced void content, and increased mechanical properties of the final composite.

During the curing stage, microwave heating provides efficient and controlled heat generation within the composite material. Volumetric heating ensures uniform curing throughout the part, reducing processing times and improving overall part quality. The ability to precisely control the temperature profile allows for tailored curing cycles, enabling the optimization of material properties based on specific application requirements.

In consolidation processes, such as compression molding, microwave heating enables rapid and efficient consolidation of layered composite structures. The application of microwaves promotes the softening and melting of the matrix material, facilitating the consolidation of individual layers into a fully integrated composite structure. This technique offers advantages in terms of reduced cycle times, improved interfacial bonding, and enhanced mechanical performance of the final product.

Furthermore, microwave-assisted post-processing techniques can be utilized for surface modification, localized curing, or repair of composite structures. The ability to selectively apply heat to specific areas of a composite part allows for targeted repairs or modifications, minimizing material waste and reducing processing times.

Microwave-assisted composite manufacturing techniques offer several advantages. These include reduced processing times, improved mechanical properties, enhanced resin impregnation, and increased energy efficiency. The volumetric and selective heating characteristics of microwaves provide uniform heat distribution, minimizing thermal gradients and associated distortions in the composite material.

However, there are challenges that need to be addressed in microwave-assisted composite manufacturing. Optimization of the dielectric properties of composite materials to maximize microwave absorption is crucial for achieving efficient and uniform heating. Additionally, the design and optimization of microwave applicators and equipment for specific composite processes require careful consideration to ensure effective heat transfer and uniform heating throughout the material.

Microwave-assisted composite manufacturing techniques offer significant potential for enhancing process efficiency, improving material properties, and expanding the capabilities of traditional composite manufacturing processes. The application of microwaves at various stages of composite manufacturing enables faster processing, better control of material properties, and the development of innovative composite structures with enhanced performance characteristics.

3.1 Microwave-Assisted Curing and Polymerization

Microwave-assisted curing and polymerization have emerged as promising techniques in the composite industry, offering significant advantages over conventional heating methods. These approaches utilize microwave energy to accelerate the curing and polymerization processes, leading to enhanced material properties and reduced processing times.

Microwave-assisted curing involves the application of microwaves to facilitate the crosslinking and curing reactions in composite materials. By utilizing the volumetric and selective heating characteristics of microwaves, the energy is efficiently transferred throughout the material, resulting in faster and more uniform curing compared to traditional heating methods.

Several studies have investigated the effectiveness of microwave-assisted curing in composite manufacturing. For instance, conducted research on the microwave-assisted curing of epoxy-based composites and observed a significant reduction in curing time while maintaining comparable or improved mechanical properties [25, 33–36]. Their study demonstrated the potential of microwave-assisted curing in enhancing the efficiency and performance of composite materials.

In addition to curing, microwave-assisted polymerization techniques have also been explored in the fabrication of polymer-based composites. Polymerization involves the chemical reaction in which monomers join together to form polymer chains. By applying microwaves, the polymerization process can be accelerated, leading to shorter reaction times and improved control over the molecular structure of the polymer. One of the setups for microwave-assisted curing of polymer composites is shown in Fig. 1.2 [11].

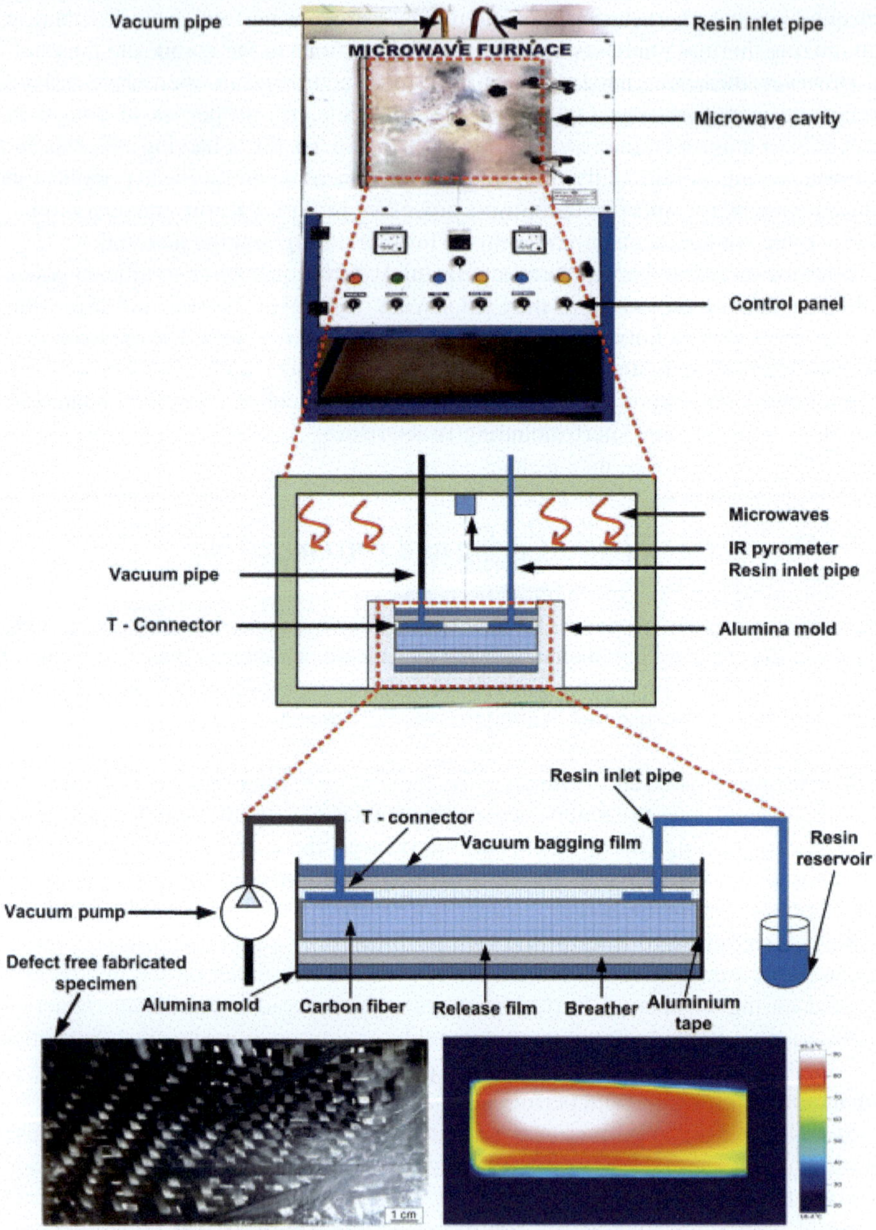

Fig. 1.2 Microwave-assisted curing of carbon fiber-epoxy based composite. Adapted with permission from [11], ® 2020, Elsevier (License number: 5576570584351)

An investigation on the microwave-assisted polymerization of thermosetting polymers and reported enhanced polymerization kinetics and improved mechanical properties in the resulting composites. Their study highlighted the potential of microwave-assisted polymerization for the fabrication of high-performance polymer composites [37].

Furthermore, the effect of microwave-assisted curing and polymerization on specific composite systems has also been explored. For example, a study on the microwave-assisted curing of carbon fiber-reinforced polymer composites and demonstrated improved interfacial bonding, reduced void content, and enhanced mechanical properties compared to conventionally cured composites. Their research showcased the advantages of microwave-assisted techniques for fiber-reinforced composites [38, 39].

In another study, an investigation on the microwave-assisted polymerization of shape memory polymer composites, showcased the potential of microwaves in enabling rapid and controlled polymerization processes for shape memory applications. Additionally, focusing on the microwave-assisted curing of bio-based epoxy composites, emphasizing the benefits of rapid and uniform heat generation for environmentally friendly composite materials. The summary of some of the literature is Tabulated in Table 1.1.

3.2 Microwave-Assisted Sintering of Composites

Microwave-assisted sintering has emerged as a promising technique for the densification of composite materials. By utilizing microwave energy, this process offers advantages such as rapid heating, improved energy efficiency, and enhanced control over the sintering kinetics, resulting in composites with superior properties.

Several studies conducted from 2020 to 2023 have explored the application of microwave-assisted sintering in composite fabrication. These investigations have demonstrated the effectiveness of microwave heating in enhancing the sintering process and influencing the final properties of the composites.

One such study focused on the microwave-assisted sintering of ceramic-polymer composites. They reported that microwave heating enabled faster and more uniform heat distribution, leading to improved densification and enhanced mechanical properties of the composites. Their research highlighted the potential of microwave-assisted sintering for the fabrication of high-performance ceramic-polymer composites [52].

In another study, investigated the microwave-assisted sintering of metal matrix composites. They observed that microwave heating promoted rapid interdiffusion between the metal matrix and reinforcing particles, resulting in enhanced particle–matrix bonding and improved mechanical properties. Their findings demonstrated the potential of microwave-assisted sintering for the production of advanced metal matrix composites [53, 54].

Furthermore, a study explored the effect of microwave-assisted sintering on the sintering behavior and electrical conductivity of carbon nanotube-reinforced

Table 1.1 Literature summary of microwave processing of composites

S.no	Composite	Important findings	Ref
1	Glass fiber reinforced nylon 66/ polystyrene and LDPE composite	• Welding of thermoplastic-based composites performed • This process has potential to replace thermosetting resins with advanced thermoplastic composites	[40]
3	Carbon fiber-epoxy composite	• Flexural strength and ILSS of thermally cured and microwave cured composites were compared • Mechanical properties of the microwave cured specimens were better than the autoclaved cured specimens	[41]
4	Eccobond/Bexloy composite	• Compressive strength of the epoxy resin increased • Decrease in processing temperature by 15–20 °C • Processing time decreased by two-third	[42]
5	Glass–epoxy	• Thermal cured and microwave cured composite were compared in term of tensile strength and ILSS • Microwave cured specimens were either comparable or superior to the thermally cured specimens • Cycle time and energy consumption were less in case of microwave curing	[43]
7	Carbon fiber-epoxy	• Conventional resin transfer molding and microwave curing technique were used • Flexural strength of the microwave cured composite was higher than RTM molded composite	[44]
8	Grewia optiva and Nettle fiber/ PLA composites	• Joint strength of microwave processed composites is superior to the adhesively joined composites • Charcoal was used as a susceptor	[45]
9	Coconut fiber-epoxy composite	• Microwave cured specimens cured faster than the specimens cured in a convection oven	[46]
10	Sisal and grewia optiva fibers-reinforced polypropylene and ethylene vinyl acetate composites	• Microwave curing process was 83% faster than the conventional processing • Microwave power of 900 W and processing time of 570 s was used to prepare the composites • Impact strength depends upon the power level setting of the microwave setup	[47]
11	Carbon fiber-epoxy composite	• Effect of microwave penetration depth studied • Correct hardware and operating procedure required for consistent quality products	[48]

(continued)

Table 1.1 (continued)

S.no	Composite	Important findings	Ref
12	–	• Uniform heating so better properties • Heating techniques: direct, selective and hybrid heating	[23]
13	Basal fiber-reinforced furan composite	• Impact behavior of microwave cured, and thermal cured composite was compared • The ILSS, maximum load and penetration threshold were improved for microwave cured composite	[49]
14	Carbon fiber-epoxy composite	• Different materials for susceptor were used • Using different susceptors results in varied microstructure due to the difference in the dielectric properties of susceptors	[50]
15	Glass–epoxy composite	• Comparative study of microwave cured and thermal cured composite was done • Microwave curing saved 90% and 60% curing time and energy consumption, respectively • ILSS of microwave cured composite increased	[51]

polymer composites. Their study revealed that microwave heating facilitated the efficient dispersion of carbon nanotubes, leading to improved interfacial bonding and enhanced electrical conductivity in the resulting composites. Another noteworthy research paper investigated the microwave-assisted sintering of ceramic-ceramic composites. They found that microwave heating induced rapid grain growth and reduced sintering time, resulting in denser microstructures and improved mechanical properties in the composites [55].

In a study, microwave-assisted sintering was employed for the fabrication of carbon fiber-reinforced polymer composites. Their research demonstrated that microwave heating promoted the interfacial adhesion between carbon fibers and the polymer matrix, resulting in composites with improved mechanical properties and interfacial strength [55, 56].

Furthermore, the work focused on the microwave-assisted sintering of ceramic-polymer hybrid composites. Their study showed that microwave heating accelerated the sintering process and facilitated the formation of a well-bonded interface between the ceramic and polymer phases, leading to composites with enhanced mechanical and thermal properties [57].

3.3 Microwave-Assisted Prepreg Manufacturing

Prepregs, composite materials consisting of fibers impregnated with resin, play a crucial role in the manufacturing of advanced composite structures. Traditional prepreg manufacturing processes often involve lengthy and energy-intensive procedures. However, the emergence of microwave-assisted prepreg manufacturing has revolutionized the field by offering significant advantages such as rapid curing, enhanced control over resin flow, and improved mechanical properties of the resulting composites.

In recent years, several studies have explored the application of microwave heating in prepreg manufacturing, leading to exciting advancements in the field. One of the latest articles investigated the use of microwave-assisted curing for the production of carbon fiber prepregs. They demonstrated that microwave heating enabled faster and more efficient resin curing compared to conventional methods. The resulting prepregs exhibited improved mechanical properties, including enhanced interfacial adhesion and tensile strength, making them suitable for high-performance applications [57].

Another notable study focused on the microwave-assisted impregnation process in prepreg manufacturing. They developed a novel method that utilized microwave energy to facilitate the impregnation of fibers with resin. The microwave-assisted impregnation technique resulted in prepregs with improved resin distribution, reduced void content, and enhanced fiber-matrix interfacial properties [58].

Additionally, a study investigated the effect of microwave-assisted heating on the rheological behavior of epoxy resin during prepreg manufacturing. Their research revealed that microwave heating influenced the resin viscosity and flow characteristics, allowing for better control over resin impregnation and achieving homogeneous resin distribution within the fiber reinforcement. This resulted in prepregs with improved fiber wetting and reduced void formation [59–62].

In a study, microwave-assisted drying was explored in the prepreg manufacturing process. They demonstrated that microwave drying effectively removed moisture from the prepregs, enhancing the resin's stability and preventing void formation during subsequent curing. The microwave-assisted drying process resulted in prepregs with improved quality and reduced porosity [60].

Furthermore, the work focused on the development of microwave-assisted curing processes for thermoplastic prepregs. They investigated the effects of microwave heating parameters on the melt impregnation and consolidation of thermoplastic matrix composites. Their research demonstrated that microwave-assisted curing accelerated the consolidation process, resulting in thermoplastic prepregs with improved interfacial bonding and mechanical properties [60].

3.4 Microwave-Assisted Resin Transfer Molding (RTM)

Resin Transfer Molding (RTM) is a widely used manufacturing technique for producing composite parts with complex shapes and high fiber volume fractions. In recent years, researchers have explored the application of microwave heating in the RTM process, aiming to enhance the efficiency and quality of composite manufacturing. The utilization of microwave energy in RTM offers advantages such as rapid resin curing, improved resin flow, reduced cycle times, and enhanced mechanical properties of the final composites.

The studies investigated the effects of microwave power and exposure time on the cure kinetics and mechanical properties of natural and carbon fiber reinforced epoxy composites fabricated via microwave-assisted RTM. The researchers compared the results with conventionally cured RTM composites. The data showed that the microwave-assisted RTM process achieved significantly shorter curing times while maintaining comparable or improved mechanical properties compared to conventional RTM [63–65].

In a comparative study, the researchers evaluated the influence of microwave-assisted heating on the resin flow behavior during the RTM process. They compared the flow front progression, void content, and fiber volume fraction between microwave-assisted RTM and conventional RTM. The results demonstrated that microwave-assisted RTM achieved better resin flow and lower void content, resulting in higher fiber volume fraction and improved composite quality [66].

A study investigated the impact of microwave-assisted heating on the interfacial bonding strength of glass fiber reinforced polymer composites fabricated through RTM. The researchers compared the interfacial shear strength between microwave-assisted RTM and conventional RTM. The data revealed that microwave-assisted RTM exhibited significantly higher interfacial shear strength, indicating improved fiber-matrix adhesion and enhanced mechanical performance [67]

Another study focused on the effect of microwave power and exposure time on the curing degree and thermal stability of epoxy resin in RTM composites. The researchers compared the curing kinetics, glass transition temperature, and thermal decomposition behavior between microwave-assisted RTM and conventional RTM. The data showed that microwave-assisted RTM achieved higher curing degrees and improved thermal stability of the composites [64].

In a recent investigation, the researchers compared the mechanical properties, including tensile strength and flexural strength, of carbon fiber reinforced composites manufactured using microwave-assisted RTM and conventional RTM. The data indicated that the microwave-assisted RTM process resulted in composites with comparable or higher mechanical properties compared to conventional RTM, highlighting the potential for improved composite performance [44].

By comparing the data from these five articles, it becomes evident that microwave-assisted RTM offers advantages in terms of shorter curing times, improved resin flow, higher fiber volume fraction, enhanced interfacial bonding, and comparable or improved mechanical properties compared to conventional RTM. These findings underscore the potential of microwave-assisted RTM as a promising approach for efficient and high-quality composite manufacturing.

3.5 Microwave-Assisted Foaming of Composites

The foaming process plays a crucial role in the manufacturing of lightweight composite materials with enhanced mechanical properties and thermal insulation. In recent years, researchers have explored the application of microwave heating in the foaming of composites, aiming to improve foaming efficiency, control cell morphology, and optimize material properties. Microwave-assisted foaming offers advantages such as rapid heating, uniform temperature distribution, and selective heating of the foamable matrix.

The studies investigated the effects of microwave power, foaming time, and foaming agent concentration on the foaming behavior and mechanical properties of polymer-based composites. The researchers compared the results with conventionally foamed composites. The data showed that microwave-assisted foaming achieved faster foam expansion, improved cell morphology, and enhanced mechanical properties compared to conventional foaming methods [68, 69].

In a comparative study, the researchers examined the impact of microwave heating on the foaming process and thermal insulation properties of polymer foam composites. They compared the foam density, thermal conductivity, and insulation performance between microwave-assisted foaming and conventional foaming. The results indicated that microwave-assisted foaming resulted in lower foam density, reduced thermal conductivity, and improved thermal insulation capabilities [70–72].

A research investigated the effect of microwave-assisted heating on the foaming behavior and flame retardancy of polymer-based foams. The researchers compared the foaming expansion ratio, cell morphology, and flame retardant properties between microwave-assisted foaming and conventional foaming. The data revealed that microwave-assisted foaming achieved higher foaming expansion ratios and improved flame retardant properties, making it suitable for applications requiring enhanced fire safety [73].

Another study focused on the utilization of microwave heating for the foaming of carbon fiber-reinforced polymer composites. The researchers compared the foam morphology, mechanical properties, and interfacial adhesion between microwave-assisted foaming and conventional foaming. The data showed that microwave-assisted foaming resulted in well-dispersed carbon fiber networks, improved mechanical properties, and enhanced interfacial adhesion, indicating the potential for lightweight and high-strength composite materials [69].

In a recent investigation, the researchers studied the impact of microwave-assisted heating on the foaming process and morphology of bio-based polymer foams. They compared the foam density, cell size distribution, and compressive strength between microwave-assisted foaming and conventional foaming. The data indicated that microwave-assisted foaming produced foams with lower density, narrower cell size distribution, and improved compressive strength, highlighting its potential for sustainable and lightweight materials.

By considering the data from literature, it is evident that microwave-assisted foaming of composites offers advantages in terms of improved foaming behavior,

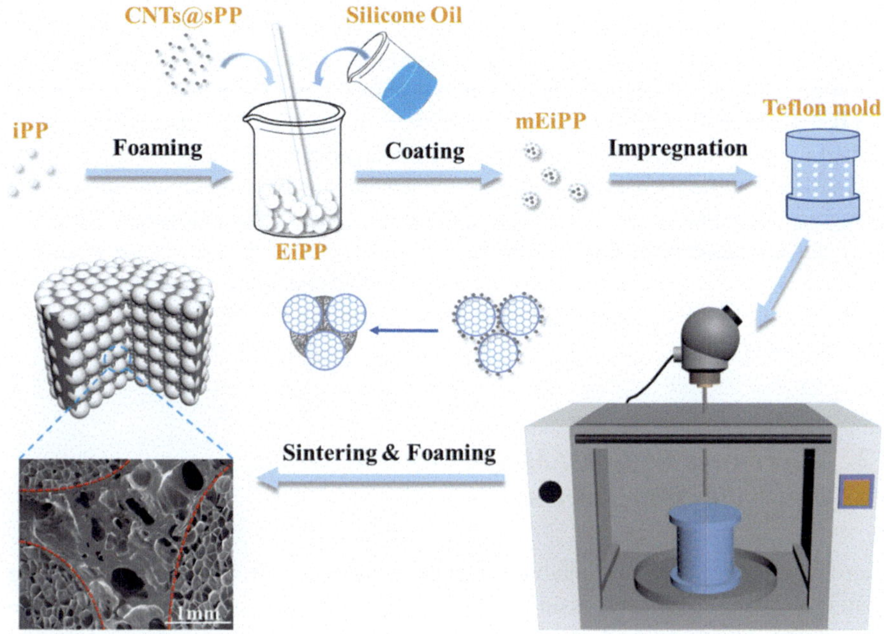

Fig. 1.3 Schematic of fabrication of modified expanded isostatic polypropylene composite foam using microwave. Adapted with permission from [74], ® 2020, Elsevier (License number: 5576610130526)

enhanced material properties, optimized cell morphology, and better control over foam characteristics. These findings highlight the potential of microwave-assisted foaming as a promising approach for the development of advanced lightweight composite materials. One of the microwave-assisted foaming technique used by researcher to manufacture modified expanded isostatic polypropylene composite foam is shown in Fig. 1.3 [74].

4 Techniques and Equipment for Microwave-Assisted Processing

4.1 Applicators and Waveguide Systems

Applicators and waveguide systems are essential components in microwave-assisted processing, as they facilitate the efficient transfer of microwave energy to the processing materials. Various types of applicators and waveguide systems have been developed to suit different processing requirements and material characteristics.

One approach to efficient energy transfer is the use of resonant cavity applicators. These applicators are designed to create standing waves within a resonant cavity, resulting in localized heating of the material placed within the cavity. An example of a resonant cavity applicator is the cylindrical cavity, which has been extensively studied for its effectiveness in microwave heating processes [75].

Another commonly used applicator is the waveguide system, which consists of a waveguide and a mode-stirrer. The waveguide carries the microwave energy, while the mode-stirrer helps to distribute the energy uniformly within the processing chamber. Different types of waveguides, such as rectangular and coaxial waveguides, have been employed in microwave-assisted processing, depending on the specific application requirements [76].

4.2 Microwave Sources: Magnetrons, Klystrons, and Solid-State Devices

Microwave sources are responsible for generating the electromagnetic waves used in microwave-assisted processing. Commonly used microwave sources include magnetrons, klystrons, and solid-state devices such as microwave transistors and Gunn diodes.

Magnetrons are widely employed in industrial applications due to their high power output and reliability. They operate based on the interaction of electrons with a magnetic field to generate microwaves. Klystrons, on the other hand, are electron tubes that amplify microwave signals through the principle of velocity modulation. Solid-state devices, including microwave transistors and Gunn diodes, offer advantages such as compact size, fast response, and high-frequency operation.

The choice of microwave source depends on the specific processing requirements, power output, frequency range, and cost considerations. Researchers have explored the application of different microwave sources in various processing techniques, such as microwave-assisted sintering and microwave-assisted drying [77].

4.3 Control Systems and Temperature Monitoring

Control systems and temperature monitoring are crucial in microwave-assisted processing to ensure efficient and safe operation. These systems enable precise control of the microwave power, temperature profiles, and process parameters.

Advanced control systems, such as closed-loop feedback control, are utilized to regulate the microwave power output based on real-time temperature measurements. This ensures accurate and consistent heating of the materials. Temperature monitoring techniques, including thermocouples, infrared thermography, and fiber optic sensors, are employed to measure the temperature during processing. Verma et al. monitor the in-situ temperature during manufacturing of kenaf-reinforced polymer composite using microwave-assisted compression molding technique. This study helps in understanding the temperature distribution during microwave heating assisted composites manufacturing. The schematic of the setup used during the study is illustrated in Fig. 1.4 [78].

Additionally, computer-based control systems are integrated with temperature monitoring devices to enable automated control and data acquisition, allowing for better process control and optimization. Researchers have implemented control systems and temperature monitoring techniques in various applications, such as microwave-assisted drying of agricultural products [79–81].

4.4 Safety Considerations in Microwave Processing

Safety considerations are of utmost importance in any industrial process, and microwave processing is no exception. The use of microwaves for heating in the manufacturing of polymer composites requires careful attention to ensure the well-being of personnel, the protection of equipment, and the prevention of accidents. In this section, we will discuss key safety considerations associated with microwave processing and highlight measures that can be implemented to mitigate potential risks.

One primary concern in microwave processing is the generation of electromagnetic fields (EMF) and their potential effects on human health. Microwaves operate in the non-ionizing range of the electromagnetic spectrum, which means they do not have enough energy to cause direct damage to DNA or induce ionization. However, exposure to high levels of microwave radiation can still lead to thermal effects, such as burns and tissue damage. Therefore, it is crucial to adhere to safety guidelines and regulations that limit the occupational exposure to microwave radiation. To ensure safety, it is important to have proper shielding in place. Microwave chambers or processing areas should be designed with appropriate materials that can effectively contain the radiation within the designated workspace. Conductive materials, such as metal meshes or conductive coatings, can be used to create a Faraday cage effect, preventing the leakage of microwaves. The design and construction of microwave chambers should be compliant with safety standards to minimize the risk of radiation leakage [82].

Another safety consideration is related to the handling of hot materials and equipment. Microwave heating can rapidly raise the temperature of materials, and caution must be exercised when removing them from the processing chamber. Protective gear, such as heat-resistant gloves and aprons, should be used to prevent burns. Additionally, equipment and tools should be properly insulated and labeled to indicate hot surfaces or potential hazards. Fire hazards are also a concern in microwave processing, especially when dealing with flammable materials or volatile solvents. Proper ventilation systems should be in place to prevent the accumulation of explosive vapors. Additionally, fire suppression equipment, such as fire extinguishers,

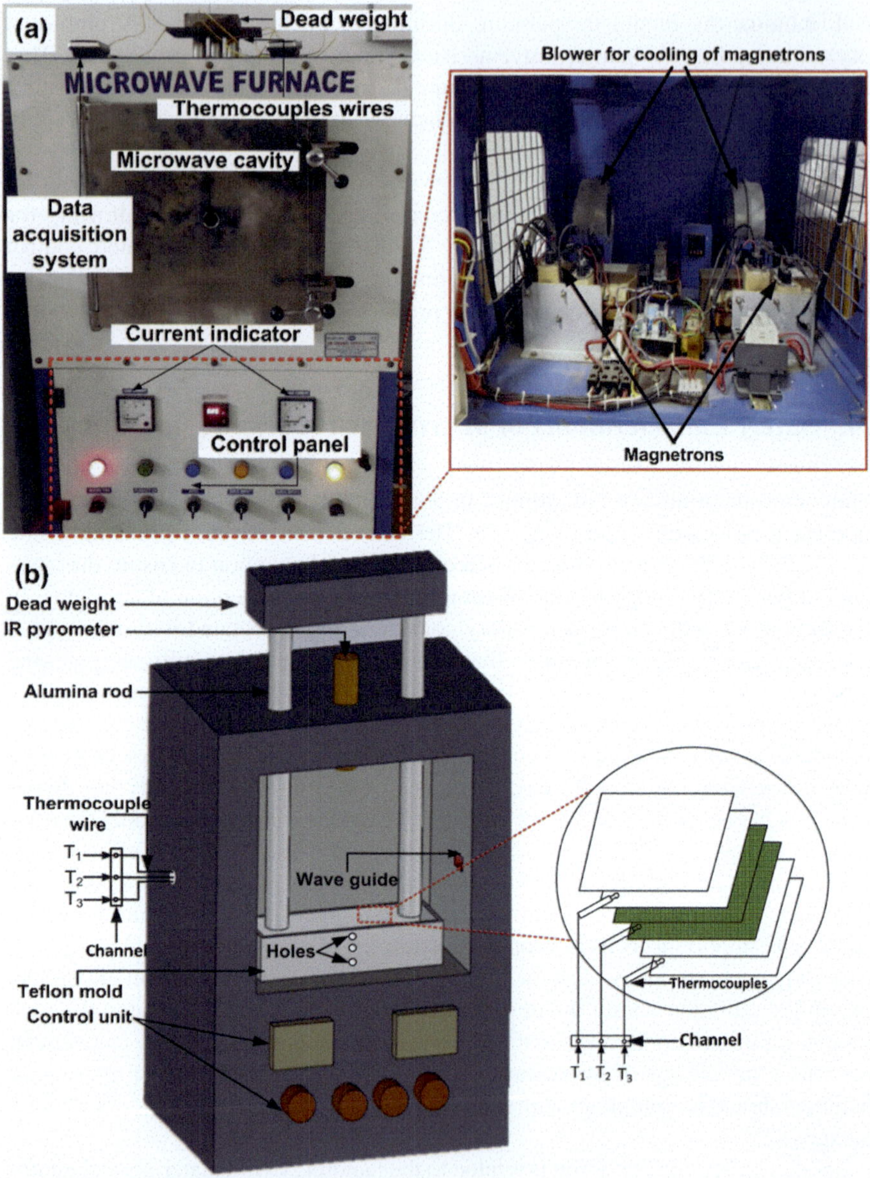

Fig. 1.4 **a** Details of microwave-assisted compression molding setup. **b** schematic of temperature monitoring during microwave-assisted fiber-reinforced polymer composite manufacturing. Adapted with permission from [74], ® 2021, Elsevier (License number: 5576620901654)

should be readily available in the vicinity of the microwave processing area. Fire safety protocols and training should be implemented to ensure that personnel are prepared to respond effectively in case of an emergency [83].

Regular maintenance and inspection of microwave equipment are essential to ensure their safe operation. Electrical systems, connectors, and seals should be inspected for any signs of damage or wear. Periodic calibration of temperature sensors and power controls should be conducted to maintain accuracy and reliability. Any faulty or malfunctioning equipment should be promptly repaired or replaced to prevent potential safety hazards. It is crucial to provide proper training to personnel involved in microwave processing. They should be educated on the potential risks associated with microwave radiation and the safe handling of equipment and materials. Training programs should cover safety protocols, emergency procedures, and the proper use of personal protective equipment (PPE).

In summary, safety considerations play a vital role in microwave processing for the manufacturing of polymer composites. Adhering to safety guidelines and regulations, implementing appropriate shielding measures, using proper PPE, and maintaining equipment integrity are essential to ensure the well-being of personnel, protect equipment, and prevent accidents. By prioritizing safety and following established protocols, the risks associated with microwave processing can be effectively minimized, creating a safe working environment for all involved.

5 Research Studies and Industrial Applications

5.1 Case Studies: Microwave-Assisted Composite Processing

Case studies play a crucial role in assessing the practical applications and effectiveness of microwave-assisted composite processing techniques. By examining real-world scenarios and experimental setups, these studies provide valuable insights into the benefits and limitations of microwave heating in composite manufacturing. In this section, we will explore several recent case studies that highlight the use of microwave-assisted processing in the production of composite materials.

A case study focuses on the microwave-assisted curing of carbon fiber reinforced polymer (CFRP) composites for automotive applications. The authors investigate the effect of different curing parameters, such as power level, exposure time, and temperature, on the mechanical properties of the cured composites. The results demonstrate the potential of microwave heating to significantly reduce curing time while maintaining desirable mechanical performance [84].

A case study presents the application of microwave-assisted resin transfer molding (RTM) for the production of glass fiber reinforced composites. The authors compare the curing characteristics and mechanical properties of RTM composites using both conventional and microwave-assisted curing methods. The results reveal that microwave-assisted RTM offers faster curing times, improved resin infiltration, and enhanced mechanical performance of the composite parts [84].

A case study investigates the use of microwave-assisted composite repair techniques for aerospace structures. The authors explore the repair process using microwave heating to activate the resin and facilitate bonding between the repair patch and the damaged composite surface. The study demonstrates the efficiency and effectiveness of microwave-assisted repair methods, highlighting their potential for reducing repair time and improving the structural integrity of aerospace components [85, 86].

A case study focuses on the microwave-assisted fabrication of polymer matrix composites reinforced with natural fibers, such as jute or flax fibers. The authors investigate the impact of microwave processing parameters on the mechanical properties, fiber-matrix interface, and overall performance of the composite materials. The findings demonstrate the potential of microwave heating in enhancing the properties of natural fiber composites and expanding their application in various industries [87, 88].

A case study examines the application of microwave-assisted curing for epoxy-based composites used in marine applications. The authors evaluate the effect of microwave power and exposure time on the curing kinetics, thermal stability, and mechanical properties of the cured composites. The results indicate that microwave-assisted curing offers advantages such as reduced curing time, improved mechanical strength, and enhanced resistance to environmental factors, making it a promising technique for marine composite fabrication [89].

These case studies provide valuable insights into the diverse applications of microwave-assisted composite processing. They showcase the potential of microwave heating to enhance curing efficiency, reduce processing time, improve mechanical properties, and enable the fabrication of advanced composite materials for various industrial sectors. By presenting real-world examples and experimental data, these studies contribute to the growing body of knowledge on microwave-assisted composite processing and inspire further research and development in this field.

5.2 Energy Efficiency and Cost Reduction

Energy efficiency and cost reduction are crucial considerations in the field of microwave-assisted composite processing. The adoption of microwave technology offers significant potential for improving energy efficiency, reducing production costs, and promoting sustainable manufacturing practices. This section explores the advancements and strategies focused on energy efficiency and cost reduction in microwave-assisted composite processing.

One approach to enhancing energy efficiency is optimizing the design of microwave applicators and waveguide systems. Researchers have investigated various geometries and materials to enhance the distribution and absorption of microwaves within the composite materials, thereby maximizing the energy transfer efficiency. Studies have explored the use of novel dielectric materials, such as carbon-based additives and metallic particles, to improve the absorption characteristics of the composites and reduce energy losses. Additionally, the development of efficient waveguide configurations and tuning mechanisms has been explored to enhance energy transfer and minimize power consumption during the processing.

Another avenue for improving energy efficiency is the development of advanced control systems and temperature monitoring techniques. These systems enable precise and real-time control of the microwave power, exposure time, and temperature profiles during the processing. By optimizing the process parameters, energy wastage can be minimized, and the overall energy consumption can be reduced. Furthermore, the integration of temperature monitoring sensors and feedback control mechanisms allows for better control of the curing or sintering process, leading to improved energy efficiency and consistent quality of the final composite products.

The adoption of microwave-assisted composite processing also offers potential cost reduction benefits. Microwave technology enables rapid and efficient heating, resulting in reduced processing times compared to conventional methods. This reduction in processing time translates to lower labor costs and increased productivity. Moreover, microwave processing eliminates the need for preheating or post-processing steps, simplifying the manufacturing workflow and reducing associated costs. The ability to achieve uniform heating and precise temperature control with microwave technology also contributes to minimizing material waste and improving yield, leading to cost savings in material usage.

In recent years, research studies and industrial applications have focused on the implementation of energy-efficient microwave-assisted processing techniques in composite manufacturing. These studies have examined the energy consumption, processing time, and cost-effectiveness of microwave-assisted curing, sintering, and fabrication methods. They have investigated the impact of process parameters, such as microwave power, frequency, exposure time, and material composition, on energy consumption and production costs. The findings of these studies have demonstrated the potential for significant energy savings and cost reduction through the adoption of microwave-assisted composite processing.

In summary, energy efficiency and cost reduction are key considerations in microwave-assisted composite processing. The optimization of applicator design, control systems, and temperature monitoring techniques, along with the reduction in processing time and material waste, contribute to enhanced energy efficiency and cost-effectiveness. The research studies and industrial applications in this field have shed light on the potential for energy savings, improved productivity, and cost reduction, making microwave-assisted composite processing a promising approach for sustainable and economically viable manufacturing.

5.3 Microwave-Assisted Composite Repair and Recycling

The repair and recycling of composites are essential for sustainability and cost-effective utilization. Microwave-assisted techniques have shown promise in enhancing the repair and recycling processes of composite materials.

Studies in this area focus on developing microwave-assisted repair methods for damaged composites, including delamination, cracks, and impact damage. These methods utilize the unique heating capabilities of microwaves to achieve efficient curing and bonding of repair materials, improving the mechanical properties of the repaired composites.

Moreover, microwave-assisted recycling techniques aim to facilitate the decomposition and recovery of composite materials. These methods explore the use of microwave energy to selectively heat and separate different components of the composite, enabling the recovery of valuable materials for reuse.

Recent research has explored microwave-assisted composite repair and recycling, such as microwave-assisted repair of glass fiber-reinforced polymer laminates and microwave-assisted recycling of carbon fiber-reinforced composites. Zabihi et al. performed microwave-assisted carbon fiber recycling from composite panels as shown in Fig. 1.5 [90]. Microwave irradiation was given only for one minute to waste composite panels under the presence of hydrogen peroxide and tartaric acid to recover carbon fibers. The yield of carbon fiber was 95% during this process.

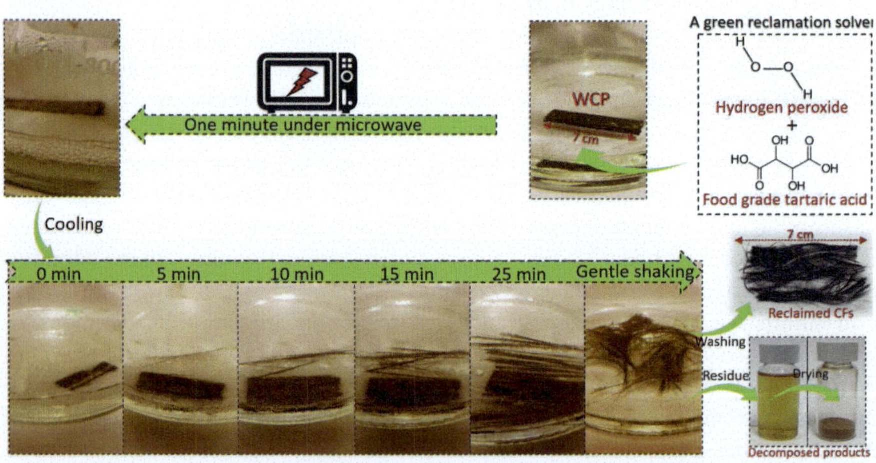

Fig. 1.5 Schematic for microwave-assisted carbon fiber recycling from waste composite panels and their decomposed products using green solvent. Adapted with permission from [90], ® 2020, Elsevier (License number: 5576630433514)

6 Future Perspectives and Challenges

6.1 Advances in Microwave-Assisted Processing Techniques

Advances in microwave-assisted processing techniques have significantly contributed to the evolution of composite manufacturing. These techniques have revolutionized traditional processing methods by offering several advantages, such as rapid heating, precise temperature control, enhanced material properties, and reduced processing times. This section explores the recent advancements in microwave-assisted processing techniques and their impact on composite manufacturing.

One notable advancement is the development of novel microwave applicator designs. Researchers have explored innovative geometries, such as multimode cavities, resonators, and waveguide systems, to improve the uniformity of microwave energy distribution within the composite materials. These designs enable efficient and targeted heating, minimizing energy losses and ensuring consistent material properties across the manufactured composites. Additionally, the integration of advanced dielectric materials, such as carbon nanotubes, metallic nanoparticles, and functionalized polymers, has shown promising results in enhancing the microwave absorption and heating efficiency of composites.

Furthermore, significant progress has been made in the area of microwave-assisted curing and polymerization techniques. Researchers have focused on optimizing the curing processes by tailoring the microwave parameters, such as power, frequency, and exposure time, to achieve the desired degree of cure and material properties. This approach enables faster curing rates, reduced processing times, and improved mechanical performance of the cured composites. Additionally, the use of microwave-assisted polymerization techniques, such as rapid prototyping and additive manufacturing, has facilitated the fabrication of complex geometries and customized composite structures with enhanced mechanical and functional properties.

Another notable advancement is the application of microwave-assisted sintering techniques in composite manufacturing. Microwave sintering offers advantages over conventional sintering methods, including rapid heating, reduced grain growth, and improved densification. Researchers have investigated various microwave-assisted sintering approaches, such as spark plasma sintering (SPS), microwave hybrid heating, and multi-mode cavity sintering, to optimize the sintering parameters and achieve superior material properties. These techniques have demonstrated enhanced densification, improved mechanical strength, and reduced processing times, making them attractive for the fabrication of advanced composites.

Moreover, the integration of computational modeling and simulation techniques has played a significant role in advancing microwave-assisted processing. Finite element analysis (FEA) and computational electromagnetics simulations have been employed to study the electromagnetic behavior, heat transfer mechanisms, and material responses during microwave processing. These simulations enable the optimization of process parameters, prediction of material behavior, and design of efficient processing strategies. The combination of experimental studies and numerical simulations has contributed to a better understanding of the underlying phenomena and the development of robust microwave-assisted processing techniques.

Recent research studies have also focused on the integration of microwave technology with other processing methods, such as ultrasonic irradiation, electric field-assisted processing, and hybrid heating approaches. These synergistic approaches have shown promising results in enhancing the processing efficiency, material properties, and overall performance of the composites. The combination of multiple energy sources enables targeted heating, improved mass transport, and enhanced curing kinetics, leading to advanced composite materials with tailored properties.

In summary, the advancements in microwave-assisted processing techniques have revolutionized the field of composite manufacturing. The development of innovative applicator designs, optimization of curing and polymerization processes, application of microwave sintering techniques, integration of computational modeling, and exploration of synergistic approaches have all contributed to the evolution of this field. These advancements have paved the way for the fabrication of high-performance composites with tailored properties, reduced processing times, and improved energy efficiency. The continuous research and development in this area hold great potential for further advancements in microwave-assisted processing techniques and their application in various industries.

6.2 Environmental and Sustainability Considerations

Environmental and sustainability considerations have become increasingly important in the field of microwave-assisted processing. As the demand for more environmentally friendly manufacturing processes grows, researchers and industry professionals have been actively exploring ways to minimize the environmental impact and promote sustainable practices in microwave-assisted processing techniques. This section discusses the key environmental and sustainability considerations associated with microwave-assisted processing and highlights recent advancements in this area.

One of the main environmental benefits of microwave-assisted processing is its energy efficiency. Compared to conventional heating methods, such as electric resistance or gas heating, microwave heating offers higher energy conversion efficiency. Microwaves directly heat the material, minimizing heat loss to the surroundings, and allowing for faster and more precise heating. This results in reduced energy consumption and lower greenhouse gas emissions, making microwave-assisted processing a more sustainable option.

Another important consideration is the potential for reducing waste and improving material utilization. Microwave-assisted processing techniques have shown promise in minimizing material waste by enabling more precise control over the heating process. With better temperature control and rapid heating, the risk of overheating, degradation, or thermal damage to the material is reduced. This allows for more efficient use of raw materials, resulting in higher yield and less waste generation. Additionally, microwave-assisted processing offers opportunities for recycling and reusing composite materials, contributing to a more circular and sustainable economy.

Furthermore, microwave-assisted processing techniques have the potential to reduce the use of harmful chemicals and solvents. Traditional manufacturing processes often rely on the use of volatile organic compounds (VOCs) or hazardous solvents for various purposes, such as resin mixing, curing, or cleaning. Microwave-assisted processing can offer alternatives to these processes by promoting solvent-free or low-VOC approaches. For example, microwave-assisted curing techniques have been developed to reduce or eliminate the need for solvent-based curing agents, leading to safer and more environmentally friendly processes.

The adoption of sustainable microwave-assisted processing techniques also requires considering the overall life cycle assessment (LCA) of the process. Researchers have started to investigate the environmental impacts associated with microwave-assisted processing, taking into account factors such as energy consumption, material sourcing, transportation, and waste management. LCA studies provide valuable insights into the environmental hotspots of the process and help identify areas for improvement in terms of resource efficiency, emissions reduction, and waste management.

Recent advancements in microwave-assisted processing have also focused on the development of eco-friendly materials. Researchers have explored the use of bio-based polymers, natural fibers, and recycled materials in composite manufacturing. By incorporating sustainable materials, microwave-assisted processing can contribute to reducing the reliance on non-renewable resources and promoting a more sustainable and environmentally conscious approach to composite production.

In summary, environmental and sustainability considerations play a vital role in the development and adoption of microwave-assisted processing techniques. The energy efficiency, waste reduction, elimination of harmful chemicals, and exploration of eco-friendly materials contribute to a more sustainable and environmentally friendly manufacturing process. Ongoing research and development efforts are aimed at further optimizing microwave-assisted processing techniques, conducting comprehensive life cycle assessments, and advancing sustainable practices in composite manufacturing. By addressing these considerations, microwave-assisted processing has the potential to make significant contributions to a greener and more sustainable future.

7 Conclusion

In conclusion, microwave-assisted processing techniques have emerged as promising approaches in the field of composite manufacturing, offering numerous benefits and opportunities for advancements. The utilization of microwaves for heating and curing processes provides advantages such as rapid and efficient heating, precise temperature control, and reduced energy consumption. It enables improved material properties, enhanced processing efficiency, and reduced production time, leading to cost savings and increased productivity.

The application of microwave-assisted processing in various stages of composite manufacturing, including prepreg manufacturing, resin transfer molding, sintering, foaming, and repair, has been extensively studied. The data presented in this book chapter highlight the recent developments and research findings in each of these areas, demonstrating the effectiveness and potential of microwave-assisted processing techniques.

The studies discussed in this chapter provide valuable insights into the benefits of microwave-assisted processing, including enhanced mechanical properties, reduced void content, improved material homogeneity, and accelerated curing and consolidation. Additionally, the utilization of microwaves offers opportunities for the development of sustainable and environmentally friendly manufacturing processes, through reduced energy consumption, waste reduction, elimination of hazardous chemicals, and exploration of eco-friendly materials.

However, challenges and limitations still exist in the widespread adoption of microwave-assisted processing. These include the optimization of process parameters, scaling up of production, equipment design and cost considerations, and the need for further research on the long-term performance and reliability of microwave-processed composites. Addressing these challenges will require collaborative efforts from researchers, industry professionals, and regulatory bodies.

Overall, microwave-assisted processing has the potential to revolutionize the composite manufacturing industry by providing faster, more efficient, and environmentally friendly processes. Further research and development are needed to explore new applications, refine process parameters, and optimize equipment design. With continued advancements in microwave technology and growing emphasis on sustainability, microwave-assisted processing is poised to play a significant role in shaping the future of composite manufacturing, offering improved product performance, reduced environmental impact, and cost-effective production.

References

1. Gupta M, Wai EW (2007) Leong, microwaves and metals. John Wiley & Sons
2. Chen M, Siochi EJ, Ward TC, McGrath JE (1993) Basic ideas of microwave processing of polymers. Polym Eng Sci 33:1092–1109. https://doi.org/10.1002/pen.760331703
3. Rangari VK, Bhuyan MS, Jeelani S (2010) Microwave processing and characterization of EPON 862/CNT nanocomposites. Mater Sci Eng B Solid State Mater Adv Technol 168:117–121. https://doi.org/10.1016/j.mseb.2010.01.013
4. Singh MK, Verma N, Zafar S (2020) Conventional processing of polymer matrix composites. In Lightweight polymer composite structures: design and manufacturing techniques, pp 21–66
5. Singh MK, Zafar S (2019) Development and mechanical characterization of microwave-cured thermoplastic based natural fibre reinforced composites. J Thermoplast Compos Mater 32:1427–1442. https://doi.org/10.1177/0892705718799832

6. Mishra RR, Sharma AK (2016) On mechanism of in-situ microwave casting of aluminium alloy 7039 and cast microstructure. Mater Des 112:97–106. https://doi.org/10.1016/J.MATDES.2016.09.041
7. Gamit D, Mishra RR, Sharma AK (2017) Joining of mild steel pipes using microwave hybrid heating at 2.45 GHz and joint characterization. J Manuf Process 27:158–168. https://doi.org/10.1016/J.JMAPRO.2017.04.028
8. Pundhir N, Zafar S, Pathak H (2019) Performance evaluation of HDPE/MWCNT and HDPE/kenaf composites. J Thermoplast Compos Mater. https://doi.org/10.1177/0892705719868278
9. Zafar S, Singh MK, Verma N (2020) Method for manufacturing thermoplastic composite from microwave-assisted compression moulding. Off J Patent Office 31228
10. Verma N, Zafar S, Pathak H (2020) Investigations on thermal damage and surface roughness of laser beam machined nano-hydroxyapatite UHMWPE composites. Manuf Lett. 25:81–87. https://doi.org/10.1016/j.mfglet.2020.08.003
11. Verma N, Kumar R, Zafar S, Pathak H (2020) Vacuum-assisted microwave curing of epoxy/carbon fiber composite: an attempt for defect reduction in processing. Manuf Lett 24:127–131. https://doi.org/10.1016/j.mfglet.2020.04.010
12. Singh MK, Zafar S (2020) Abrasive wear mechanism of microwave-assisted compression molded kenaf/HDPE composite. J Tribol 142:1–11. https://doi.org/10.1115/1.4046858
13. Verma N, Singh MK, Zafar S (2020) Development of porous bio-nano-composites using microwave processing. In Biofibers and biopolymers for biocomposites, pp 209–228. https://doi.org/10.1007/978-3-030-40301-0_10
14. Arora G, Pathak H, Zafar S (2019) Fabrication and characterization of microwave cured high-density polyethylene/carbon nanotube and polypropylene/carbon nanotube composites. J Compos Mater 53:2091–2104. https://doi.org/10.1177/0021998318822705
15. Verma N, Zafar S, Talha M (2019) Influence of nano-hydroxyapatite on mechanical behavior of microwave processed polycaprolactone composite foams. Mater Res Express 6:085336. https://doi.org/10.1088/2053-1591/ab260d
16. Singh MK, Chauhan D, Gupta MK, Diwedi A (2015) Optimization of process parameters of aluminum alloy (Al-6082 T-6) machined on CNC lathe machine for low surface roughness. J Mater Sci Eng 04:2169–22. https://doi.org/10.4172/2169-0022.1000202
17. Singh MK, Zafar S, Talha M (2019) Development and characterisation of poly-L-lactide-based foams fabricated through microwave-assisted compression moulding. J Cell Plast 55:523–541. https://doi.org/10.1177/0021955X19850728
18. Singh MK, Zafar S (2018) Influence of microwave power on mechanical properties of microwave-cured polyethylene/coir composites. J Nat Fibers 17:845–860. https://doi.org/10.1080/15440478.2018.1534192
19. Arora G, Pathak H (2020) Experimental and numerical approach to study mechanical and fracture properties of high-density polyethylene carbon nanotubes composite. Mater Today Commun 22:100829. https://doi.org/10.1016/j.mtcomm.2019.100829
20. Arora G, Pathak H (2021) Nanoindentation characterization of polymer nanocomposites for elastic and viscoelastic properties: experimental and mathematical approach. Compos C: Open Access 4:100103. https://doi.org/10.1016/j.jcomc.2020.100103
21. Padmanabhan S, Gupta A, Arora G, Pathak H, Burela RG, Bhatnagar AS (2020) Meso–macro-scale computational analysis of boron nitride nanotube-reinforced aluminium and epoxy nanocomposites: a case study on crack propagation. In: Proceedings of the institution of mechanical engineers, Part L: journal of materials: design and applications, pp 1–16. https://doi.org/10.1177/1464420720961426
22. Bhatnagar AS, Gupta A, Arora G, Padmanabhan S, Burela RG (2021) Mean-field homogenization coupled low-velocity impact analysis of nanofibre reinforced composites. Mater Today Commun 102089. https://doi.org/10.1016/j.mtcomm.2021.102089
23. Mishra RR, Sharma AK (2016) Microwave-material interaction phenomena: heating mechanisms, challenges and opportunities in material processing. Compos Part A Appl Sci Manuf 81:78–97. https://doi.org/10.1016/j.compositesa.2015.10.035

24. Thostenson ET, Chou T (1999) Microwave processing : fundamentals and applications 30:1055–1071
25. Fotiou I, Baltopoulos A, Vavouliotis A, Kostopoulos V (2013) Microwave curing of epoxy polymers reinforced with carbon nanotubes. J Appl Polym Sci 129:2754–2764. https://doi.org/10.1002/app.39003
26. Hassan MN, Mahmoud MM, Link G, El-Fattah AA, Kandil S (2016) Sintering of naturally derived hydroxyapatite using high frequency microwave processing. J Alloys Compd 682:107–114. https://doi.org/10.1016/j.jallcom.2016.04.266
27. Smith VK, Pollard JF, Ward TC, Graybeal JD, Ma JJ, Quirk RP (1987) Dielectric properties of polymers at microwave frequencies. Pam Chem Soc 28:326–327
28. Chang J, Liang G, Gu A, Cai S, Yuan L (2012) The production of carbon nanotube/epoxy composites with a very high dielectric constant and low dielectric loss by microwave curing. Carbon NY 50:689–698. https://doi.org/10.1016/J.CARBON.2011.09.029
29. Drozdov AD, Christiansen J (2020) Modeling dielectric permittivity of polymer composites at microwave frequencies. Mater Res Bull 126:110818. https://doi.org/10.1016/j.materresbull.2020.110818
30. Al-Saleh MH, Jawad SA, El Ghanem HM (2014) Electrical and dielectric behaviors of dry-mixed CNT/UHMWPE nanocomposites. High Perform Polym 26:205–211. https://doi.org/10.1177/0954008313507590
31. Azlan AA, Badrun MKA, Ali MT, Awang MZ (2013) A study of dielectric constant and loss tangent of Leucaena Leucocephala wood plastic composite (WPC) substrate. In: IEEE Symposium on Wireless Technology and Applications, ISWTA, pp 176–181. https://doi.org/10.1109/ISWTA.2013.6688765
32. Singh MK, Zafar S (2021) Wettability, absorption and degradation behavior of microwave-assisted compression molded kenaf/HDPE composite tank under various environments. Polymer Degradation and Stability 185 (2021): 109500. https://doi.org/10.1016/j.polymdegradstab.2021.109500
33. Ghadami F, Dadfar MR, Kazazi M (2016) Hot-cured epoxy-nanoparticulate-filled nanocomposites: fracture toughness behavior. Eng Fract Mech 162:193–200. https://doi.org/10.1016/j.engfracmech.2016.05.016
34. Qing Y, Zhou W, Luo F, Zhu D (2010) Epoxy-silicone filled with multi-walled carbon nanotubes and carbonyl iron particles as a microwave absorber. Carbon NY 48:4074–4080. https://doi.org/10.1016/J.CARBON.2010.07.014
35. Zhi CY, Bando Y, Terao T, Tang C, Golberg D (2010) Dielectric and thermal properties of epoxy/boron nitride nanotube composites. Pure Appl Chem 82:2175–2183. https://doi.org/10.1351/PAC-CON-09-11-41/MACHINEREADABLECITATION/RIS
36. Kausar A, Rafique I, Muhammad B (2016) Review of applications of polymer/carbon nanotubes and Epoxy/CNT composites. Polym Plast Technol Eng 55:1167–1191. https://doi.org/10.1080/03602559.2016.1163588
37. Li Y, Li N, Gao J (2014) Tooling design and microwave curing technologies for the manufacturing of fiber-reinforced polymer composites in aerospace applications. Int J Adv Manuf Technol 70:591–606. https://doi.org/10.1007/S00170-013-5268-3/METRICS
38. Mgbemena CO, Li D, Lin M, Daniel P, Katnam KB, Kumar VT, Nezhad HY (2018) Accelerated microwave curing of fibre-reinforced thermoset. Compos A. https://doi.org/10.1016/j.compositesa.2018.09.012
39. Li Y, Li N, Zhou J, Cheng Q (2019) Microwave curing of multidirectional carbon fiber reinforced polymer composites. Compos Struct 212:83–93. https://doi.org/10.1016/J.COMPSTRUCT.2019.01.027
40. Ku HS, Siores E, Ball JAR (1999) Microwave facilities for welding thermoplastic composites and preliminary results. J Microw Power Electromagn Energy 34:195–205. https://doi.org/10.1080/08327823.1999.11688406
41. Nightingale C, Day RJ (2002) Flexural and interlaminar shear strength properties of carbon fibre/epoxy composites cured thermally and with microwave radiation. Compos Part A Appl Sci Manuf 33:1021–1030. https://doi.org/10.1016/S1359-835X(02)00031-3

42. Zhou S, Hawley MC (2003) A study of microwave reaction rate enhancement effect in adhesive bonding of polymers and composites. Compos Struct 61:303–309. https://doi.org/10.1016/S0263-8223(03)00061-8
43. Rao RMVGK, Rao S (2006) Studies on tensile and interlaminar shear strength properties of thermally cured and microwave cured glass-epoxy composites. J Reinforce Plastics Compos 25:783–795. https://doi.org/10.1177/0731684406063542
44. Papargyris DA, Day RJ, Nesbitt A, Bakavos D (2008) Comparison of the mechanical and physical properties of a carbon fibre epoxy composite manufactured by resin transfer moulding using conventional and microwave heating. Compos Sci Technol 68:1854–1861. https://doi.org/10.1016/j.compscitech.2008.01.010
45. Bajpai PK, Singh I, Madaan J (2012) Joining of natural fiber reinforced composites using microwave energy: experimental and finite element study. Mater Des 35:596–602. https://doi.org/10.1016/j.matdes.2011.10.007
46. Reem SM, Ansari M, Saleh M (2021) A study on mechanical, thermal and morphological properties of natural fibre/epoxy composite. J Purity Utility Reaction Environ 1:237–266
47. Ali S, Bajpai PK, Singh I, Sharma AK (2014) Curing of natural fibre-reinforced thermoplastic composites using microwave energy. J Reinf Plast Compos 33:993–999. https://doi.org/10.1177/0731684414523326
48. Kwak M, Robinson P, Bismarck A, Wise R (2015) Microwave curing of carbon–epoxy composites: penetration depth and material characterisation. Compos Part A Appl Sci Manuf 75:18–27. https://doi.org/10.1016/J.COMPOSITESA.2015.04.007
49. Vikas G, Sudheer M (2017) A review on properties of basalt fiber reinforced polymer composites. Am J Mater Sci 7:156–165. https://doi.org/10.5923/j.materials.20170705.07
50. Li Y, Cheng L, Zhou J (2018) Curing multidirectional carbon fiber reinforced polymer composites with indirect microwave heating. Int J Adv Manuf Technol 97:1137–1147. https://doi.org/10.1007/s00170-018-1974-1
51. Rao S, Chiranjeevi MC, Rajendra Prakash M (2018) Vacuum-assisted microwave processing of glass-epoxy composite laminates using novel microwave absorbing molds. Polym Compos 39:1152–1160. https://doi.org/10.1002/pc.24044
52. Bajakke PA, Malik VR, Mugali P, Deshpande AS (2021) Microwave processing of engineering materials BT—coatings: materials, processes, characterization and optimization. In: Kumar K, Babu BS, Davim JP (Eds.). Springer International Publishing, Cham, pp 31–55. https://doi.org/10.1007/978-3-030-62163-6_2
53. Alem SAA, Latifi R, Angizi S, Hassanaghaei F, Aghaahmadi M, Ghasali E, Rajabi M (2020) Microwave sintering of ceramic reinforced metal matrix composites and their properties: a review. Mater Manuf Processes 35:303–327. https://doi.org/10.1080/10426914.2020.1718698
54. Manohar G, Maity SR, Pandey KM (2022) Microstructural and mechanical properties of microwave sintered AA7075/Graphite/SiC hybrid composite fabricated by powder metallurgy techniques. SILICON 14:5179–5189. https://doi.org/10.1007/s12633-021-01299-7
55. Mehta A, Vasudev H, Jeyaprakash N (2023) Role of sustainable manufacturing approach: microwave processing of materials. Int J Interact Design Manuf (IJIDeM). https://doi.org/10.1007/s12008-023-01318-4
56. Lunetto V, Galati M, Settineri L, Iuliano L (2023) Sustainability in the manufacturing of composite materials: A literature review and directions for future research. J Manuf Process 85:858–874. https://doi.org/10.1016/j.jmapro.2022.12.020
57. Borrell A, Salvador MD (2018) Advanced ceramic materials sintered by microwave technology. In: Liu M (Ed.) IntechOpen, Rijeka, p. Ch. 1. https://doi.org/10.5772/intechopen.78831
58. Bogdal D, Penczek P, Pielichowski J, Prociak A (2003) Microwave assisted synthesis, crosslinking, and processing of polymeric materials BT—Liquid Chromatography/FTIR Microspectroscopy/Microwave Assisted Synthesis. Springer Berlin Heidelberg, Berlin, Heidelberg, pp 194–263. https://doi.org/10.1007/b11051
59. Li N, Li Y, Hang X, Gao J (2014) Analysis and optimization of temperature distribution in carbon fiber reinforced composite materials during microwave curing process. J Mater Process Technol 214:544–550. https://doi.org/10.1016/j.jmatprotec.2013.10.012

60. Ghosh SK, Das NC (2023) Application of radiation curing on properties and performance of polymers and polymer composites BT—applications of high energy radiations: synthesis and processing of polymeric materials. In: Chowdhury SR (Ed.) Springer Nature Singapore, Singapore, pp 1–39. https://doi.org/10.1007/978-981-19-9048-9_1
61. Pastarnokienė L, Jonikaitė-Švėgždienė J, Lapinskaitė N, Kulbokaitė R, Bočkuvienė A, Kochanė T, Makuška R (2023) The effect of reactive diluents on curing of epoxy resins and properties of the cured epoxy coatings. J Coat Technol Res. https://doi.org/10.1007/s11998-022-00737-4
62. Almushaikeh AM, Alaswad SO, Alsuhybani MS, AlOtaibi BM, Alarifi IM, Alqahtani NB, Aldosari SM, Alsaleh SS, Haidyrah AS, Alolyan AA, Alshammari BA (2023) Manufacturing of carbon fiber reinforced thermoplastics and its recovery of carbon fiber: a review. Polym Test 122:108029. https://doi.org/10.1016/j.polymertesting.2023.108029
63. Haider I, Gul IH, Umer MA, Baig MM (2023) Silica-fiber-reinforced composites for microelectronic applications: effects of curing routes. Materials 16. https://doi.org/10.3390/ma1605 1790
64. Kumar R, Rani M, Zafar S (2020) Microwave-based manufacturing of epoxy/fiber composites BT—handbook of epoxy/fiber composites. In: Mavinkere Rangappa S, Parameswaranpillai J, Siengchin S, Thomas S (Eds.). Springer Singapore, Singapore, pp 1–40. https://doi.org/10.1007/978-981-15-8141-0_54-1
65. Ramdani N, Derradji M, Mokhnache EO (2022) Natural fiber reinforced polybenzoxazine composites: a review. Mater. Today Commun. 31:103645. https://doi.org/10.1016/j.mtcomm.2022.103645
66. Chen D, Arakawa K, Xu C (2015) Reduction of void content of vacuum-assisted resin transfer molded composites by infusion pressure control. Polym Compos 36:1629–1637. https://doi.org/10.1002/pc.23071
67. Maenz S, Mühlstädt M, Jandt KD, Bossert J (2014) Mechanical properties of microwave cured glass fibre epoxy composites prepared by resin transfer moulding. J Compos Mater 49:2839–2847. https://doi.org/10.1177/0021998314557295
68. Muñoz-Pascual S, Saiz-Arroyo C, Vuluga Z, Corobea MC, Rodriguez-Perez MA (2020) Foams with enhanced ductility and impact behavior based on polypropylene composites. Polymers 12. https://doi.org/10.3390/polym12040943
69. Zubair M, Ferrari R, Alagha O, Mu'azu ND, Blaisi NI, Ateeq IS, Manzar MS (2020) Microwave foaming of materials: an emerging field. Polymers (Basel) 12. https://doi.org/10.3390/polym1 2112477
70. Shahbazi M, Aghvami-Panah M, Panahi-Sarmad M, Seraji AA, Zeraatkar A, Ghaffarian Anbaran R, Xiao X (2022) Fabricating bimodal microcellular structure in polystyrene/carbon nanotube/glass-fiber hybrid nanocomposite foam by microwave-assisted heating: a proof-of-concept study. J Appl Polym Sci 139:52125. https://doi.org/10.1002/app.52125
71. Song SA, Chung YS, Kim SS (2014) The mechanical and thermal characteristics of phenolic foams reinforced with carbon nanoparticles. Compos Sci Technol 103:85–93. https://doi.org/10.1016/j.compscitech.2014.08.013
72. Verma N, Zafar S, Talha M (2020) Application of microwave energy for rapid fabrication of nano-hydroxyapatite reinforced polycaprolactone composite foam. Manuf Lett 23:9–13. https://doi.org/10.1016/j.mfglet.2019.11.006
73. Kuhnigk J, Standau T, Dörr D, Brütting C, Altstädt V, Ruckdäschel H (2022) Progress in the development of bead foams—A review. J Cell Plast 58:707–735. https://doi.org/10.1177/002 1955X221087603
74. Ding M, Xu D, Wang Q (2022) Microwave-assisted foaming and sintering high-strength and antistatic polypropylene bead foams by constructing conductive 3D skeleton structure. Compos Part A Appl Sci Manuf 163:107196. https://doi.org/10.1016/j.compositesa.2022.107196
75. Prapuchanay E, Puangngernmak N, Chalermwisutkul S (2012) A single mode cylindrical microwave cavity for heating applications of liquid material. In 2012 9th international conference on electrical engineering/electronics, computer, telecommunications and information technology, pp 1–4. https://doi.org/10.1109/ECTICon.2012.6254353

76. Eghbal Sarabi F, Ghorbani M, Stankiewicz A, Nigar H (2020) Coaxial traveling-wave microwave reactors: Design challenges and solutions. Chem Eng Res Design 153:677–683. https://doi.org/10.1016/j.cherd.2019.11.022
77. Gaukel V, Siebert T, Erle U (2017) 8—Microwave-assisted drying. In Regier M, Knoerzer K, H.B.T.-T.M.P. of F. (Second E. Schubert (Eds.), Woodhead Publishing Series in Food Science, Technology and Nutrition, Woodhead Publishing, pp 152–178. https://doi.org/10.1016/B978-0-08-100528-6.00008-5
78. Verma N, Singh MK, Zafar S, Pathak H (2021) Comparative study of in-situ temperature measurement during microwave-assisted compression-molding and conventionally compression-molding process. CIRP J Manuf Sci Technol 35:336–345. https://doi.org/10.1016/j.cirpj.2021.07.005
79. Askari GR, Emam-Djomeh Z, Mousavi SM (2009) An investigation of the effects of drying methods and conditions on drying characteristics and quality attributes of agricultural products during hot air and hot air/microwave-assisted dehydration. Drying Technol 27:831–841. https://doi.org/10.1080/07373930902988106
80. Zhang M, Jiang H, Lim R-X (2010) Recent developments in microwave-assisted drying of vegetables, fruits, and aquatic products—drying kinetics and quality considerations. Drying Technol 28:1307–1316. https://doi.org/10.1080/07373937.2010.524591
81. Rattanadecho P, Makul N (2016) Microwave-assisted drying: a review of the state-of-the-art. Drying Technol 34:1–38. https://doi.org/10.1080/07373937.2014.957764
82. Zhi W-J, Wang L-F, Hu X-J (2017) Recent advances in the effects of microwave radiation on brains. Mil Med Res 4:29. https://doi.org/10.1186/s40779-017-0139-0
83. Ozdemir F, Kargi A (2011) Electromagnetic waves and human health. In: Zhurbenko V (Ed.). IntechOpen, Rijeka, 2011, Ch. 22. https://doi.org/10.5772/16343
84. Zabihi O, Ahmadi M, Liu C, Mahmoodi R, Li Q, Naebe M (2020) Development of a low cost and green microwave assisted approach towards the circular carbon fibre composites. Compos B Eng 184:107750. https://doi.org/10.1016/j.compositesb.2020.107750
85. Li N, Li Y, Jelonnek J, Link G, Gao J (2017) A new process control method for microwave curing of carbon fibre reinforced composites in aerospace applications. Compos B Eng 122:61–70. https://doi.org/10.1016/j.compositesb.2017.04.009
86. Kim M, Kim H, Lee W (2015) Repair of aircraft structures using composite patches bonded through induction heating. Adv Compos Mater 24:307–323. https://doi.org/10.1080/09243046.2014.899553
87. Naik TP, Singh I, Sharma AK (2022) Processing of polymer matrix composites using microwave energy: a review. Compos Part A Appl Sci Manuf 156:106870. https://doi.org/10.1016/j.compositesa.2022.106870
88. Singh I, Bajpai PK, Malik D, Madaan J, Bhatnagar N (2012) Microwave joining of natural fiber reinforced green composites. Adv Mat Res 410:102–105. https://doi.org/10.4028/www.scientific.net/AMR.410.102
89. Bélanger JMR, Paré JRJ, Poon O, Fairbridge C, Ng S, Mutyala S, Hawkins R (2007) Remarks on various applications of microwave energy, journal of microwave power and electromagnetic. Energy 42:24–44. https://doi.org/10.1080/08327823.2007.11688597
90. Zabihi O, Ahmadi M, Liu C, Mahmoodi R, Li Q, Naebe M (2020) Development of a low cost and green microwave assisted approach towards the circular carbon fibre composites. Compos B Eng 184:107750. https://doi.org/10.1016/j.compositesb.2020.107750

Application of Microwave Heating in Polymer Composites Manufacturing

Papiya Bhowmik

Abstract Polymer composites have transformed industries with their exceptional properties, but conventional manufacturing processes face challenges such as long processing times, high energy consumption, and environmental concerns. Microwave heating has emerged as a promising technology for polymer composites manufacturing. It utilizes electromagnetic waves to generate heat within materials, offering advantages such as rapid and selective heating, high energy efficiency, and improved control. The benefits of microwave heating include reduced processing time, expedited curing and consolidation processes, and enhanced mechanical properties through improved dispersion and bonding. Challenges remain in optimizing processing parameters, designing microwave-absorbing materials, and ensuring safety. Microwave heating offers rapid and selective heating, enhanced composite properties, and reduced processing time. Ongoing research aims to unlock its full potential in advanced polymer composites production. Microwave heating has the potential to shape the future of polymer composites in various industries by improving efficiency and performance.

Keywords Microwave · Mechanical properties · Electrical properties · Dielectric · Curing · Composites

P. Bhowmik (✉)
Department of Mechanical Engineering, Indian Institute of Technology, Ropar, Rupnagar, Punjab 140001, India
e-mail: papiya.nita@gmail.com; papiya.bhowmik@lnu.se

Research Scientist, Department of Forestry and Wood Technology, Linnaeus University, Växjö, Sweden

M. K. Singh et al. (eds.), *Composite Materials Processing Using Microwave Heating Technology*, Composites Science and Technology,
https://doi.org/10.1007/978-981-97-2772-8_2

1 Introduction

Polymer composites have revolutionized various industries, including aerospace, automotive, construction, and sports, by offering exceptional mechanical, thermal, and electrical properties that surpass those of traditional materials [1]. These composites are composed of a polymer matrix reinforced with strong fibers or particles, resulting in enhanced strength, stiffness, and other desired characteristics. However, the conventional manufacturing processes of polymer composites often entail certain challenges, such as prolonged processing times, high energy consumption, and environmental considerations. As a result, researchers have been actively exploring alternative and innovative manufacturing techniques to address these issues and further enhance the properties of the final composite products.

In recent years, microwave heating has emerged as a promising technology that has garnered significant attention in the field of polymer composites manufacturing [2–19]. Microwave heating is a form of dielectric heating that utilizes electromagnetic waves to generate heat within materials by stimulating molecular vibrations and inducing frictional heating. By exploiting the unique properties of microwaves, this heating technique offers several advantages over traditional heating methods, including rapid and selective heating, high energy efficiency, and improved control over the heating process. These advantages make microwave heating an attractive option for the fabrication of polymer composites with enhanced performance and reduced processing times.

The fundamental principle behind microwave heating lies in the interaction between the electric field of the microwave radiation and the polar molecules present in the polymer matrix. When exposed to microwaves, the polar molecules attempt to align themselves with the alternating electric field. As a result, they continuously reorient themselves, leading to rapid molecular motion and, consequently, the conversion of electromagnetic energy into heat. This phenomenon allows for efficient and uniform heating of the composite material [16].

The benefits of microwave heating in polymer composites manufacturing are multifaceted. Firstly, the rapid heating rate associated with microwave heating significantly reduces the processing time compared to conventional methods. The ability to achieve higher temperatures in a shorter period allows for expedited curing and consolidation processes, enabling faster production cycles and increased manufacturing throughput. Additionally, microwave heating promotes the uniform dispersion of filler materials, such as fibers or particles, within the polymer matrix. This uniform dispersion enhances the interfacial bonding between the matrix and reinforcements, resulting in improved mechanical properties, such as strength, stiffness, and impact resistance [20, 21].

Another advantage of microwave heating is its selective heating capability. Unlike conventional heating methods that heat the entire composite material, microwave heating can selectively target specific regions or components within the composite. This selective heating enables localized processing, such as curing only the desired areas or promoting specific reactions, while minimizing energy consumption and reducing the risk of thermal damage to the surrounding regions. Furthermore, microwave heating can mitigate issues associated with void formation and outgassing, as the rapid and efficient heating minimizes the entrapment of gases and ensures better consolidation of the composite structure.

The impact of microwave heating on the properties of polymer composites extends beyond the realm of mechanical performance. Microwave irradiation can influence the thermal and electrical conductivity of the composites, as well as their electromagnetic interference shielding effectiveness [22–26]. Through the careful selection of processing parameters, such as power, frequency, and exposure time, the thermal conductivity of the composites can be enhanced, enabling improved heat dissipation and thermal management. Similarly, the electrical conductivity can be tailored by incorporating conductive fillers and optimizing the microwave heating conditions, enabling applications in electrical and electronic fields. Moreover, microwave heating can facilitate the dispersion and alignment of conductive fillers, leading to enhanced electrical and thermal conductivities within the composite structure.

Despite the numerous advantages offered by microwave heating, there are still challenges and limitations that need to be addressed. One primary challenge lies in the optimization of processing parameters to achieve the desired properties of the composites. Factors such as microwave frequency, power level, exposure time, and sample geometry need to be carefully controlled to ensure uniform heating and prevent potential thermal damage. Additionally, the design and development of microwave-absorbing materials and structures are essential for maximizing the heating efficiency and penetration depth of the microwave energy. Furthermore, the safety and health aspects related to microwave radiation exposure and equipment design must be thoroughly considered to ensure the well-being of operators and to comply with regulatory standards.

Looking ahead, the application of microwave heating in polymer composites manufacturing holds promising potential. Researchers are actively exploring the utilization of microwave heating in the fabrication of more complex and high-performance composites, including nanocomposites, biocomposites, and hybrid composites [27]. The integration of microwave heating with other advanced manufacturing techniques, such as 3D printing or additive manufacturing, offers opportunities for the production of customized and functionalized composite structures with precise control over material properties. Furthermore, the development of predictive models and simulation tools can assist in optimizing the processing parameters and designing composite structures with desired properties, thereby accelerating the adoption of microwave heating in industrial applications.

In conclusion, microwave heating has emerged as a promising and versatile technology for polymer composites manufacturing. Its ability to provide rapid and selective heating, along with the potential to enhance the mechanical, thermal, and electrical properties of composites, makes it an attractive alternative to conventional heating methods. While challenges and limitations remain, ongoing research and development efforts are expected to overcome these hurdles and unlock the full potential of microwave heating in the production of advanced polymer composites.

With its potential to reduce processing time, improve energy efficiency, and enhance composite performance, microwave heating is poised to play a significant role in shaping the future of polymer composites in various industries.

2 Basic Principles of Microwave Heating

Microwave heating is based on the principle of dielectric heating, which involves the interaction of electromagnetic waves with polar molecules in a material. The energy of the electromagnetic waves is converted into heat as the polar molecules rotate and align with the alternating electric field of the waves. The rate of heating and the amount of heat generated depend on several factors, such as the frequency and power of the electromagnetic wave, the dielectric properties of the material, and the penetration depth of the wave. In this section, we will provide a more detailed explanation of the basic principle of microwave heating, along with mathematical expressions that describe the heating process.

Electromagnetic waves are composed of an electric field and a magnetic field that oscillate perpendicular to each other and to the direction of propagation. The frequency of the wave, denoted by f, is the number of oscillations per unit time, measured in Hertz (Hz). The wavelength of the wave, denoted by λ, is the distance between two consecutive points of maximum amplitude, and is related to the frequency by the equation $\lambda = c/f$, where c is the speed of light in vacuum (3 × 10^8 m/s) [2].

In a material, the electric field of the electromagnetic wave causes the polar molecules to rotate and align with the field direction. The magnitude of the rotation, denoted by θ, is proportional to the electric field strength, denoted by E, and the dipole moment of the molecules, denoted by μ. The dipole moment is a measure of the separation and magnitude of the positive and negative charges in the molecule, and is related to the polarizability of the molecule. The relationship between the rotation and the field strength is given by the equation [2]:

$$\theta = (\mu \mathrm{E})/\mathrm{kT} \tag{1}$$

where k is the Boltzmann constant (1.38×10^{-23} J/K) and T is the temperature in Kelvin. This equation shows that the rotation of the polar molecules increases with the electric field strength and the temperature.

As the polar molecules rotate, they generate friction and heat, which can raise the temperature of the material. The heat generated per unit volume of the material, denoted by Q, is proportional to the square of the electric field strength, the frequency of the wave, and the dielectric loss factor of the material, denoted by ε”. The dielectric loss factor represents the ability of the material to dissipate electromagnetic energy as heat, and is related to the conductivity and relaxation properties of the material. The relationship between the heat generated and the field strength and frequency is given by the equation [2]:

$$Q = \left(2\pi f\varepsilon'' E^2\right)/\rho \tag{2}$$

where ρ is the density of the material.

The penetration depth of the electromagnetic wave, denoted by δ, is the distance over which the wave amplitude decreases by a factor of 1/e. The penetration depth is related to the frequency and the dielectric properties of the material, and determines the depth of heating and the uniformity of heating. The relationship between the penetration depth and the frequency and dielectric properties is given by the equation [2]:

$$\delta = \left(2/\pi f\varepsilon'\right)^{0.5} \tag{3}$$

where ε' is the real part of the dielectric constant of the material, which represents its ability to store electrical energy.

The heating rate, denoted by dT/dt, is the rate at which the temperature of the material increases with time. The heating rate depends on the rate of energy absorption and the rate of energy dissipation, which are related to the field strength, the frequency, the dielectric properties, and the heat capacity of the material. The relationship between the heating rate and these factors can be described by complex mathematical expressions, such as the heat transfer equation, the electromagnetic wave equation, and the dielectric relaxation equation [2].

Microwave heating is a complex process that involves the interaction of electromagnetic waves with materials and depends on a variety of factors that can affect the heating efficiency, uniformity, and quality. Some of the key factors that can influence microwave heating include the frequency and power of the wave, the dielectric properties of the material, the geometry and size of the material, the mode of energy transfer, and the control parameters of the microwave system [2].

One of the advantages of microwave heating over conventional heating methods is the ability to selectively heat certain parts of a material while leaving others relatively unaffected. This is because the heating is primarily determined by the dielectric properties of the material, which can vary depending on the composition, structure, and morphology of the material. For example, materials with high water content or high ionic conductivity tend to absorb microwave energy more efficiently than materials with low water content or low ionic conductivity. This can be exploited in various applications, such as drying, sterilization, and sintering [2].

In the context of polymer composites manufacturing, microwave heating can offer several benefits, such as reduced cycle time, improved mechanical properties, and better control of the curing process. Polymer composites are typically composed of a matrix material, such as epoxy or polyester resin, and reinforcement materials, such as carbon fibers or glass fibers. The curing of the resin involves a chemical reaction that converts the liquid resin into a solid polymer network, and can be accelerated by heating. Microwave heating can provide a faster and more uniform heating of the material, which can reduce the time and energy required for curing, and can also enhance the interfacial bonding between the matrix and the reinforcement materials. Additionally, microwave heating can help reduce the formation of defects, such as voids and resin-rich areas, which can affect the mechanical properties of the composites.

In conclusion, the basic principle of microwave heating is based on the interaction of electromagnetic waves with polar molecules in a material, which generates friction and heat. The heating efficiency, uniformity, and quality depend on various factors, such as the frequency and power of the wave, the dielectric properties of the material, and the penetration depth of the wave. Microwave heating can offer several advantages over conventional heating methods, such as selective and faster heating, better control of the curing process, and improved mechanical properties of the material. In the next sections, we will explore the application of microwave heating in polymer composites manufacturing, and review some of the recent research in this field.

However, microwave heating also has some limitations. The heating rate depends on the absorption properties of the material, which can be affected by the material composition, shape, and size. The penetration depth of the microwave energy also depends on the frequency and material properties, which can limit the heating efficiency of thick or complex-shaped materials. The microwave heating equipment can also be expensive and require specialized expertise and safety precautions.

3 Application of Microwave Heating in Polymer Composites

Microwave heating has emerged as a promising technology for the manufacturing of polymer composites, due to its ability to provide faster and more uniform heating, better control of the curing process, and improved mechanical properties of the material. In this section, we will explore the various applications of microwave heating in polymer composite manufacturing, and review some of the recent research in this field.

3.1 Curing of Polymer Composites

The most common application of microwave heating in polymer composite manufacturing is the curing of the resin matrix. The curing process involves a chemical reaction that converts the liquid resin into a solid polymer network, and can be accelerated by heating. Conventional heating methods, such as convection and infrared heating, have been used for curing polymer composites, but they often suffer from slow heating rates, poor temperature control, and uneven heating.

Microwave heating offers several advantages over conventional heating methods for the curing of polymer composites. For example, microwave heating can provide faster and more uniform heating of the material, which can reduce the time and energy required for curing, and can also enhance the interfacial bonding between the matrix and the reinforcement materials. Additionally, microwave heating can help reduce the formation of defects, such as voids and resin-rich areas, which can affect the mechanical properties of the composites.

Several studies have investigated the use of microwave heating for the curing of polymer composites, and have demonstrated significant improvements in the curing efficiency and mechanical properties of the material. Li et al. [28] prepares the CFRP composites with multidirectional fiber orientations using vacuum-assisted resin transfer molding (VARTM) and compare the curing behavior and mechanical properties of samples cured with both microwave and conventional methods. The mechanical properties of the cured samples are evaluated using various tests, such as tensile, compression, and interlaminar shear tests.

The results of the study show that microwave curing offers several advantages over conventional curing for multidirectional CFRP composites. The authors found that the use of microwave-assisted curing significantly reduces the curing time, resulting in a more uniform degree of cure throughout the composite. Additionally, microwave curing is found to improve the interlaminar shear strength of the composites compared to conventional curing, while maintaining their other mechanical properties.

Overall, the study suggests that microwave curing can offer significant benefits for the manufacture of multidirectional CFRP composites, particularly in terms of reducing curing time and improving interlaminar shear strength. The authors note that further research is needed to optimize the microwave curing process for this specific type of composite.

Another study by Li et al. [29] discuss the advantages and disadvantages of microwave curing compared to conventional curing methods, and explore the design and development of microwave tooling systems for FRP composite production. They highlight the importance of optimizing the microwave tooling design and processing parameters, including the power and frequency of the microwave system, to ensure uniform and effective curing of the FRP composites.

The study also includes experimental studies on the use of microwave curing for the manufacture of FRP composites, particularly for aerospace applications. The authors investigate the effect of processing parameters on the mechanical properties of the cured composites, including their strength, stiffness, and fatigue resistance. They also compare the mechanical properties of FRP composites cured using microwave technology to those cured using conventional methods.

The results of the study show that microwave curing can offer significant benefits for the manufacture of FRP composites, particularly in terms of reducing curing time and improving the mechanical properties of the cured composites. However, the authors note that further research is needed to optimize the microwave tooling design and processing parameters for specific applications, and to better understand the effects of microwave radiation on the curing process and the resulting material properties.

Overall, the paper provides valuable insights into the potential of microwave curing for the manufacture of FRP composites in aerospace applications, and highlights the need for further research and development in this field.

Feher et al. [30] describes the development of a new technology called HEPHAISTOS for industrial composite fabrication using microwaves. This technology uses microwaves to heat and cure composite materials quickly and uniformly, resulting in a more efficient and cost-effective manufacturing process.

The study explains how traditional composite fabrication processes involve the use of ovens or autoclaves to cure the materials, which can be time-consuming and expensive. HEPHAISTOS technology, on the other hand, uses a microwave-assisted process to cure the composite materials in a fraction of the time required by traditional methods.

The study discusses the various parameters that can be controlled using HEPHAISTOS technology, such as the temperature, power level, and curing time, to optimize the curing process for different types of composite materials. It also highlights the advantages of using this technology, such as reduced energy consumption, improved quality of the final product, and increased manufacturing efficiency.

In the study by Zhang et al. [31] waste CFRP composite panels were subjected to each of the three upcycling technologies, and the resulting products were characterized using various analytical techniques. The study evaluated the quality of the carbon fibers recovered through upcycling and compared them with virgin carbon fibers to determine their suitability for various applications.

The study found that all three upcycling technologies resulted in the recovery of high-quality carbon fibers with similar mechanical properties to virgin carbon fibers. However, the study also found that the upcycling process can negatively affect the mechanical properties of the matrix material, which could limit the potential applications of upcycled CFRP composites. The study concludes that upcycling of CFRP composites using pyrolysis, solvolysis, and hydrothermal treatment can provide a sustainable and economically viable alternative to landfill disposal. The study also suggests that the recovered carbon fibers could potentially be used in various applications such as reinforced plastics, conductive materials, and energy storage devices. Overall, the study demonstrates the potential of upcycling technologies for the sustainable and environmentally friendly management of waste CFRP composites, while also highlighting the need for further research to optimize the upcycling process and evaluate the mechanical properties of the upcycled composite materials.

3.2 Effects of Microwave Heating on Polymer Composites

Microwave heating can have significant effects on the properties of polymer composites, such as thermal, mechanical, and electrical properties. The effects depend on the processing parameters, material composition, and filler properties.

3.2.1 Dielectric Properties

The study proposes an offline test method to measure the dielectric properties of polymer composites during microwave curing, which is an energy-efficient alternative to conventional autoclave curing for composite processing. The dielectric properties are important for energy conversion efficiency during microwave curing but existing measurement methods are not suitable for this process. To address this, the authors establish a kinetics model based on differential scanning calorimetry to describe the entire curing process of the composites under microwave curing [15].

To reflect the dielectric properties during microwave curing, the authors prepare several specially designed samples with different degrees of curing. These samples are used in the offline test method. Through extensive experimental research, the method is proven to be feasible in terms of test accuracy and efficiency. Using this method, the authors accurately determine the anisotropic complex permittivity of a carbon fiber/epoxy composite during microwave curing. Statistical analysis reveals that both the dielectric constant and dielectric loss of the composite increase initially, reach a maximum at the point of maximum reaction rate, and then decrease during the microwave curing process. The corresponding mechanism behind these changes is also systematically investigated in the study. The findings highlight the temporal variations in the dielectric properties of the carbon fiber/epoxy composite and provide insights into the underlying mechanisms driving these changes [15].

A study examined the impact of dielectric properties on various aspects of rubber sheet heating through microwave heating. Both experimental and numerical simulation studies were conducted to validate the simulation method used. The results indicated that dielectric constant and dielectric loss tangent significantly influenced the ability of rubber to absorb microwave energy and convert it into heat [32].

Different dielectric properties resulted in distinct electric field strength distributions within the rubber sheets during microwave heating. Consequently, varying amounts of electric field energy were converted into heat, leading to diverse temperature distributions across the rubber sheets. The temperature distributions were found to be uneven, and the temperature difference increased with prolonged heating time. However, the growth rate of the temperature difference decreased as the heating time increased [32].

Moreover, the dielectric constant or loss tangent of the rubber had a notable impact on microwave heating efficiency and temperature rise rate. Higher dielectric constant or loss tangent values corresponded to increased microwave heating efficiency and faster temperature rise. However, these conditions also resulted in more uneven temperature distributions within the rubber sheets. Additionally, as the heating time increased, the average body temperature of the rubber sheet increased, but the growth rate of the average body temperature slowed down. The heat flux density of the rubber sheet was significantly influenced by the dielectric properties. Interestingly, the maximum temperature gradient value was observed in the central area of the rubber sheet rather than in the hot spot area [32].

The dielectric properties of rubber sheets played a crucial role in microwave heating processes, affecting temperature distributions, heating efficiency, and heat flux density. Understanding these effects can contribute to the optimization of microwave heating processes for rubber sheets and facilitate the development of more efficient and uniform heating methods [32].

3.2.2 Electrical Properties

An accurate measurement of the electrical conductivity of carbon/carbon (C/C) composites in the higher range of microwave frequencies is presented. The C/C composite investigated is an 8 mm thick laminate made from a 2D woven fabric, which is machined to create a rectangular waveguide. By acquiring transmission and reflection parameters across a wide frequency band of 23 to 30 GHz, the power loss in the waveguide is evaluated, enabling the extraction of the electrical conductivity of the C/C composite. The obtained results indicate that the conductivity values at microwave frequencies are approximately 10 times lower than those measured at very low frequencies shown in Fig. 2.1. Furthermore, the effects of temperature on the conductivity are examined, revealing variations in conductivity with changing temperature. A precise measurement method is developed for determining the electrical conductivity of C/C composites at higher microwave frequencies. The findings highlight significant disparities between the conductivity values obtained at microwave frequencies compared to very low frequencies. Additionally, the investigation sheds light on the behavior of conductivity in response to temperature fluctuations [33].

Epoxy nanocomposites with varying carbon nanofiber contents up to 1 wt.% were manufactured using room temperature and refrigerated curing conditions. The mechanical and electrical properties of these composites were analyzed. The flexural modulus and hardness of the refrigerated samples exhibited a significant increase compared to the room temperature cured samples, primarily due to the prevention of nanofiber aggregation during the curing process. This prevention of aggregation

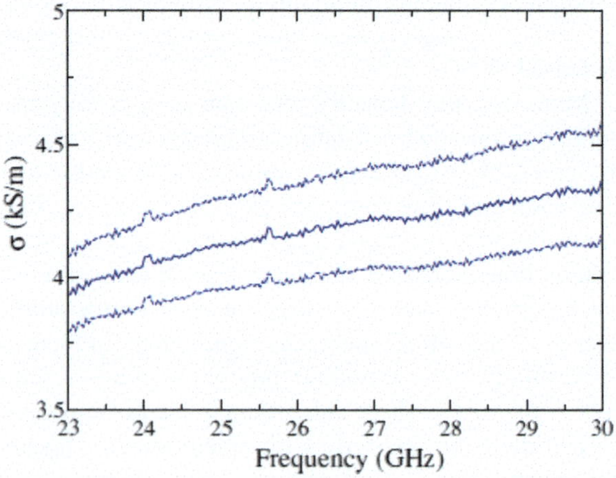

Fig. 2.1 Electrical conductivity of the employed C/C composite. Dashed lines represent the expected value variation range due to the VNA residual uncertainty [33]

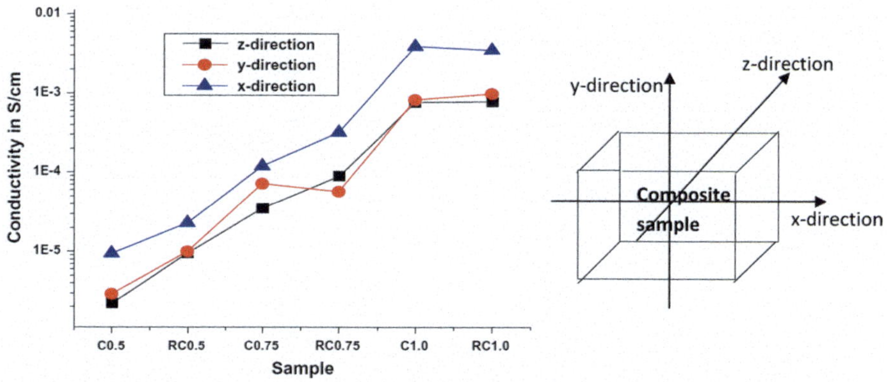

Fig. 2.2 Electrical conductivity of CNF/epoxy nanocomposites in three different directions along with schematic of the sample indicating directions [34]

enhanced the stress transfer and reinforcement between the epoxy matrix and carbon nanofibers, which was confirmed by the increase and shifting in the G-band observed in the Raman spectra of these samples [34].

The electrical conductivity of the composites improved by 3 to 6 orders of magnitude after the incorporation of carbon nanofibers into the insulating epoxy matrix as shown in Fig. 2.2. The room temperature cured samples demonstrated higher conductivity, which was attributed to the formation of a network structure through the aggregation of nanofibers along the fiber alignment direction. This phenomenon was validated through electron microscopic studies [34].

The fabrication of epoxy nanocomposites with carbon nanofibers under different curing conditions revealed significant improvements in mechanical and electrical properties. Refrigerated curing prevented nanofiber aggregation, resulting in enhanced mechanical properties. The infusion of carbon nanofibers drastically improved the electrical conductivity of the composites, with room temperature cured samples exhibiting higher conductivity attributed to the formation of a network structure through nanofiber aggregation [34].

The ultra-thinning of large tow carbon fiber (CF) has emerged as a key advancement in the industrialization of carbon fiber-reinforced polymer (CFRP) composites. However, conventional epoxy resins alone are unable to meet the demands of high-performance composites. To address this, the impact of carbon nanotube (CNT) reinforced epoxy resin on ultra-thin CFRP composites was investigated [35].

The study revealed that the epoxy resin matrix achieved its highest compressive strength, bending strength, bending modulus, and tensile strength when the CNT content was 0.2 wt%. Notably, the incorporation of CNTs significantly enhanced the transverse fiber bundle test tensile strength, interlaminar shear strength, and compressive strength of ultra-thin CF reinforced modified epoxy composites as shown in Fig. 2.3 [35].

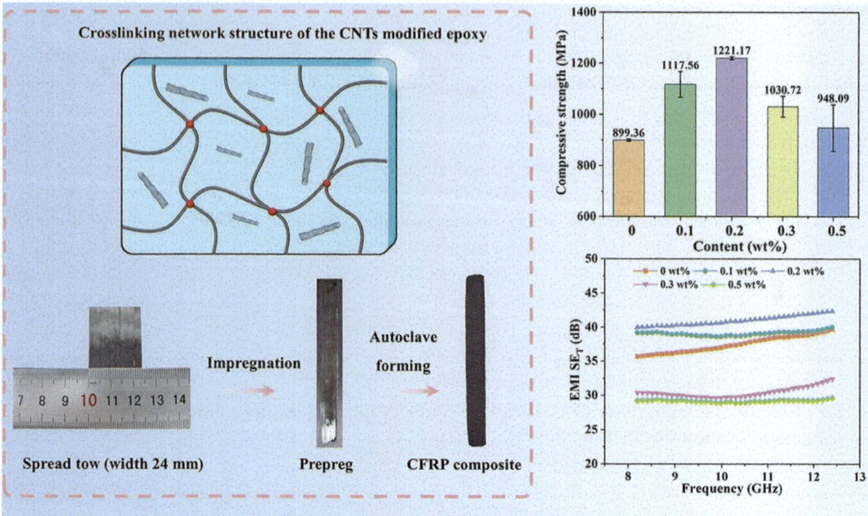

Fig. 2.3 Ultra-thin carbon fiber reinforced carbon nanotubes modified epoxy composites with superior mechanical and electrical properties for the aerospace field [35]

Moreover, the composite exhibited excellent thermal conductivity, electrical conductivity, and electromagnetic interference shielding effectiveness. These properties make it suitable for a wide range of applications, including aerospace and energy equipment. This work provides valuable insights into the development of ultra-thin CFRP composites that integrate structural and functional properties. By reinforcing the epoxy resin matrix with CNTs, significant improvements were achieved in mechanical properties, along with enhanced thermal conductivity, electrical conductivity, and electromagnetic interference shielding effectiveness. These findings open up potential applications in various industries, highlighting the versatility and potential of these advanced composites [35].

3.2.3 Mechanical Properties

Microwave curing technologies offer numerous advantages over traditional autoclave curing methods in the manufacturing of aerospace fiber-reinforced polymer composites. However, the direct penetration and heating of multidirectional carbon fiber reinforced polymer composites using microwaves pose a significant challenge, limiting their widespread industrial application. To address this issue, this paper proposes an indirect microwave curing method.

The indirect microwave curing method focuses on optimizing the microwave absorption performance of a heating medium, utilizing dielectric properties and reflection loss evaluated through transmission line theory. Based on these optimizations, a microwave-susceptible mold is meticulously designed and manufactured.

Subsequently, successful curing of multidirectional carbon fiber/epoxy composites is achieved using indirect microwave heating. The curing process is monitored through infrared thermal imaging, and the final products are analyzed using differential scanning calorimetry. A comparative analysis reveals that the indirect microwave curing method reduces the curing cycle by 42.1% and energy consumption by 75.9% compared to traditional thermal curing methods as shown in Fig. 2.4 [36]. Furthermore, additional characterization experiments demonstrate that the mechanical properties of specimens cured using the indirect microwave method are slightly superior to those cured thermally. This work introduces an innovative solution to the challenges of microwave curing in multidirectional carbon fiber reinforced polymer composites. The proposed indirect microwave curing method effectively reduces curing time, energy consumption, and demonstrates comparable or improved mechanical properties compared to traditional thermal curing methods [36].

Cf/SiC (carbon fiber reinforced silicon carbide matrix) composites were fabricated using chemical vapor infiltration, and the impact of SiC content on the mechanical and electromagnetic properties of the materials was systematically investigated. The inclusion of a higher volume fraction of SiC matrix was found to facilitate load transfer to the carbon fiber. As the SiC content increased from 21.5 to 42.2 vol%, the total porosity decreased from 38.5 to 17.8 vol%, resulting in improved flexural strength (38 ± 4 to 375 ± 10 MPa) and fracture toughness (6.2 ± 0.7 to 21 ± 0.3 MPa m1/2) of the Cf/SiC composites as shown in Fig. 2.5 [37]. However, the increase in SiC content had a negative effect on the electromagnetic interference (EMI) shielding effectiveness of the materials. Over the frequency range of 8.2–12.4 GHz, the EMI shielding effectiveness decreased from 43 ± 1.4 to 31 ± 1.1 dB with increasing SiC

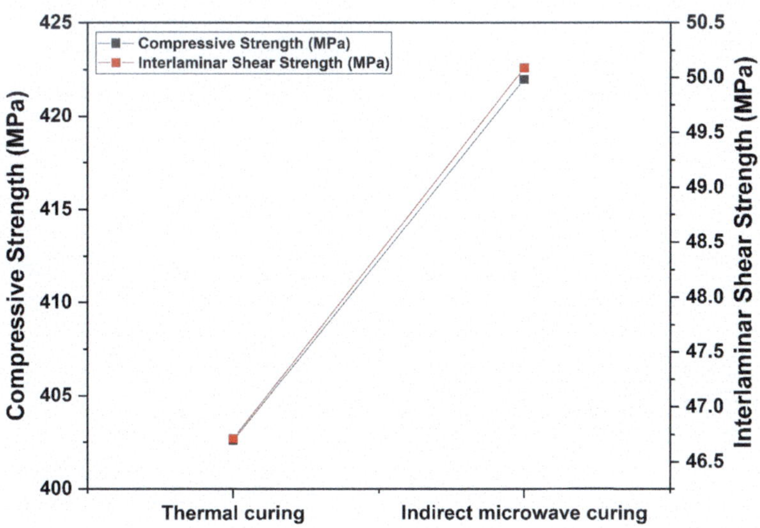

Fig. 2.4 Mechanical properties of the thermally cured and indirect microwave cured multidirectional carbon fiber/epoxy composite [36]

content. This decrease in shielding effectiveness was primarily attributed to a decline in absorption loss. The electrical conductivity of the Cf/SiC composites decreased with the increase in SiC content, resulting in a significant drop in conductive loss, which played a crucial role in reducing the absorption of electromagnetic waves. The systematic study revealed that higher SiC content in Cf/SiC composites led to reduced porosity and improved mechanical properties. However, it also resulted in a decrease in EMI shielding effectiveness due to the decline in absorption loss, which was primarily influenced by the diminishing conductive loss caused by the decrease in electrical conductivity of the composites [37].

Sample S1 exhibits a relatively clean and brittle fracture surface, with minimal fiber pull-out. This suggests that the majority of the fibers experienced shearing forces, indicating a lack of effective load transfer. Moreover, the fiber bundles in S1 appear disorganized, with the individual fibers scattered and separated from each other. In contrast, sample S2 displays improved compactness compared to S1, indicating a higher SiC content and a better ability of the matrix to transfer loads. The

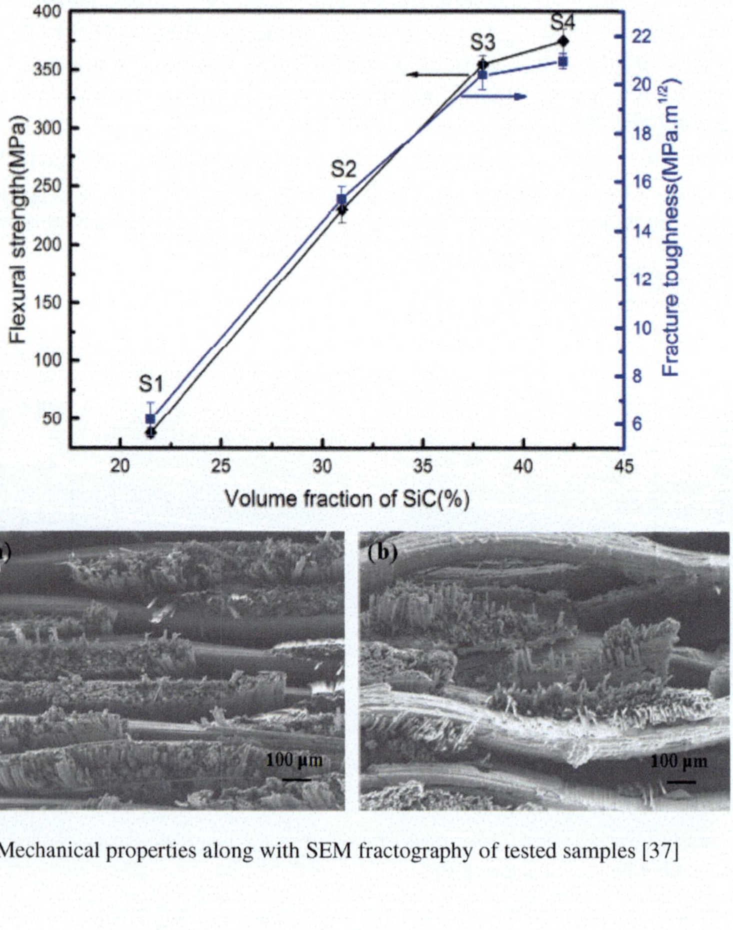

Fig. 2.5 Mechanical properties along with SEM fractography of tested samples [37]

fracture surface of S2 demonstrates enhanced fiber bundle integrity and compactness. It shows an uneven fracture surface with a significant number of fibers being pulled out, indicating the highest effective fiber fraction for load bearing among all the samples [37].

In another study [38], a series of short carbon fiber (SCF)/Al_2O_3 composites with MgO as a sintering additive were fabricated using a pressureless sintering process. The aim was to investigate the influence of SCF content on the mechanical, dielectric, and microwave absorbing properties of the composite. The results indicate that the addition of MgO enhances the density, hardness, and flexural strength of the alumina ceramic. However, as the Csf content increases, these mechanical properties of the (SCF)/Al_2O_3–MgO composite tend to decrease. In the frequency range of 8.2–12.4 GHz, both the real and imaginary parts of the complex permittivity increase with increasing SCF content. This behavior is attributed to the increasing electron polarization and associated polarization relaxation in the composite. When the SCF content is 0.3 wt%, the composite exhibits a reflection loss of less than −10 dB and a minimum value of −27 dB with a coating thickness of 1.4 mm. These results suggest that the (SCF)/Al_2O_3–MgO composite is a promising material for microwave absorption, offering favorable mechanical properties as shown in Fig. 2.6. [38].

An investigation on the impact of microwave curing processes, governed by different temperature profiles, on the mechanical properties of carbon fiber-reinforced composite materials were studied [39]. Microwave curing offers faster manufacturing

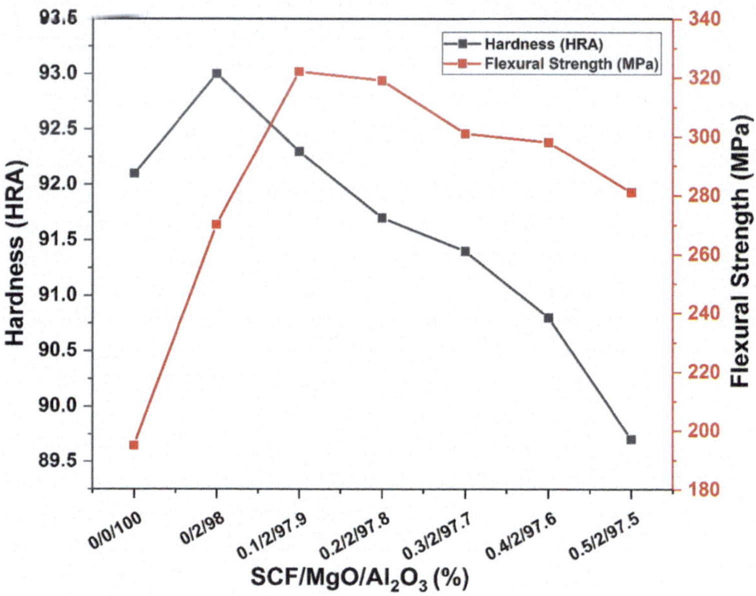

Fig. 2.6 Mechanical properties of (SCF)/Al_2O_3–MgO composite [38]

compared to conventional thermal curing, but the effects of shortened temperature profiles on the final cured parts remain unclear.

Experimental investigations were conducted, and the results revealed that microwave-cured composites exhibited similar curing kinetics to conventionally thermally cured ones. Furthermore, microwave-cured composites demonstrated superior interlaminar shear and flexural strengths compared to thermally cured composites, although slightly lower tensile and compressive strengths were observed. When considering the effects of heating rates within certain ranges, it was found that higher heating rates in the temperature profiles enhanced compressive and flexural strengths. However, tensile and interlaminar strengths were moderately compromised. On the other hand, reducing the holding time in the temperature profiles moderately decreased the mechanical performances of the composites, except for interlaminar shear strength [39].

Micrographs of the fracture surfaces following interlaminar shear tests provided insights into the fiber-matrix interfacial bonding. They demonstrated that microwave curing had a positive effect on enhancing the bonding, although this effect was slightly compromised when increasing the heating rates. Overall, these findings offer valuable information for understanding the trade-offs between reducing manufacturing time and preserving the mechanical properties of microwave-cured composites. The study highlights the potential of microwave curing in enhancing specific mechanical properties, while also providing insights into the effects of different temperature profiles on composite performance [39].

A study explores the incorporation of microwave heating into the resin transfer molding technique to enhance composite manufacturing. By utilizing microwave heating, a remarkable 50% reduction in cure cycle time was achieved compared to conventional resin transfer molding methods. The mechanical and physical properties of carbon fiber/epoxy composites produced through microwave curing were compared to those of conventionally cured composites [40].

Mechanical testing demonstrated similar flexural moduli and flexural strength values for both types of composites when the data were normalized to a common fiber volume fraction. However, an intriguing finding was that the interlaminar shear strength (ILSS) of the microwave-cured composites exhibited a notable 9% increase. This enhancement in ILSS was attributed to a reduction in resin viscosity during the initial stages of the curing process, leading to improved fiber wetting and reduced fiber pull-out. Scanning electron microscopy supported these observations, revealing better fiber wetting and less fiber pull-out in the microwave-cured composites [40].

Both microwave-cured and conventionally cured composites exhibited minimal void content, with levels below 2%. Dynamic mechanical thermal analysis indicated comparable glass transition temperatures between the two composite types. However, a notable 15 °C shift in the position of the β-transition peak was observed, suggesting a deviation in the cross-linking pathway followed during the curing process between the thermally and microwave-cured composites [40].

Another study introduces a microwave heating-assisted additive manufacturing method for continuous carbon fiber reinforced plastic (CCFRP) that enables faster and more energy-efficient production compared to traditional methods. The research investigates the permittivity of printed specimens with varying fiber volume fractions and establishes a micro-scale microwave radiation and heat transfer model to analyze the relationship between the filament's skin–core temperature difference, diameter, and printing speed. The main findings of the study are as follows [41]:

- Microwave heating-assisted additive manufacturing enables higher printing speeds and improved energy efficiency compared to conventional methods. This is attributed to the faster and more efficient heat transfer provided by microwave heating.
- The dielectric properties of printed specimens, influenced by the fiber volume fraction, were investigated. These properties are crucial in establishing the microwave radiation and heat transfer model.

- The simulated skin–core temperature difference of moving CCFRP filaments during conventional thermal and microwave heating provides insights into the relationship between temperature difference, filament diameter, and printing speed as shown in Fig. 2.7 [41]. This understanding aids in optimizing the printing process for achieving desired temperature profiles.
- The study compares the non-isothermal crystallization behavior and mechanical strengths of thermal and microwave-printed specimens. It reveals variations in these properties due to the different heating mechanisms employed, and provides analysis on the reasons behind the observed differences.
- The microwave heating-assisted additive manufacturing method for CCFRP offers enhanced production speed and energy efficiency. The investigation of dielectric properties, along with the developed microwave radiation and heat transfer model, aids in understanding the relationship between temperature profiles and printing parameters. Furthermore, the study analyzes the non-isothermal crystallization behavior and mechanical strengths of printed specimens, providing valuable insights into the impact of different heating methods on the final product [41].

4 Challenges and Future Perspectives

In order to fully harness the potential of microwave heating in polymer composites manufacturing, several challenges and limitations must be addressed. One key aspect is the optimization of processing parameters. The selection of the appropriate frequency, power level, and exposure time is critical to achieve the desired properties and avoid issues such as overheating or insufficient curing. This requires a thorough understanding of the material system and the interaction between microwaves and the composite components.

Another area of focus is the development of microwave-absorbing materials and structures. Enhancing the absorption efficiency and penetration depth of microwave

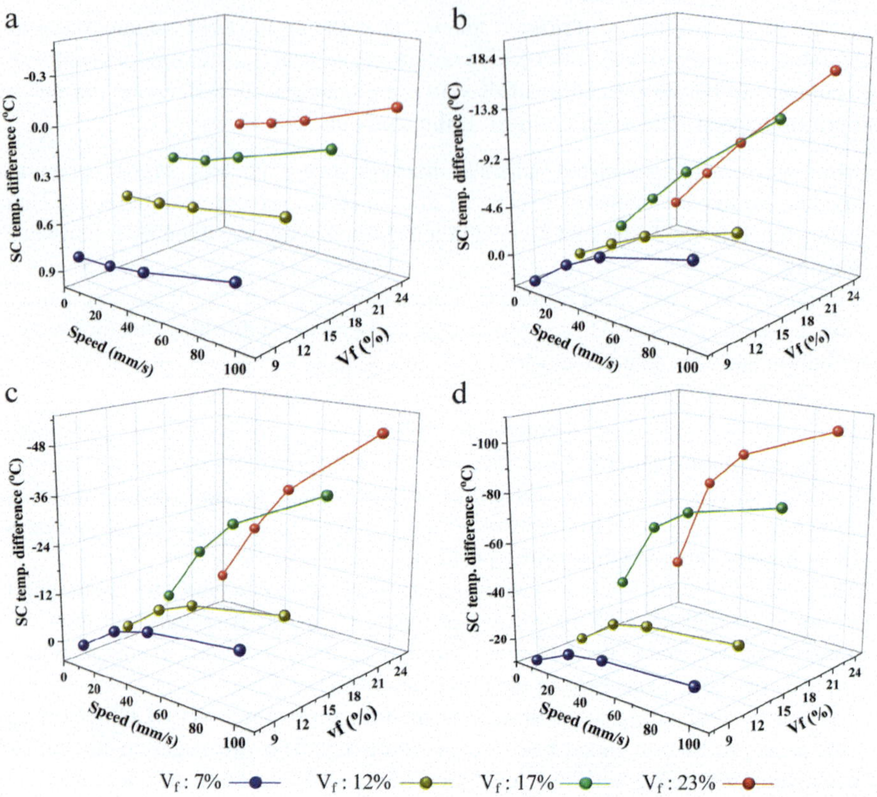

Fig. 2.7 SC temperature difference of four CCFRP filaments with different diameters printed by microwave, **a** 0.45 mm, **b** 1.75 mm, **c** 2.85 mm, **d** 5.48 mm [41]

energy within the composites can improve the overall heating efficiency and uniformity. This could involve the integration of microwave-absorbing additives or the design of composite structures that enhance the interaction with microwaves.

Safety and health considerations related to microwave radiation exposure and equipment design are paramount. Proper shielding, monitoring, and control measures should be implemented to ensure the safety of operators and minimize any potential risks associated with microwave radiation. Additionally, the design of microwave applicators and processing equipment should be optimized for efficient and uniform energy distribution.

Looking towards the future, the application of microwave heating in polymer composites manufacturing can be expanded to more complex and high-performance composites. This includes nanocomposites, where the incorporation of nanoparticles can further enhance the material properties. Additionally, exploring the potential of microwave heating in biocomposites and hybrid composites could open up new avenues for innovative and sustainable materials.

Integrating microwave heating with other advanced manufacturing techniques, such as 3D printing, holds promise for the fabrication of customized and functionalized composite structures. The combination of microwave heating with additive manufacturing processes can enable the production of intricate geometries with tailored properties, offering significant opportunities for advanced applications.

To facilitate process optimization and design, the development of predictive models and simulation tools is crucial. These tools can aid in understanding the complex interactions between microwaves and the composite materials, optimizing processing parameters, and predicting the final properties of the composite structures.

Addressing the challenges and limitations of microwave heating in polymer composites manufacturing, along with exploring new applications and technologies, will pave the way for the widespread adoption of this promising technique. Continued research and development in this field hold the potential to revolutionize the manufacturing of polymer composites and unlock new opportunities for advanced materials with tailored properties.

5 Summary

In summary, microwave heating is a promising and innovative manufacturing technique for polymer composites. It has several advantages over conventional heating methods, such as faster heating rate, selective heating, and improved properties of the composites. Microwave heating can be applied in various stages of the manufacturing process, such as mixing, molding, and curing, to achieve different goals and optimize the process. The effects of microwave heating on the properties of polymer composites depend on the processing parameters, material composition, and filler properties. Microwave heating can improve the thermal, mechanical, and electrical properties of the composites by promoting cross-linking and curing reactions, improving dispersion and alignment of the fillers, and achieving a more homogeneous and uniform structure.

However, there are still challenges and limitations that need to be addressed, such as the optimization of processing parameters, development of microwave-absorbing materials and structures, and safety and health issues related to microwave radiation exposure. In the future, the application of microwave heating in polymer composites manufacturing can be extended to more complex and high-performance composites, and integrated with other advanced manufacturing techniques to enable the fabrication of customized and functionalized composite structures. The development of predictive models and simulation tools can also help optimize the process and design the composite structures with desired properties.

Overall, the application of microwave heating in polymer composites manufacturing represents a promising and innovative approach to overcome the limitations of conventional heating methods and achieve improved properties and performance of the composites. This chapter aims to provide an overview of the current state-of-the-art of microwave heating in polymer composites manufacturing, and highlight the challenges and future perspectives of this technique.

References

1. Silvestre J, Silvestre N, De Brito J (2016). Mechanics of advanced materials and structures polymer nanocomposites for structural applications: recent trends and new perspectives polymer nanocomposites for structural applications: recent trends and new perspectives. https://doi.org/10.1080/15376494.2015.1068406
2. Mishra RR, Sharma AK (2016) Microwave-material interaction phenomena: heating mechanisms, challenges and opportunities in material processing. Compos Part A Appl Sci Manuf 81:78–97. https://doi.org/10.1016/j.compositesa.2015.10.035
3. Ghosh SK, Das NC (2023) Application of radiation curing on properties and performance of polymers and polymer composites BT—applications of high energy radiations: synthesis and processing of polymeric materials. In: Chowdhury SR, editor. Singapore: Springer Nature Singapore, pp 1–39. https://doi.org/10.1007/978-981-19-9048-9_1
4. Tewari R, Singh MK, Zafar S (2021) Utilization of forest and plastic wastes for composite manufacturing using microwave-assisted compression molding for low load applications. J Polym Res 28. https://doi.org/10.1007/s10965-021-02778-6
5. Singh MK, Zafar S (2020) Effect of layering sequence on mechanical properties of woven kenaf/jute fabric hybrid laminated microwave-processed composites. J Ind Text 00:1–22. https://doi.org/10.1177/1528083720911219
6. Singh MK, Verma N, Zafar S (2019) Optimization of process parameters of microwave processed PLLA/coir composites for enhanced mechanical behavior. J Phys Conf Ser 1240:012038. https://doi.org/10.1088/1742-6596/1240/1/012038
7. Singh MK, Verma N, Pundhir N, Zafar S, Pathak H (2020) Optimization of microwave power and reinforcement in microwave-cured coir/HDPE composites. Adv Mech Eng 159–71. https://doi.org/10.1038/scientificamerican06051909-354supp
8. Singh MK, Zafar S, Talha M (2019) Development and characterisation of poly-L-lactide-based foams fabricated through microwave-assisted compression moulding. J Cell Plast 55:523–541. https://doi.org/10.1177/0021955X19850728
9. Arora G, Pathak H (2020) Experimental and numerical approach to study mechanical and fracture properties of high-density polyethylene carbon nanotubes composite. Mater Today Commun 22:100829. https://doi.org/10.1016/j.mtcomm.2019.100829
10. Arora G, Pathak H (2021) Nanoindentation characterization of polymer nanocomposites for elastic and viscoelastic properties: Experimental and mathematical approach. Compos Part C Open Access 4:100103. https://doi.org/10.1016/j.jcomc.2020.100103
11. Arora G, Pathak H, Zafar S (2019) Fabrication and characterization of microwave cured high-density polyethylene/carbon nanotube and polypropylene/carbon nanotube composites. J Compos Mater 53:2091–2104. https://doi.org/10.1177/0021998318822705
12. Pastarnokienė L, Jonikaitė-Švėgždienė J, Lapinskaitė N, Kulbokaitė R, Bočkuvienė A, Kochanė T et al (2023) The effect of reactive diluents on curing of epoxy resins and properties of the cured epoxy coatings. J Coatings Technol Res. https://doi.org/10.1007/s11998-022-00737-4
13. Markandan K, Lai CQ (2023) Fabrication, properties and applications of polymer composites additively manufactured with filler alignment control: a review. Compos Part B Eng 256:110661. https://doi.org/10.1016/j.compositesb.2023.110661
14. Mehta A, Vasudev H, Jeyaprakash N (2023) Role of sustainable manufacturing approach: microwave processing of materials. Int J Interact Des Manuf. https://doi.org/10.1007/s12008-023-01318-4

15. Zhang L, Li Y, Zhou J (2018) Anisotropic dielectric properties of carbon fiber reinforced polymer composites during microwave curing. Appl Compos Mater 25:1339–1356. https://doi.org/10.1007/s10443-017-9669-6
16. Naik TP, Singh I, Sharma AK (2022) Processing of polymer matrix composites using microwave energy: a review. Compos Part A Appl Sci Manuf 156:106870. https://doi.org/10.1016/j.compositesa.2022.106870
17. Singh MK, Zafar S (2021) Wettability , absorption and degradation behavior of microwave-assisted compression molded kenaf/HDPE composite tank under various environments 185
18. Singh MK, Zafar S (2020) Abrasive wear mechanism of microwave-assisted compression molded kenaf/HDPE composite. J Tribol 142:1–11. https://doi.org/10.1115/1.4046858
19. Singh MK, Zafar S (2019) Development and mechanical characterization of microwave-cured thermoplastic based natural fibre reinforced composites. J Thermoplast Compos Mater 32:1427–1442. https://doi.org/10.1177/0892705718799832
20. Manohar G, Maity SR, Pandey KM (2022) Microstructural and mechanical properties of microwave sintered AA7075/graphite/SiC hybrid composite fabricated by powder metallurgy techniques. Silicon 14:5179–5189. https://doi.org/10.1007/s12633-021-01299-7
21. Lunetto V, Galati M, Settineri L, Iuliano L (2023) Sustainability in the manufacturing of composite materials: a literature review and directions for future research. J Manuf Process 85:858–74. https://doi.org/10.1016/j.jmapro.2022.12.020
22. Zabihi O, Ahmadi M, Liu C, Mahmoodi R, Li Q, Naebe M (2020) Development of a low cost and green microwave assisted approach towards the circular carbon fibre composites. Compos Part B Eng 184:107750. https://doi.org/10.1016/j.compositesb.2020.107750
23. Singh I, Bajpai PK, Malik D, Madaan J, Bhatnagar N (2012) Microwave joining of natural fiber reinforced green composites. Adv Mater Res 410:102–105. https://doi.org/10.4028/www.scientific.net/AMR.410.102
24. Li N, Li Y, Jelonnek J, Link G, Gao J (2017) A new process control method for microwave curing of carbon fibre reinforced composites in aerospace applications. Compos Part B Eng 122:61–70. https://doi.org/10.1016/j.compositesb.2017.04.009
25. Kim M, Kim H, Lee W (2015) Repair of aircraft structures using composite patches bonded through induction heating. Adv Compos Mater 24:307–323. https://doi.org/10.1080/09243046.2014.899553
26. Bélanger JMR, Paré JRJ, Poon O, Fairbridge C, Ng S, Mutyala S et al (2007) Remarks on various applications of microwave energy. J Microw Power Electromagn Energy 42:24–44. https://doi.org/10.1080/08327823.2007.11688597
27. Arora G, Pathak H (2022) Fracture and elastoplastic behavior of polymer-carbon nanotube composites under thermomechanical environment: an integrated dual-scale modeling and experimental study. J Mater Eng Perform 31:7120–7137. https://doi.org/10.1007/s11665-022-06743-2
28. Li Y, Li N, Zhou J, Cheng Q (2019) Microwave curing of multidirectional carbon fiber reinforced polymer composites. Compos Struct 212:83–93. https://doi.org/10.1016/J.COMPSTRUCT.2019.01.027
29. Li Y, Li N, Gao J (2014) Tooling design and microwave curing technologies for the manufacturing of fiber-reinforced polymer composites in aerospace applications. Int J Adv Manuf Technol 70:591–606. https://doi.org/10.1007/S00170-013-5268-3/METRICS
30. Feher LE, Thumm MK (2004) Microwave innovation for industrial composite fabrication - The HEPHAISTOS technology. IEEE Trans Plasma Sci 32:73–79. https://doi.org/10.1109/TPS.2004.823983
31. Zhang L, Liu W, Jiang H, Zhang X, Shang Y, Jiang C et al (2023) Upcycling of carbon fiber-reinforced polymer composites. Compos Sci Technol 231:109824. https://doi.org/10.1016/J.COMPSCITECH.2022.109824
32. Chen H, Li T, Wang Z, Ye R, Li Q (2020) Effect of dielectric properties on heat transfer characteristics of rubber materials via microwave heating. Int J Therm Sci 148:106162. https://doi.org/10.1016/j.ijthermalsci.2019.106162

33. Maenz S, Mühlstädt M, Jandt KD, Bossert J (2014) Mechanical properties of microwave cured glass fibre epoxy composites prepared by resin transfer moulding. J Compos Mater 49:2839–2847. https://doi.org/10.1177/0021998314557295
34. Bal S (2010) Experimental study of mechanical and electrical properties of carbon nanofiber/epoxy composites. Mater Des 31:2406–13. https://doi.org/10.1016/j.matdes.2009.11.058
35. Zhang C, Ling Y, Zhang X, Liang M, Zou H (2022) Ultra-thin carbon fiber reinforced carbon nanotubes modified epoxy composites with superior mechanical and electrical properties for the aerospace field. Compos Part A Appl Sci Manuf 163:107197. https://doi.org/10.1016/j.compositesa.2022.107197
36. Li Y, Cheng L, Zhou J (2018) Curing multidirectional carbon fiber reinforced polymer composites with indirect microwave heating. Int J Adv Manuf Technol 97:1137–1147. https://doi.org/10.1007/s00170-018-1974-1
37. Chen L, Yin X, Fan X, Chen M, Ma X, Cheng L, et al. Mechanical and electromagnetic shielding properties of carbon fiber reinforced silicon carbide matrix composites. Carbon NY 95:10–9. https://doi.org/10.1016/j.carbon.2015.08.011
38. Huang S, Zhou W, Luo F, Wei P, Zhu D (2014) Mechanical and dielectric properties of short carbon fiber reinforced Al_2O_3 composites with MgO additive. Ceram Int 40:2785–91. https://doi.org/10.1016/j.ceramint.2013.10.038
39. Hang X, Li Y, Hao X, Li N, Wen Y (2015) Effects of temperature profiles of microwave curing processes on mechanical properties of carbon fibre–reinforced composites. Proc Inst Mech Eng Part B J Eng Manuf 231:1332–1340. https://doi.org/10.1177/0954405415596142
40. Papargyris DA, Day RJ, Nesbitt A, Bakavos D. Comparison of the mechanical and physical properties of a carbon fibre epoxy composite manufactured by resin transfer moulding using conventional and microwave heating. Compos Sci Technol 68:1854–61. https://doi.org/10.1016/j.compscitech.2008.01.010
41. Li N, Link G, Jelonnek J, Morais MVC, Henning F (2021) Microwave additive manufacturing of continuous carbon fibers reinforced thermoplastic composites: characterization, analysis, and properties. Addit Manuf 44:102035. https://doi.org/10.1016/j.addma.2021.102035

Chronological Development in Microwave Heating Technology: Design and Applications

Neeraj Kumar Bhoi, Harpreet Singh, and Pramod K. Jain

Abstract The present chapter will concentrate on the many development methods that have been created for the processing of materials ever since the introduction of microwave radiation. This will serve as a reference for the advanced manufacturing technologies for the fabrication of immaculate net shape components and the qualities connected with those products for various applications. The development of sustainable product generation, as well as the influence on the environmental elements of a range of production methods using microwave heating technology has been discussed.

Keywords Microwave sintering · Hybrid hearing · Mechanical behavior · Microstructure characterizations · Sustainable processing · Simulation study

1 Introduction

Heating is a crucial manufacturing step that entails bringing the consolidated powder or bulk metallic material to the melting point. Traditional heating reduces thermal energy from the outside surface of the material to the inner surface by conduction, convection, and radiation. Each material reacts with microwave radiation at a distinct skin depth. The primary problem with conventional heating approaches

N. K. Bhoi (✉)
School of Automation (Mechatronics Engineering), Banasthali Vidyapith, Banasthali, Newai, Rajasthan 304022, India
e-mail: neerajbitd@gmail.com

N. K. Bhoi · H. Singh
Department of Mechanical Engineering, Dr. B. R. Ambedkar National Institute of Technology Jalandhar, Jalandhar 144011, Punjab, India

P. K. Jain
Department of Mechanical and Industrial Engineering, Indian Institute of Technology, Roorkee, UttaraKhand, India

M. K. Singh et al. (eds.), *Composite Materials Processing Using Microwave Heating Technology*, Composites Science and Technology,
https://doi.org/10.1007/978-981-97-2772-8_3

is the occurrence of thermal shock, interlayer diffusion, and thermal expansion coefficient mismatch, all of which impede the desired functional performance in the workplace. Microwave processing of the material utilizes electromagnetic radiation to heat the material. The target material's molecules absorb the microwave energy, causing them to vibrate and oscillate in a configuration that is oriented on the mean location of the molecule [28, 39]. The production of heat is a direct consequence of the vibrations of the molecule. In this manner the heating of the material takes place with the help of microwave energy. The material's core generates heat, which is then transferred to its surface. Therefore, microwave energy is very quick, requires little energy to create, and is simple to use for the purposes of material processing. Unique benefits and process capabilities of microwave synthesis enhance material performance by decreasing the quantity of residual and unwanted product produced during synthesis [7, 11]. The new advent technique solves a number of problems in the present microwave synthesis technology and is anticipated to be suitable for batch production of custom-built vehicles and military components. Material processing technologies that reduced the complete product cycle time were developed after World War II in response to the industrial sector's desire for comparatively quick processing procedures with higher product quality. Since its invention in 1948, the microwave has found widespread application in the kitchen, especially for heating food and liquids. In 1960, manufacture of polymers using microwave radiation began because of the precision and speed with which the thick polymeric material could be heated. The first successful experimentation with microwave sintering was done by Roy et al. for compacted metallic particles composed of Fe–Ni and Fe-Cu by using a 2.45 GHz microwave field [34]. Microwave sintering produces a product with greater mechanical properties than conventional sintering. Following this, a number of studies reported using microwave processing methods to treat a variety of polymeric, metallic, and nonmetallic components, stating several advantages and capabilities over conventional processing procedures [11, 14].

This chapter explores the interaction of microwave radiation with various metallic and composite materials during sintering. With microwave radiation, the heating mechanism, process capabilities, and various interaction behaviors of metallic materials are described. The various progress in microwave heating technology in context with the expanding demand of the manufacturing industries has been examined and comprehensive frameworks have been presented. This will help the researcher identify potential parameters and factor responsible for designing and working with microwave energy for end use. Microwave sintering progress constraints and potential prospects are studied.

2 Interactions Between Microwaves and Materials

Microwave interaction with the material depends on various factors such as magnetic properties, electrical properties, penetration dept and thermal conductivity. The penetration depth of the material is defined as surface distance from which the current density fall to 0.3679 of its's surface value. The empirical relationship for penetration depth of the material is given by Eq. 1.

$$d = \sqrt{\frac{\rho}{\pi f \mu_0 \mu'}} \quad (1)$$

where ρ refers to the resistivity of the material, f refers to the frequency of the alternating magnetic field, μ_0 refers to the permeability of free space ($4\pi \times 10^{-7}$ H/m) and μ' refers to the relative permeability of the material (which is assumed to be equal to 1 for most metals except for magnetic materials where the permeability is higher). The penetration depth of the material is in direct relation with frequency, permeability, and resistivity of the material[14]. A schematic of materials' interactions with microwaves, potential changes as a function of exposure duration, temperature, phases of microwave hybrid heating, and the resulting macrostructure of the compacted material after sintering is given in Fig. 3.1.

The interaction behavior of the microwave with different material are classified in following different categories:

(a) **Opaque material**: The major amount of microwave energy is reflected from the surface of the material. Bulk metallic material such as Aluminum, titanium etc. falls in this category.
(b) **Transparent material**: In this category of material all the incident microwave energy is passed through the material without interacting with the molecules. Teflon, quart is the typical example in this category.

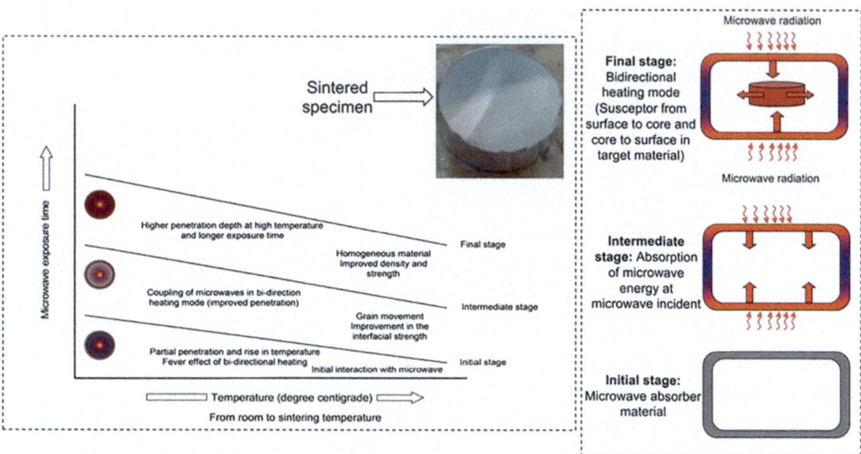

Fig. 3.1 Interaction of material under microwave and possible evolution with respect to exposure time and temperature and different stages in microwave hybrid heating and final macrostructure of the compacted material after sintering. [12] *Taken from Bhoi,* et al., *J Inorg Organomet Polym 32, 1319–1333 (2022). With kind permission from Springer Nature* (Bhoi et al.)

(c) **Absorbing material**: In this category, the material absorbs and converts microwave radiation to heat. These materials are known as susceptor materials. In this group are often silicon carbide, graphite, charcoal, boron carbide, boron nitride, carbon nanotubes, and water, among other substances.
(d) **Mixed absorbing materials**: This category contains the various metals and composites. This phenomenon is shown by polymer, ceramic, and metallic composite materials. Few composite materials exhibit partial absorption, in which a portion of the incoming microwave radiation is both transmitted and absorbed.

3 Assessment of Potential Heating Configurations

Depending on the kind of material that is exposed to microwave radiation, there are two distinct methods for heating the material. In radiate mode, the substance that absorbs microwave radiation may be heated directly. However, the material with a shallower penetration depth requires an additional heating arrangement to accelerate and provide microwave radiation coupling. In this category, the arrangement falls under susceptor material. In general, susceptor materials have a significant amount of skin depth and a significant amount of dielectric loss factor. Microwave absorption is the process that is involved in the amplified and quick heating that occurs in susceptor material. In metals, the ability of the material to conduct electricity and ions is primarily responsible for microwave absorption. This is because the conductivity of the material enables electromagnetic radiation to penetrate the material's body. Because of this, a number of microscopic processes take occur, including molecular vibration, the collision of rapidly moving electrons, and magnetic spin, which ultimately leads to the heating of the susceptor material [19]. There will be no evidence of heating since the bulk metallic material is a microwave reflector even when it is at room temperature. The bulk metal is shielded by the susceptor material, and the heat is transferred to the metals through conduction (at room temperature). In general, charge, such as metals, does not absorb microwaves up to a specific temperature, which is referred to as the critical temperature. The microwave absorption process begins after the charge has been brought up to the critical temperature by the conduction of heat from the susceptor material. Because conventional (conduction) heating and microwave heating take occur concurrently now, the rate of heating and temperature reached may both be increased in a much shorter amount of time. Because of this, the susceptor was a very crucial player in the process of making the charge absorb microwaves by bringing it to the critical temperature [8, 34]. The susceptor arrangement is grouped in the following subdomain.

3.1 Microwave Susceping Crucible

The most easy and effective method for increasing the heating efficiency of microwave processing is to make use of a crucible that has a propensity to absorb electromagnetic energy. It functions very well as a medium for the processing of bulk materials in preparation for sintering and melting. Graphite, silicon carbide, and boron carbide are examples of the class of material that has been used in the fabrication of microwave-receptive crucibles for the purpose of sintering and melting various types of metallic elements.

3.2 Powder and Bulk Susceptor Materials

Gupta and Wong used a 2.45 GHz frequency in a home microwave oven to sinter aluminum, magnesium, and lead-free Viromet 349 solder alloy utilizing SiC powder as a susceptor medium. Because of the improved coupling with microwave and susceptor aided heating to the metal compact, the resulting microstructure is almost completely dense and has very little porosity. The sintering method that was explored in their research was one in which microwaves supported the mechanism from the inside–outside direction, and SiC-susceptors assisted the process from the outside–inside direction. The procedure of heating the sintered material with radiant heat results in the material having superior mechanical characteristics [19].

Wong and Gupta devised a susceptor aided quick sintering technique for the manufacturing of Mg/Cu nanocomposite material while operating in the same configuration of heating mode. In the process of developing the composite, the sintering temperature was adjusted to 640 °C using a 900 W microwave oven operating at 2.45 GHz. To attain the sintering temperature during processing, a total of 25 min is required with no holding time in between [44].

Bhoi and his team have established the feasibility of using SiC bulk material in the process of developing an Al-ZnO nanorod composite material. In their research, it took a total of 11 min to reach the required sintering temperature of 630 °C in household microwave oven. The holding time during the sintering temperature was not maintained, and the whole process was carried out in a normal working environment. Further, with the similar setup, the same group has developed Al-Y_2O_3 composite material with superior material performance compared to conventionally processed material [3, 5, 8, 12]. A comparative assessment of the different susceptor material utilized in the different work has been listed in Table 3.1.

Table 3.1 Different material manufacturing operation and types of susceptor material utilized

Materials/processes	Susceptor material	Processing techniques and temperature	Processing time and mode	Reported by
Al-Si_3N_4 composites fabrication	Bulk Silicon Carbide (SiC)	Microwave heating at 0.9 kW and 2.45 GHz 550 °C followed hot extrusion	30 min at ambient environment	[27]
Al–Mg–Si–Cu fabrication		Microwave heating at 6 kW and 2.45 GHz frequency	1 h under vacuum level of 10^{-6} Torr	[31]
Al-$Cu_{50}Ti_{50}$ composite fabrication		Microwave heating at 0.9 kW and 2.45 GHz 550 °C	30 min at ambient environment	[33]
Al-ZnO composite fabrication		Microwave hybrid sintering at 0.9 kW and 2.45 GHz frequency 630 °C	11 min at ambient environment	[5]
Al-Y_2O_3 composite fabrication				[12]
Al-ZnO-Y_2O_3 composite fabrication				[8]
Mg-Cu nanocomposite fabrication	Powder SiC	Microwave hybrid sintering at 0.9 kW and 2.45 GHz frequency 640 °C	25 min at ambient environment	[44]
AA7075/B_4C/ZrC hybrid nano composite fabrication	Bulk SiC	Microwave hybrid sintering at 1.4 kW and 2.45 GHz frequency 500 °C	heating rate of 10 °C/min for 30 min	[26]
Mild steel coating with Al	Bulk SiC	Microwave hybrid heating at 0.9 kW	NR	[22]
SS-304 material with WC-12Co cladding	Charcoal	Microwave heating with 1.1 kW at 2.45 GHz	520 s	[45]
Copper casting	Bulk SiC	Microwave heating with 2.45 GHz and 1400 W	–	[29]
Inconel 625 joining	Charcoal	Microwave heating with 2.45 GHz and 900 W	21 min	[2]
Cast iron joining	Charcoal	Microwave heating with 2.45 GHz and 900 W	6 min 40 s	[41]
UO_2 sintering	Bulk SiC and 3Y-TZP	Microwave heating with 2.45 GHz and 3 kW	–	[15]
Ni-SiC composite casting	Charcoal	Microwave heating with 2.45 GHz and 900 W	–	[40]

3.3 *Effect of Susceptor Material Shape and Size*

The size of the susceptor material has a substantial effect on the heating rate of the material. The needed shape of the material's susceptor depends on the kind of application. In recent research conducted by Patel et al., the thickness of the susceptor material was shown to be significantly impacted by microwave power, exposure time, and duration. In their work, the thickness of silicon carbide was subjected to changing microwave power and duration. Silicon carbide (SiC) with a thickness of 25 mm and a microwave power of 900 W was seen to heat more efficiently in a home oven, according to the research [32].

Based on a finding somewhat like this one, Croquesel et al. investigated the use of two susceptors, namely SiC and 3Y-TZP, for the sintering of uranium oxide-based nuclear fuels. They made use of microwave power that was 3 kW in capacity and 2.45 GHz in frequency level. During the sintering process, a more uniform electrical field generation was observed as a direct result of the 3Y-TZP susceptor material having a lower electrical conductivity and a greater permittivity. When it came to the temperature of the target zone during the sintering process, similar findings were made. During the process of heating UO_2 pellets with SiC and 3Y-TZP susceptors material, Fig. 3.2 displays the progression of the data that was obtained (temperature vs time, incident power vs time, absorbed power vs time, and reflected power vs time). When compared to a SiC susceptor, the temperature absorption ability of the 3Y-TZP material is much quicker under comparable operational settings [15].

4 Novel Manufacturing Processes Using Microwave Energy

With the goal of achieving zero defects in manufacturing, this section focuses on the most important process parameters that determine the ultimate output of the microwave-processed material.

4.1 *Microwave and 3D Printing*

The ever-increasing need for intricate form production as well as quick product development, the technology of 3D printing has emerged as a game-changing instrument for the field of industrial practice. Metal printing operations are difficult to carry out because they need complex equipment and a specific atmosphere. This makes the actualization of products more difficult. Powder bed fusion, selective laser melting operation and directed energy deposition are the common method for 3D printing of the metallic material. Recently, induction conduction based metallic printing came into existence with joule heating concept for metal deposition [36].

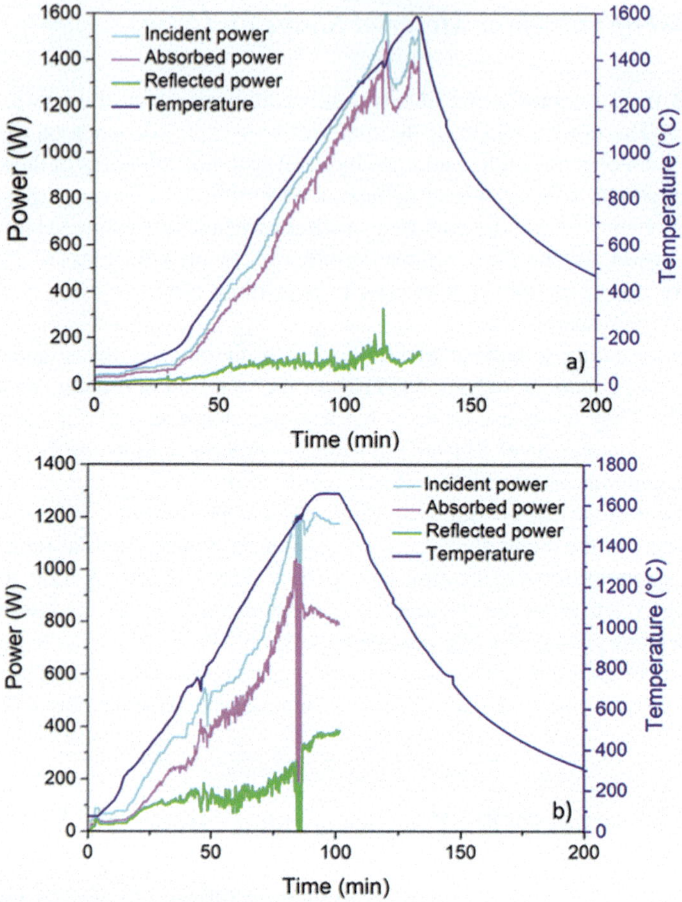

Fig. 3.2 Temperature and incident, absorbed and reflected powers versus time for the pellets sintered with a) SiC and b) 3Y-TZP susceptors. *Adapted from Croquesel* et al., *Materials & Design, Volume 204, 2021, 109,638, ISSN 0264–1275, Open access article under a Creative Commons license requires no permission* [15]

For the layer-by-layer deposition of the metallic powder material, Shelef and Jerby demonstrated the advantages of the microwave energy's volumetric heating as well as its selective heating. This was attained using microwave energy. For the deposition of the bronze base powder material, they employ a procedure known as deliberate localized microwave heating (LMH). The LMH was designed in an atmosphere including a shielding gas composed of nitrogen, and the incident average microwave power was 180 W. The 3D printing process for creating metallic material utilizing LMH is shown here by Fig. 3.3 as a schematic example. The use of uncomplicated magnetron-based technology for the additive manufacture of metallic stuff in three dimensions opens a whole new world of possibilities to produce complex structures across large areas [35].

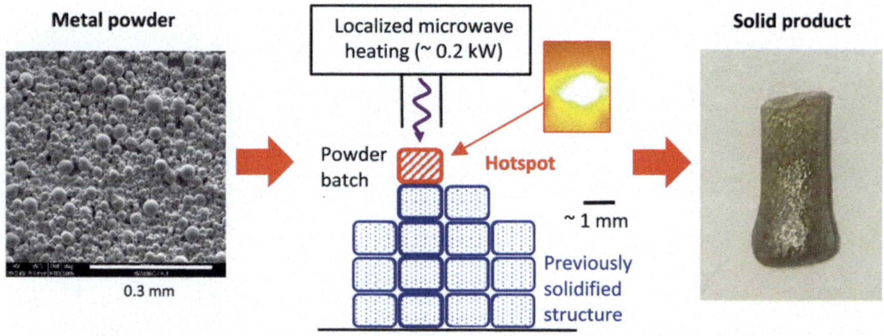

Fig. 3.3 Graphical representation of the Incremental solidification of metal powders by transistor-based microwave applicator. *Taken from Shelef and Jerby, Materials & Design, Volume 185, 2020, 108,234. Require no permission Under a Creative Commons* license. [35]

4.2 Microwave Sintering for Material Development

Sintering involves heating the green, compacted component to between 75 and 90% of the bulk material's melting temperature under varying circumstances (i.e., atmosphere, argon, helium, etc.). Microwave sintering (MWS) is a volumetric heating method that facilitates the formation of a crystalline microstructure and enhanced material properties. During MWS, the green compressed part is placed into the microwave chamber. A detailed schematic of powder metallurgy and microwave sintering has been given in Fig. 3.4. To prevent direct microwave radiation, contact between the microwave cavity and the compacted component, a transparent material such as alumina, fiber-frax, or mullite tube is introduced between the microwave cavity and the component. Microwave sintering's results are affected by several factors, including frequency of processing, interaction between microwaves and materials, and penetration depth. When using microwave heating, the core of the material will be hotter than its surface because of the interaction of microwave radiation with the core molecule. This leads in the following benefits over conventional heating methods:

1. Minimal processing time.
2. Improved diffusion.
3. Deeper penetration.
4. Reduced average grain size.
5. Increases the density factor.
6. Higher mechanical and tribological qualities.
7. Quicker heating rate.

Bhoi et al. used a microwave hybrid sintering technique to generate Al-based metal matrix composite materials. For processing with powder metallurgy assisted method, they use various reinforcing elements, such as ZnO nanorods and Y_2O_3 nanoparticles. With the microwave hybrid sintering technology, processing time

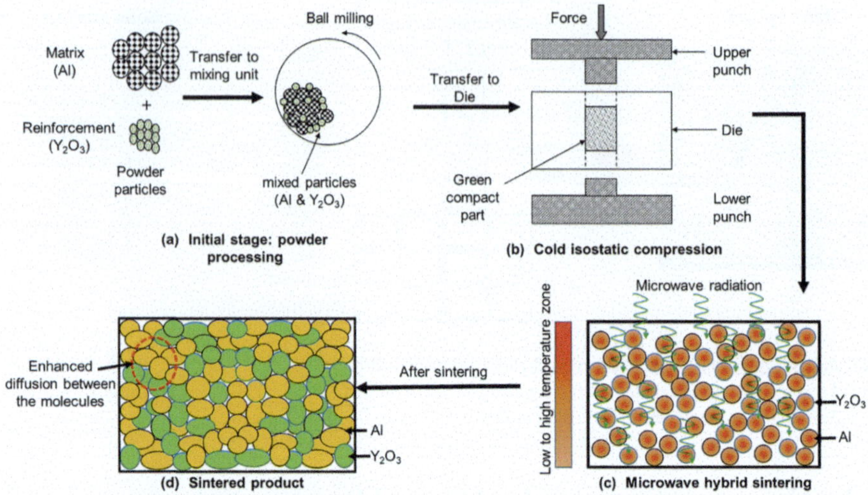

Fig. 3.4 Schematic illustration of steps involved in powder metallurgy process (a) initial mixing stage of matrix and reinforcement material, (b) cold isostatic compression of mixed powder material using die and punch system, (c) microwave hybrid sintering of compacted part and (d) final sintered product exhibit diffusion between the matrix and reinforcement material. [12] *Taken from Bhoi* et al., *J Inorg Organomet Polym 32, 1319–1333 (2022). With kind permission from Springer Nature* (Bhoi et al.)

(~95%) and energy use (~85%) are lowered in compared to conventional heating methods. In addition, the material integrity and final product quality are unaffected by microwave hybrid sintering under normal operating conditions. The material performance attained by microwave processing is superior to standard powder metallurgy and casting processes(Bhoi et al. 2021b). On a similar investigation, Kansal et al. used microwave hybrid sintering in conjunction with rapid manufacturing technologies to build a porous zinc scaffold [20].

The influence of sintering mode and temperature on the sintering of Al-6711 was studied under conditions of high vacuum (10×10^{-6} torr) and temperatures of 570, 580, 610, and 630 °C by [31]. Observations indicate that with higher vacuum sintering, mechanical characteristics increase considerably, with 19% more ductility. T6 age hardening increases the sample's ultimate and yield strengths from 115 and 100 MPa to 271 and 256 MPa, respectively. With an increase in temperature from 590 to 630 °C, the porosity of a traditional sintered sample with fine particle size is drastically decreased. Due to the quick heating and shorter processing time for diffusion of alloying elements, inhomogeneous grain with a greater proportion and coarser grain was found in microwave-sintered samples [31].

Al-TiC metal matrix composites created using conventional, spark plasma sintering (SPS), and microwave sintering techniques by Ghasali et al. in the year 2017. Al and TiC powder and 15 wt% TiC were blended with aluminum in ethanol, dried at 70 °C, and uniaxial pressed at 250 MPa. The conventional sintering temperature is 700 °C at a rate of 10 °C per minute for one hour. The microwave sintering temperature was 600 °C with an average heating rate of 30 °C/min. SPS procedure at 400 °C with a starting pressure of 10 MPa to 40 MPa. Total sintering times for SPS, microwave, and traditional heating approaches are around 10, 21, and 120 min, respectively. Compared to microwave and traditional sintering, SPS results in the greatest amount of shrinkage, and after 9 min of sintering, no significant changes occur. Microstructural study of the surface reveals that a mismatch in the thermal expansion coefficients of Al and TiC particles produces more particle separation at higher temperatures during both conventional and microwave sintering. An interfacial layer of TiO_2 forms around TiC particles due to longer sintering times and higher temperatures in conventional methods. Improvements in micro-hardness, bending strength, and density were all achieved using SPS sintering [17].

Powder metallurgy and microwave sintering are used to create a composite out of Al and FeCoNi1.5CrCu in a study by Wang et al. The effect of microwave sintering time on the mechanical properties and microstructural behavior were studied. The microstructure behavior of the produced composite material as a function of sintering time is shown schematically in Fig. 3.5 below. When sintering composites, it has been shown that more time spent in the furnace results in stronger molecular bonds between the reinforcing phase and the matrix. The existence of porosity in the formed composite was seen in a similar fashion, but over a longer period. The particle size distribution in the resulting composite strongly depends on the sintering time. A minimal size was observed after 40 min of sintering, with the maximum size after 60 min of sintering. When sintered for 40 min, the composite material had similar properties across the board, including increased compressive strength, relative density, and micro-hardness value. Hardness and elastic modulus values of 3893.2 MPa and 100.82 GPa, respectively, were found to be the highest among samples prepared in 40 min. Interface wettability and interface compatibility of the HEA reinforcement and matrix are excellent. Consequently, the composites' strength and hardness are greatly enhanced [43].

4.3 Microwave Drilling

Drilling is the very basic machining operation in the production process. The conventional drilling operations for metallic, plastic and composites provide a quick and easy way, but with considerable burr, dust effusion, heat generation, and noise. Some sophisticated drilling methods, including water jet, abrasive waste jet, plasma jet, and laser jet, are used for difficult-to-machine materials and intricate product form machining [10]. The microwave-based cutting and drilling process was originally introduced to rock excavation and stone crushing. Microwave drilling is advocated due to the many benefits of microwave processing and the notion of selective and volumetric heating [36, 37].

The fundamental idea behind microwave drilling is to focus the radiation on a very tiny region that is far smaller than the wavelength that is being used. In this stage of the machining process, the target region is moved mechanically using an electrode or

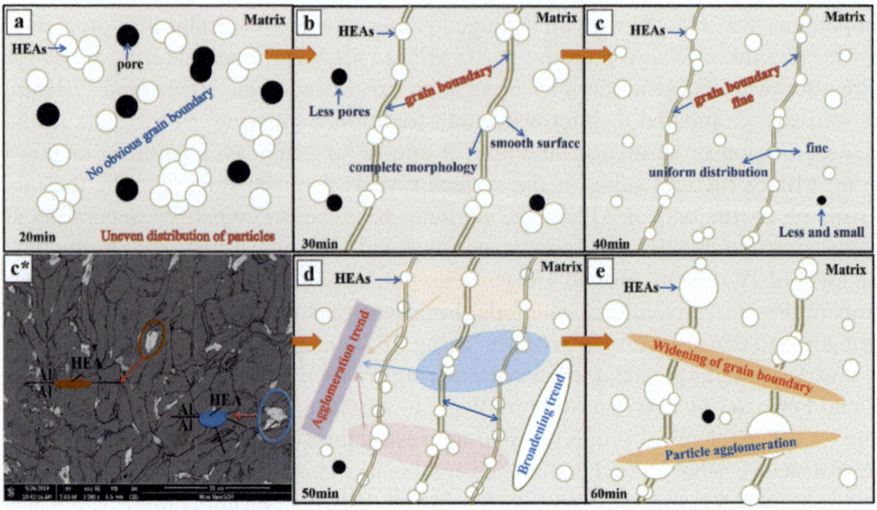

Fig. 3.5 Schematic representation of the mechanism of the changes in the reinforcement particles and grain boundaries with sintering time. *Taken from Wang* et al., *Materials Science and Engineering: A, Volume 801, 2021, 140,406, With kind permission from Elsevier.* [43]

a drill bit as the moving mechanical element. Melting and pressing are used in order to accomplish the task of drilling in the microwave. To prevent any local melting of the tool material that may occur during a microwave drilling operation, the use of an electrode or drill bit composed of tungsten is often recommended. Insulating material is placed all around the tool material to prevent heat from escaping into the surrounding environment.

Microwave drilling of cortical bone from chicken femora was achieved in vitro by Eshet et al., The primary objective of the drilling is to assess the feasibility of the microwave machining operation for orthopedic purposes. With a microwave power of 200 W and a frequency of 2.45 GHz, the drilling was completed. The constant drilling speed for penetration was ~1 mm/s. When comparing microwave drilling to traditional drilling, they found no discernible difference in material integrity. Microwave drilling produced a hole free of fragmentation and fracture [16]. In the subsequent year Kumar and Sharma came up with the idea to drill holes in borosilicate glass using a variety of dielectric media and microwave radiation. Stainless steel, graphite, and tungsten carbide were the three distinct types of concentrator media that were used in the testing setup that was carried out using a microwave power setting of 700 W. Coconut oil, transformer oil, and electrical discharge machining (EDM) oil were the three types of media used in the die-electric process. The choice of the medium with a low dielectric constant was made to produce a microwave medium that is transparent, as well as to confine the plasma and reduce the temperature differential surrounding the hole. Therefore, there is less formation of heat damage and cracks when drilling activities are being carried out. Because it has a high thermal stability, tungsten carbide delivers superior drilling performance when compared to the other

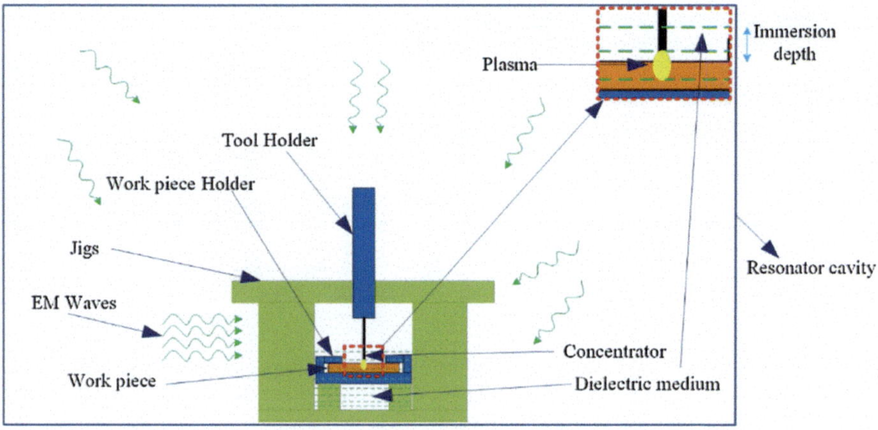

Fig. 3.6 Schematic diagram of the microwave drilling set up (inset: enlarged view of the machining zone). *Taken from Kumar and Sharma, Journal of Manufacturing Processes, Volume 33, 2018, Pages 184–193, With kind permission from Elsevier.* [23]

two concentrates produced by the preferred concentrator [23]. A typical arrangement of their drilling setup is given in Fig. 3.6.

4.4 Microwave Melting/Casting Operation

Casting is a method that is used in every industry that deals with material processing. The conventional approach of heating the material and casting it has several limitations, including a drawn-out process, a high level of energy consumption, and inconsistent heating of the material. Utilizing a more recent method known as microwave casting is one way to mitigate the drawbacks that are associated with using the conventional casting process. Irradiation of castings using microwaves is a key scientific discovery that is gaining attention in the industry [30]. When compared to conventional casting methods, microwave material casting offers several benefits, such as time and energy savings, volumetric heating, shorter processing time, enhanced microstructure characteristics, improved product performance, and improved mechanical and tribological behavior. Casting may be allocated down into two distinct categories: in-situ casting and ex-situ casting. A line diagram for the microwave assisted casting setup is given in Fig. 3.7.

Mishra et al. [48] examined the interaction of microwaves with charge, the involvement of the oxide layer during melting, and the characterization of in-situ Microwave cast copper. In-situ Microwave casting is accomplished via hybrid microwave heating at 2.45 GHz and 1400 W in a graphite mold. The oxide layer is essential for heating bulk copper during microwave hybrid heating. The grain structure is uniform, equiaxed, and compact (preheating of mold). The porosity of the in-situ castings is

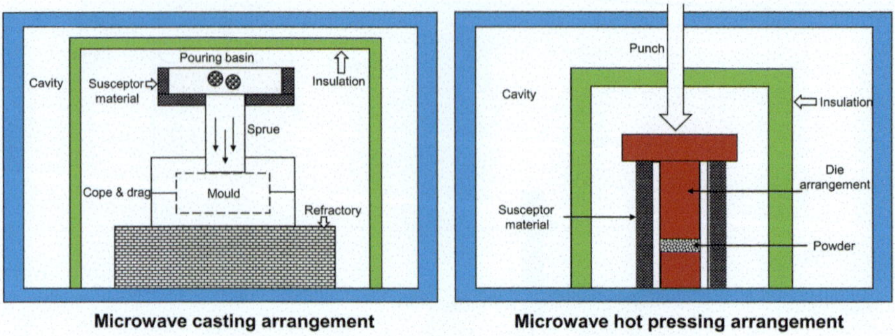

Fig. 3.7 Working schematic representation of the microwave casting and microwave hot pressing arrangements. [8] *Taken from Bhoi* et al., *Composite Materials, Elsevier, 2021, Pages 3–23, ISBN 9,780,128,205,129, with kind permission from Elsevier.* (Bhoi)

2–5%. The average microwave hardness of the cast product was found to be 93 $\pm$ 20 HV. During the solidification process, the qualities of a cast product may be modified as required, allowing for more control flexibility [48]. Subsequently, [39] examined the microstructures, hardness, and progress toward a cleaner manufacturing sector by processing in-situ microwave casted metal-ceramic composite (Ni-SiC) using microwave hybrid heating and solidifying. Microwave hybrid heating at 2.45 GHz with 900 W in a graphite mold yields an in-situ Microwave cast of Ni-SiC composite. The MHH enables conventional charge heating up to the critical temperature, beyond which it was discovered that the rate of heating increased (bi-directional heating), enabling microwave heating to interact directly with charge. The microstructure of the cast consists of evenly dispersed matrix (Ni) grains and reinforcement (SiC) grains throughout the matrix phase. The highest micro hardness achievable for a given microstructure is 1085 $\pm$ 210 HV for composites produced via microwave heating. Similarly, Singh et al. examined the effect of microwave heating and melting on the microstructures, hardness, and strength of Ni–Al powder composite castings. It has been observed that an increase in microwave heating results in an increase in intermetallic formation, which impacts the overall strength and hardness. Intermetallic compounds such as aluminide, nickel silicide, and chromium carbides were identified in nickel reinforced with 10% alumina. Due to the volumetric heating, alumina is evenly distributed with equiaxed grain growth. The micro hardness of 10% alumina-reinforced Ni was 2.4 times higher than that of pure Ni. There is a greater amount of plastic deformation in the matrix region and a lesser amount in the reinforcement region. For alumina-reinforced composites with 10% alumina, the wear rate is decreased by 72% [40].

4.5 *Microwave Curing and Hot-Compaction*

Curing is a highly important and time-consuming step for composite produced components. For instance, a composite panel for an airplane wing required a 24 h autoclave with a 5000 kW energy usage. The production process might be significantly enhanced if the composite curing could be decreased using cutting-edge heating technology. For the quick synthesis of composite material products, the microwave curing technique has proved a game-changer. The polymer-based composite material is partly absorbent of microwave radiation, making it an efficient curing procedure. In similar investigations, Li et al., used 16 separate microwave sources operating at 2.45 GHz 50 MHz distributed evenly across the working zone to further cure a composite panel made of carbon fiber. In terms of energy use and processing time, a comparative analysis revealed that microwave curing was more effective. In comparison to the conventional thermal curing procedure, there is a total reduction in energy usage of 76% and a time reduction of 42%. The compression and ILSS of the microwave-cured material are found to be 4.8 and 7.2% higher side, in contrast to the energy and time savings [24]. In light of this, Li et al., created a method for curing multidirectional carbo fibers with various arrangements for electrical field traversal with regard to ply orientations. Composite panels underwent the curing process using a microwave curing system (octagon chamber, 2.45 GHz, 20 kW power, 16 microwave sources, and waveguide slots antennas). Their observation suggests a vertical penetrating microwave method for effective curing can be used [25]. In research by Abedinzadeh et al., it was shown that microwave hot pressing may be used to create metal composites. They processed Al-Al_2O_3 composites utilizing hot pressing and microwave technology. Hydraulic equipment with a 2.45 GHz frequency and a 6-kW maximum power output at 30 MPa working pressure was employed for microwave hot pressing. When treating composite material conventionally, 400 °C and 300 MPa compaction pressure were used. In the composite form, material that has been microwave-processed shows improved load bearing capability with decreased porosity. The hot-pressing approach for composite materials that uses microwave assistance results in improved reinforcing diffusion [1]. An experimental arrangement for microwave assisted hot pressing has been shown in Fig. 3.7. Furthermore, Singh and Zafar make use of a distinctive fixture for the hot compaction procedure with microwave assistance for developing polymer-based composite materials. With a compaction pressure of 0.02 MPa and a temperature of 148 °C, a kenaf/HDPE composite was created in a 140 × 57 mm alumina mold. For composite materials made of polymers, the idea of simultaneously heating and compacting improves the responsiveness of the material and reduces the processing time needed. The setup for the hot compaction technique with microwave assistance that has been developed to fabricate composite materials is shown in Fig. 3.8. The temperature control during processing was done with the help non-contact infrared thermometer [38].

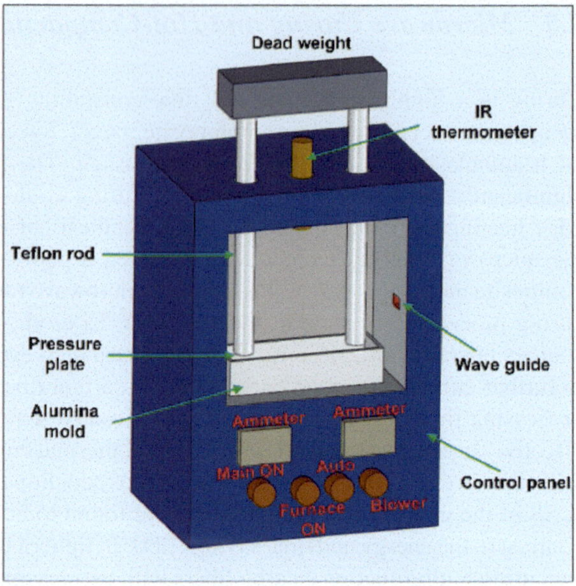

Fig. 3.8 Microwave-assisted compression molding (MACM) setup for polymer composite manufacturing. *Taken from Singh and Zafar, Polymer Degradation and Stability, Volume 185, 2021, 109,500, With kind permission from Elsevier* [38]

4.6 Microwave Cladding Operation

Depositing a metallic coating onto a substrate may be done using a few distinct methods. Laser cladding and numerous types of welding, including metal inert gas welding, submerged arc welding, and shielded metal arc welding, are examples of such technologies. All of these techniques have been investigated extensively; nevertheless, they each have substantial downsides, such as a higher cost badge and thermal distortion brought on by fast heating and cooling, which may potentially result in the production of fractures in the clads. Because of this, it is necessary to devise an innovative strategy that might potentially overcome these shortcomings [18]. The microwave cladding process is a relatively new metal processing technique that offers several advantages. Some of these advantages include a shorter response time, a better densification component, a homogeneous grain structure, improved mechanical and physical properties, an environmentally conscious process, minimal scrap created, cost-effective process, and, most importantly, volumetric heating. There is a possibility that porosity will diminish because of volumetric heating, as will the creation of fractures, as well as low levels of thermal distortion and residual stresses. A schematic representation of the microwave cladding arrangement is given in Fig. 3.9.

Cladding of WC10Co2Ni composition powder material on austenitic stainless steel was successfully established by Gupta et al. via microwave irradiation (SS-316). The cladding has a thickness of 2 mm and a porosity of less than 1%. By effective diffusion of components, the characterization of the clad sample demonstrates a strong metallurgical link between the clad and the substrate. The average micro-Vickers hardness value for 30 s and 50 g load is 1064.99 HV. The wear test

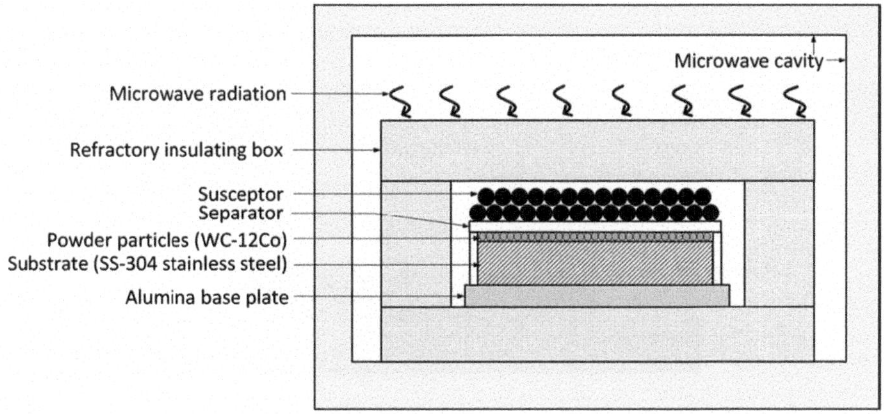

Fig. 3.9 Schematic of MHH for developing the WC-12Co microwave clads. *Taken from Zafar, and Sharma, Vacuum, Volume 131, 2016, Pages 213–222. With kind permission from Elsevier* [45]

is performed at three different sliding speeds (0.5 m/s, 1 m/s, and 1.5 m/s) over 2000 m exhibit a good resistance for the clad surface [18]. Functional graded (FG) coating is a novel microwave-energy-based coating technique. Variable compositions of coated powder materials have been layered one on top of the other in this method. The outcome of FG coating is superior to that of coating with a single layer In a work conducted by Kaushal et al., Ni with Cr_3C_2 is utilised to coat SS-306 strain-free steel. The coating composition of Cr_3C_2 varies from 0 to 30% (wt%) with Nickel powder, and over the base material, a total of four clad layers are created. Each cladding layer has a thickness of 0.5 mm. The production of Ni_3C, Ni_3Si, and Ni_3Cr_2 results in a 2.5-fold improvement in hardness (576 HV) over the basic material, indicating the engineering performance potential of the coated material [21]. In a similar manner, Zafar and Sharma deposited SS-304 material with WC-12Co cladding. The cladding processes were carried out using charcoal as the susceptor medium in a home microwave oven. With homogeneous surface properties, the clad layer adhered very well and had increased wear resistance. The production of complicated intermetallic during the cladding process aided in the improvement of the SS-304 material's performance [45]. More comprehensive reading can be found in open literature for microwave-based cladding and coating operations.

4.7 *Microwave Joining Operation*

Joining techniques are used in different industrial scenarios to combine several components into the required complex structure. For instance, if the geometry is too complex to be created by conventional manufacturing or machining techniques. Microwave joining is a state-of-the-art joining method that utilizes electromagnetic

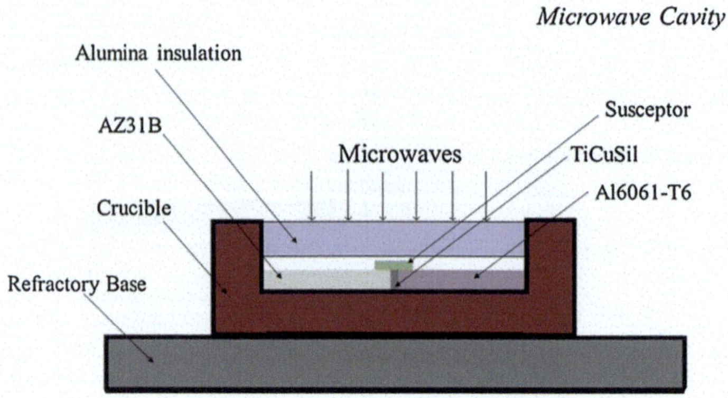

Fig. 3.10 Schematic diagram of setup for joining by microwave hybrid heating. *Adapted from Tamang and Aravindan, Journal of Manufacturing Processes, Volume 28, Part 1, 2017, Pages 94–100. With kind permission from Elsevier* [42]

radiation to join two dissimilar or similar metallic materials, often without the need of filler metal [2]. Joining can be classified in 3 different categories based upon the type of application of filler metal and application of heat and pressure. A representative illustration for the microwave joining process has been given in Fig. 3.10.

- **Autogenous joining:** In this procedure, the combining of two comparable types of metallic or plastic material is accomplished using heat or pressure. No additional filler metal is used in this procedure to join the two metallic components together.
- **Homogeneous joining:** Similar type of metal is joint with the help of homogeneous joining process. When performing the homogeneous welding process, one should choose a filler material type with a base material composition that is comparable to that of the filler material.
- **Heterogeneous joining:** Heterogeneous welding is used for joining metals that are not compatible. It is necessary to choose a filler material that is compatible with the dissimilar base metal when combining two metals that are different from one another. The filler material's compositional proportion must be identical to or comparable to that of the base metal.

In the early 1990's, the microwave joining procedure was introduced for ceramic materials. Later, as technology advanced, several researchers from all over the world used microwave joining techniques to investigate various comparable and dissimilar metals. In contrast to the conventional joining method, microwave joining uses a unique setup to combine two distinct metallic elements. This novel configuration includes insulating components, susceptor materials, and filler materials between the joining metal zones. In the illustration, a unique arrangement of this kind is shown in Figure. A schematic illustration of the different stages in the microwave joining process is given in Fig. 3.11. Badiger et al., discussed the joining of inconel-625 alloy through microwave heating by using inconel-625 powder as interface layer material in multimode applicator of frequency 2.45 GHz and at exposure power of

900 W. During joining the susceptor material used to transfer the heat is charcoal powder and very thin sheet of alumina is used as separator the separate the work piece and the susceptor material and then it is heated up to exposure time of 21 min. The metallurgical characterization is performed using a scanning electron microscope and x-ray diffraction, which demonstrates complete melting of inconel-625 powder particles and very good metallurgical bonding with the base metal. X-ray diffraction analysis (XRD) revealed the dominance of inconel-625 powder particles in the joint zone and the formation of intermetallic carbides during joining formation, resulting in increased hardness at the joint zone [2]. Singh et al. describe the joining of cast iron using microwave heating using nickel powder as the interface layer material in a multimode applicator with a frequency of 2.45 GHz and an exposure power of 900 W, with charcoal powder as the susceptor material utilized to transmit the heat. As a separator between the work piece and the susceptor material, a very thin sheet of graphite is utilized, and it is heated for an exposure duration of 6 min and 40 s. The scanning electron microscope and XRD are used to do the metallurgical characterization, which reveals the full melting of nickel powder particles and a very strong metallurgical connection with the base metal. The dominance of nickel powder particles in the joint zone and formation of intermetallic carbides during the formation of joining causes increased hardness at the joint zone. The ultimate tensile strength obtained are 176.5 MPa which is 90% strength of the Base metal [41].

5 Challenges and Opportunities

Microwave heating provides a diverse range of opportunities to materials and design professionals. Microwave material processing is anticipated to be used more often in the near future for sophisticated manufacturing operations that need efficiency and sustainability. Casting composite materials, developing tissue generation-based materials, performing painless drilling operations, printing metallic parts, and processing polymer-based materials are all examples of applications that might benefit greatly from using microwave-based operations. Even though microwave processing has some features that are exclusive to metal processing and that it cuts down on the amount of time and energy required for processing, the following issues still need to be reexamined for proper deployment and product development in modern industrial production.

- The curing of the polymer-based composite material is still in the stage of development for efficient heating since curing is a crucial operation for the homogeneity and density of the polymer-based composite material.
- The mathematical and simulation model development for processing ceramic and metallic composite material using microwave heating technology is still in its inception.

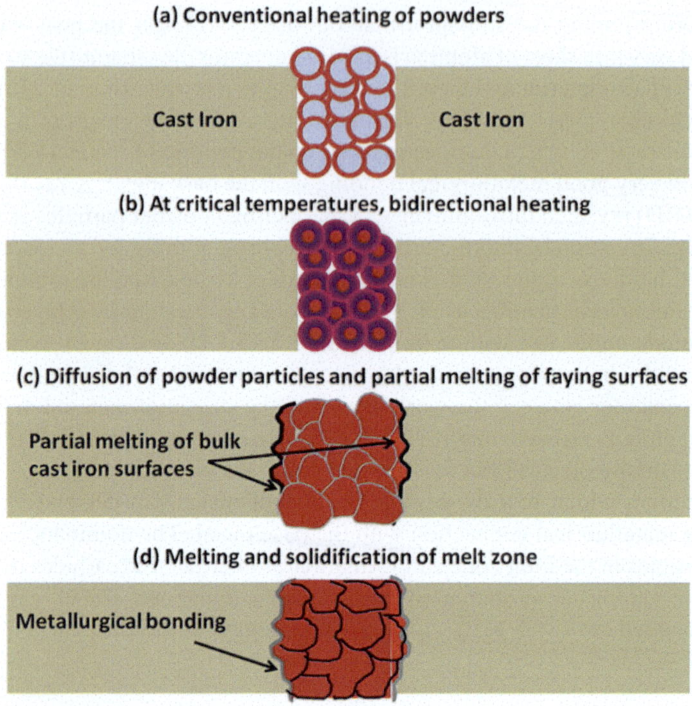

Fig. 3.11 a–d Mechanism of microwave joining of bulk materials through hybrid heating concept. *Taken from Singh* et al., *Engineering Science and Technology, an International Journal, Volume 22, Issue 2, 2019, Pages 569–577, With Kind permission from Elsevier* [41]

- Considering realistic models (3D printing of metallic materials) and real-world application, innovative tool development for efficient microwave energy consumption is still in its early stages.
- Developing an intensive strategy and model is necessary to design an innovative methodology for micro-machining advanced functionally graded composite materials using microwave heating technology.

6 Conclusions

In this chapter, several different applications of microwave heating technology have been suggested in terms of their respective potentials. Because it is harmless to the environment, cost-effective, saves time, and improves the quality of the materials being processed, microwave radiation is gaining popularity as a means of processing composite materials, biomaterials, and sophisticated machining procedures. This is due to the fact that it improves the quality of the materials being processed. Explorations in the core area of microwave heating technology, as well as optimization of process parameters for diverse areas of application, will increase the existing state of knowledge in the field and provide a sustainable solution to issues that are encountered by industrial operations.

References

1. Abedinzadeh R, Safavi SM, Karimzadeh F (2016) A study of pressureless microwave sintering, microwave-assisted hot press sintering and conventional hot pressing on properties of aluminium/alumina nanocomposite. J Mech Sci Technol 30:1967–1972. https://doi.org/10.1007/s12206-016-0402-4
2. Badiger RI, Narendranath S, Srinath MS (2015) Joining of Inconel-625 alloy through microwave hybrid heating and its characterization. J Manuf Process 18:117–123. https://doi.org/10.1016/j.jmapro.2015.02.002
3. Bhoi NK, Patel DK, Singh H et al (2022) Multi-physics simulation study of microwave hybrid sintering of aluminium and mechanical characteristics. Proc Instit Mech Eng E: J Process Mech Eng. https://doi.org/10.1177/09544089221074829
4. Bhoi NK, Singh H, Pratap S (2021a) Futuristic synthesis strategies for aluminum-based metal-matrix composites. In: Low I-M, Dong Y (eds) Composite Materials: Manufacturing, Properties and Applications, 1st edn, pp 3–23. Elsevier Ltd.
5. Bhoi NK, Singh H, Pratap S (2020) Synthesis and characterization of zinc oxide reinforced aluminum metal matrix composite produced by microwave sintering. J Compos Mater 54:3625–3636. https://doi.org/10.1177/0021998320918646
6. Bhoi NK, Singh H, Pratap S et al (2022) Zinc oxide nanorods effect in micro structural and mechanical characteristics of aluminium composite material. Mater Sci Eng B Solid State Mater Adv Technol 278:115641. https://doi.org/10.1016/j.mseb.2022.115641
7. Bhoi NK, Singh H, Pratap S (2019a) A study on microwave susceptor material for hybrid heating. J Phys Conf Ser 1240.https://doi.org/10.1088/1742-6596/1240/1/012097
8. Bhoi NK, Singh H, Pratap S et al (2021) Investigation on the combined effect of ZnO nanorods and Y_2O_3 nanoparticles on the microstructural and mechanical response of aluminium. Adv Compos Mater 31:289–310. https://doi.org/10.1080/09243046.2021.1993555
9. Bhoi, N.K., Singh, H., Pratap, S. (2022). Reciprocating Wear Behaviour of Al–SiC Composite Processed with MWS. In: Kumar, R., Chauhan, V.S., Talha, M., Pathak, H. (eds) Machines, Mechanism and Robotics. Lecture Notes in Mechanical Engineering. Springer, Singapore. https://doi.org/10.1007/978-981-16-0550-5_64
10. Bhoi, N.K., Singh, H., Pratap, S. (2020). Strategies for Controlling the Accuracy and Reliability of Abrasive Water Jet Machining. In: Gupta, V., Varde, P., Kankar, P., Joshi, N. (eds) Reliability and Risk Assessment in Engineering. Lecture Notes in Mechanical Engineering. Springer, Singapore. https://doi.org/10.1007/978-981-15-3746-2_13
11. Bhoi NK, Singh H, Pratap S, Jain PK (2019b) Microwave material processing: A clean, green, and sustainable approach. In: Kumar K, Zindani D, Davim P (eds) Sustainable Engineering Products and Manufacturing Technologies, pp 3–23. Academy Press Elsevier
12. Bhoi NK, Singh H, Pratap S, Jain PK (2022d) Aluminum Yttrium Oxide Metal Matrix Composite Synthesized by Microwave Hybrid Sintering: Processing, Microstructure and Mechanical Response. J Inorg Organomet Polym Mater. https://doi.org/10.1007/s10904-021-02195-8
13. Bhoi NK, Singh H, Saurabh P (2021c) Reciprocating wear behaviour of Al-SiC composite processed with MWS. In: Kumar R, Chauhan VS, Talha M, Pathak H (eds) Machines, Mechanism and Robotics. Lecture Notes in Mechanical Engineering. Springer, Singapore, pp 653–660

14. Chan S, Olevsky EA, Diego S et al (2021) Microwave sintering of complex shapes: from multiphysics simulation to improvements of process scalability. J Am Ceram Soc 104:3105–3118. https://doi.org/10.1111/jace.15892
15. Croquesel J, Meunier C, Petit C, et al. (2021) Design of an instrumented microwave multimode cavity for sintering of nuclear ceramics. Mater Des 204.https://doi.org/10.1016/j.matdes.2021.109638
16. Eshet Y, Mann RR, Anaton A et al (2006) Microwave drilling of bones. IEEE Trans Biomed Eng 53:1174–1182. https://doi.org/10.1109/TBME.2006.873562
17. Ghasali E, Fazili A, Alizadeh M, et al. (2017) Evaluation of microstructure and mechanical properties of Al-TiC metal matrix composite prepared by conventional, microwave and spark plasma sintering methods. Materials 10.https://doi.org/10.3390/ma10111255
18. Gupta D, Sharma AK (2011) Investigation on sliding wear performance of WC10Co2Ni cladding developed through microwave irradiation. Wear 271:1642–1650. https://doi.org/10.1016/j.wear.2010.12.037
19. Gupta M, Wong WLE (2005) Enhancing overall mechanical performance of metallic materials using two-directional microwave assisted rapid sintering. Scr Mater 52:479–483. https://doi.org/10.1016/j.scriptamat.2004.11.006
20. Kansal A, Dvivedi A, Kumar P (2022) Development and performance study of biomedical porous zinc scaffold manufactured by using additive manufacturing and microwave sintering. Mater Manuf Processes. https://doi.org/10.1080/10426914.2022.2089896
21. Kaushal S, Gupta D, Bhowmick H (2018) An approach for functionally graded cladding of composite material on austenitic stainless steel substrate through microwave heating. J Compos Mater 52:301–312. https://doi.org/10.1177/0021998317705977
22. Kumar A, Bhoi NK, Singh H (2020) Corrosion behavior of microwave clad material under different acidic environment. In: Manufacturing engineering, lecture notes on multidisciplinary industrial engineering, pp 603–612. Springer Singapore
23. Kumar G, Sharma AK (2018) Role of dielectric fluid and concentrator material in microwave drilling of borosilicate glass. J Manuf Process 33:184–193. https://doi.org/10.1016/j.jmapro.2018.05.010
24. Li Y, Cheng L, Zhou J (2018) Curing multidirectional carbon fiber reinforced polymer composites with indirect microwave heating. Int J Adv Manuf Technol 97:1137–1147. https://doi.org/10.1007/s00170-018-1974-1
25. Li Y, Li N, Zhou J, Cheng Q (2019) Microwave curing of multidirectional carbon fiber reinforced polymer composites. Compos Struct 212:83–93. https://doi.org/10.1016/j.compstruct.2019.01.027
26. Manohar G, Pandey KM, Maity SR (2021) Effect of microwave sintering on the microstructure and mechanical properties of AA7075/B4C/ZrC hybrid nano composite fabricated by powder metallurgy techniques. Ceram Int 47:32610–32618. https://doi.org/10.1016/j.ceramint.2021.08.156
27. Matli PR, Ubaid F, Shakoor RA et al (2017) Improved properties of Al-Si3N4 nanocomposites fabricated through a microwave sintering and hot extrusion process. RSC Adv 7:34401–34410. https://doi.org/10.1039/c7ra04148a
28. Mishra RR, Sharma AK (2016) On mechanism of in-situ microwave casting of aluminium alloy 7039 and cast microstructure. Mater Des 112:97–106. https://doi.org/10.1016/j.matdes.2016.09.041
29. Mishra RR, Sharma AK (2018) Experimental investigation on in-situ microwave casting of copper. In: IOP conference series: materials science and engineering. Institute of Physics Publishing
30. Mishra RR, Sharma AK (2017) Structure-property correlation in Al–Zn–Mg alloy cast developed through in-situ microwave casting. Mater Sci Eng A 688:532–544. https://doi.org/10.1016/j.msea.2017.02.021
31. Padmavathi C, Upadhyaya A, Agrawal D (2011) Effect of microwave and conventional heating on sintering behavior and properties of Al–Mg–Si–Cu alloy. Mater Chem Phys 130:449–457. https://doi.org/10.1016/j.matchemphys.2011.07.008

32. Patel DK, Bhoi NK, Singh H (2021) Microwave heating capabilities of different susceptor material: experimental and simulation study. Silicon. https://doi.org/10.1007/s12633-021-01426-4
33. Reddy MP, Ubaid F, Shakoor RA, Mohamed AMA (2018) Microstructure and mechanical behavior of microwave sintered Cu50Ti50 amorphous alloy reinforced al metal matrix composites. JOM 70:817–822. https://doi.org/10.1007/s11837-018-2831-2
34. Roy R, Agrawal D, Cheng J, Gedevanlshvili S (1999) Full sintering of powdered-metal bodies in a microwave field. Nature 399:668–670. https://doi.org/10.1038/21390
35. Shelef A, Jerby E (2020) Incremental solidification (toward 3D-printing) of metal powders by transistor-based microwave applicator. Mater Des 185. https://doi.org/10.1016/j.matdes.2019.108234
36. Shoshani Y, Jerby E (2021a) Local melting and cutting of iron bulks by a synergic microwave-DC thermal skin effect. Appl Phys Lett 118.https://doi.org/10.1063/5.0050045
37. Shoshani Y, Levin T, Jerby E (2021) Concrete cutting by a solid-state, localized microwave-heating applicator. IEEE Trans Microw Theory Tech 69:4237–4245. https://doi.org/10.1109/TMTT.2021.3081098
38. Singh MK, Zafar S (2021) Wettability, absorption and degradation behavior of microwave-assisted compression molded kenaf/HDPE composite tank under various environments. Polym Degrad Stab 185. https://doi.org/10.1016/j.polymdegradstab.2021.109500
39. Singh S, Gupta D, Jain V (2018) Processing of Ni-WC-8Co MMC casting through microwave melting. Mater Manuf Processes 33:26–34. https://doi.org/10.1080/10426914.2017.1291954
40. Singh S, Gupta D, Jain V (2016) Novel microwave composite casting process: theory, feasibility and characterization. Mater Des 111:51–59. https://doi.org/10.1016/j.matdes.2016.08.071
41. Singh S, Singh P, Gupta D et al (2019) Development and characterization of microwave processed cast iron joint. Eng Sci Technol Int J 22:569–577. https://doi.org/10.1016/j.jestch.2018.10.012
42. Tamang S, Aravindan S (2017) An investigation on joining of Al6061-T6 to AZ31B by microwave hybrid heating using active braze alloy as an interlayer. J Manuf Process 28:94–100. https://doi.org/10.1016/j.jmapro.2017.05.027
43. Wang H, Ren W, Li G, et al. (2021) Microstructure and properties of FeCoNi1.5CrCu/2024Al composites prepared by microwave sintering. Mater Sci Eng A 801. https://doi.org/10.1016/j.msea.2020.140406
44. Wong WLE, Gupta M (2007) Development of Mg/Cu nanocomposites using microwave assisted rapid sintering. Compos Sci Technol 67:1541–1552. https://doi.org/10.1016/j.compscitech.2006.07.015
45. Zafar S, Sharma AK (2016) Microstructure and wear performance of heat treated WC-12Co microwave clad. Vacuum 131:213–222. https://doi.org/10.1016/j.vacuum.2016.06.021

Further reading on microwave processing:

46. Jerby E, Dikhtiar V (2000) Method and device for drilling, cutting, nailing and joining solid non-conductive materials using microwave radiation, US Patent 6,114,676 (Sep. 5 2000)
47. Chandrasekaran S, Basak T, Ramanathan S (2011) Experimental and theoretical investigation on microwave melting of metals. J Mater Process Technol 211(3):482–487
48. Mishra RR, Sharma AK (2018) Microwave heating characteristics of bulk metallic materials and role of oxides. J Mater Sci 53(24):16567–16584
49. Mishra RR, Sharma AK (2016) A review of research trends in microwave processing of metal-based materials and opportunities in microwave metal casting. Crit Rev Solid State Mater Sci 41(3):217–255
50. Nawathe S, Wong W, Gupta M (2009) Using microwaves to synthesize pure aluminum and metastable Al/Cu nanocomposites with superior properties. J Mater Process Technol 209(10):4890–4895

51. Roy R, Cheng J, Agrawal D (2002) Microwave processing in pure H fields and pure E fields. US Patent 6365885, 2nd April 2002
52. Takayama S, Saito Y, Sato M, Nagasaka T, Muroga T, Ninomiya Y (2006) Sintering behavior of metal powders involving microwave-enhanced chemical reaction. Jpn J Appl Phys 45:1816–1822

Recycling of Thermoset Composites Waste: End of Life Solution

Tejas Pramod Naik, Inderdeep Singh, Apurbba Kumar Sharma, and HaNa Yu

Abstract Companies are under enormous pressure to implement a reuse or recycling solutions for manufactured composite components at the end of their useful lives. This includes waste-free manufacturing techniques or a method to reuse the manufacturing waste. The recycling dilemma for composite components is intrinsically problematic since composite consists of two different materials (matrix and reinforcement). To date, the most popular recycling processes for composite waste recycling are mechanical process (chopping or shredding), pyrolysis (flame-based), solvolysis (chemical-based), hydrolysis (water-based), and thermolysis (heat-based). Most of these techniques results in short fibers with compromised mechanical properties, cost escalation, and use of harsh chemicals. Environmental pressure, strict rules and regulations laid by the governments to protect the environment and for overcoming the drawbacks of the existing recycling methods, novel methods which uses microwaves, eco-friendly chemicals to recycle polymer composites are being developed. This chapter will present a brief on various recycling methods currently available for thermoset composites along with the past work carried out by various researchers worldwide. At the end the chapter focuses on the current advancements that have been carried out in recycling of composite waste by various composite industries. The further research direction has also been highlighted.

Keywords Recycling · Polymer composites · Thermoset · Carbon fiber · Glass fiber · Microwave recycling

---

T. P. Naik (✉) · H. Yu
Department of Mechanical Engineering, University of Bath, Claverton Down, Bath BA2 7AY, UK
e-mail: tpnn20@bath.ac.uk

H. Yu
e-mail: hy539@bath.ac.uk

I. Singh · A. K. Sharma
Department of Mechanical and Industrial Engineering, Indian Institute of Technology (IIT) Roorkee, Roorkee, Uttarakhand 247667, India
e-mail: inderdeep.singh@me.iitr.ac.in

A. K. Sharma
e-mail: aks@me.iitr.ac.in

M. K. Singh et al. (eds.), *Composite Materials Processing Using Microwave Heating Technology*, Composites Science and Technology,
https://doi.org/10.1007/978-981-97-2772-8_4

1 Introduction

Composite materials are widely employed in the renewable energy, automotive, aerospace, transportation, and construction industries due to their low weight, superior quality, durability, high strength, and minimal maintenance [28, 29]. The advantage composites offer over traditional materials has led to their adoption in a wide range of technical applications [11, 37]. Disposal and recycling of composites waste at the end of life (EOL) is a growing concern. In the year 2021, about 85% of plastic waste has gone to landfill in United States creating economic and ecological losses [31]. The plastic waste problem is not only in USA, it can be seen in different parts of the world (Europe, Caribbean, Asia and Latin America) [16]. The amount of waste plastics being generated by industrial, and household operations is increasing on a daily basis. Recycled plastic accounted for just 9% of the projected 359 million tonnes of plastic produced globally in 2018. The remaining plastic trash was either burnt (12%) or landfilled (79%) [14]. A greater proportion of waste dumped in landfills has a negative impact on the ecology, as does the continual growth in the production of virgin polymer. Therefore, continued production might lead to the depletion of fossil fuels, increased environmental emissions during processing, and incineration at the end of their useful life. Many nations are presently transitioning to the circular economy idea in order to prevent all these issues [12, 16]. In order to solve this problem of plastic waste, researchers, scientists and engineers worldwide have come up with the terms "reuse" and "recycle" of the plastic or composite waste at its EOL (Fig. 4.1). Recycling and reusing of the waste also satisfy the "United Nations Sustainable Development Goals" (SDGs). Due to technological and economic restrictions, there are limited commercial recycling processes for polymeric materials, although R&D activities are increasing. It has become increasingly challenging to adequately handle polymer waste in recent years. Traditional disposal techniques, such as landfill disposal, burying, or burning, pollute water and environment by releasing dust particles, toxic gases, and fumes. Traditional approaches are no longer employed to conserve the environment, and new methods of recycling are preferred to handle polymer waste. For thermosetting based composites, the maximum waste is seen from automobile and aerospace industries utilizing carbon fibers (CFs) and from wind turbine industries utilizing mainly glass fibers (GFs). These fibers are incorporated in the polymer matrices because of their exceptional mechanical properties, high strength/weight ratio, corrosion resistance, light weight, etc. [8, 34, 38, 48]. According to Albers et al., a sizeable proportion of deployed wind turbines will reach the end of their typical lifetime between 2020 and 2034, producing 0.2 million tonnes of waste [1]. Aerospace and automobile industries are also contributing continuously towards the waste generation for many decades now. Thus, to cater such a huge waste generated and to reclaim the fibers (with less damage to their mechanical properties), there is a need towards exploration of composite recycling techniques and potential uses for recycled materials. Chemical, Thermal, and Mechanical processes can recover CFs/GFs from waste composites. All of these technologies have downsides, including deterioration of

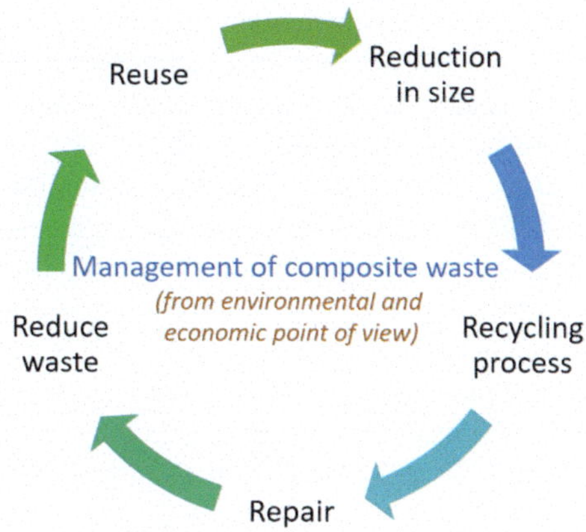

Fig. 4.1 A schematic illustration depicting the strategies to handle composites waste from an environmental and economic viewpoint

recycled fibers (rFs) mechanical properties, long cycle time, high energy consumption, high cost and generation of very harmful gases that degrade the environment [32, 34, 35]. However, at the moment, handling and reprocessing of fiber reinforced polymer composites (FRPCs) while adhering to environmental restrictions, legislative/regulatory requirements, manufacturing costs, financial projections, and asset management are crucial global goals. These demand the use of science and technology for sustainability. On this note, recently a new and novel method called as "microwave-assisted chemical" recycling is gaining attention due to its eco-friendly processing with very less degradation to the mechanical properties. Also, recently in year (2022) the official theme of JEC World (leading international composites show) held in Paris was sustainability [52]. The products showcased in the show were either made up of recycled polymers or they were made up of natural fibers. It was found that all the major composite industries are now focussing on sustainability by addressing the problems of recycling the product at its EOL. This chapter will provide an overview of the numerous recycling technologies now accessible for thermoset composites, as well as previous research conducted by various researchers worldwide. At the end of the chapter, the present accomplishments made in the recycling of composite waste by various composite companies, as well as future research directions are discussed.

2 Types of Recycling Methods

The three primary recycling categories are mechanical, thermal, and chemical [48]. Each recycling method is further divided into subcategories as shown in Fig. 4.2. A new and hybrid, microwave-assisted chemical method [29] is now gaining interest for recycling of the composites waste. Every recycling method demands for three fundamental steps, i.e., collection, sorting and cleaning before being recycled using any of the methods. Finally, the recycled raw material produced after recycling is ready to become finished product again. Thermal and chemical processes, as opposed to mechanical process, can recover long continuous fibers with a better prospect of reusing them. However, using high temperatures (450–550 °C) in thermal processes results in fibers that only retain 70–75% of their strength and cause toxic gas emissions as well as fiber surface contamination from char deposition [45]. Supercritical fluids have emerged during the past ten years as environment friendly options for the chemical recycling of fiber reinforced plastic (FRP) waste [15, 36]. However, these processes are energy-intensive since the fluids must reach a supercritical state at extremely high temperature (400–500 °C) and pressure values (10–30 MPa). Recovering long, continuous fibers with greater strength is guaranteed by chemical recycling processes, which also guarantee the removal of all resin. However, from an environmental and financial standpoint, the use of hazardous chemicals (such as nitric acid) [4, 5] and/or the application of high operation conditions (supercritical fluids) restricts their potential for large-scale use. In light of these problems, recently a sonochemical approach for recycling FRP that utilised a moderate nitric acid/ H_2O_2 combination and ultrasound at high frequency has been developed. Due to the N_2O emission during reaction and the challenge in recovering the waste acids, the approach may still not be regarded as environmentally safe [7]. With recent success, environment friendly chemical recycling techniques have been developed which can recover both the fibers and the matrix material with the least amount of environmental harm [39].

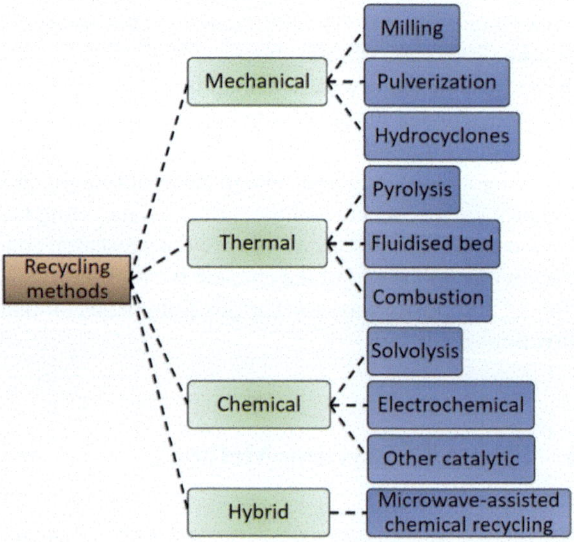

Fig. 4.2 Various waste recycling techniques for composite materials

2.1 Mechanical Recycling

2.1.1 Milling

The most mature recycling process in terms of technology is mechanical recycling, which uses many stages (Fig. 4.3) to minimise the size of the composite waste [48]. Cutting and crushing (milling) of leftovers or discarded composite components into small pieces, sometimes referred to as generated scrap, is the first step in the mechanical recycling process [16]. Then, by straining the generated scrap, fibrous and powders materials are recovered. Sometimes throughout the production process, these recycled resources are combined with virgin materials. Mechanical recycling is a more effective way to recycle glass fiber reinforced composites (GFRCs) and carbon fiber reinforced composites (CFRCs) than landfilling or incineration. The main materials recycled using this technique are GFRCs and thermoplastics (PE, PVC, and PP) [16, 38, 41]. Although mechanical recycling is a cheap procedure, it requires a larger initial investment (shredder machine) and maintenance of the equipment. Mechanical recycling by-products can be used as fillers in short fiber composites like SMCs, construction, etc. The mechanical milling procedure is safe to use and can be carried out at ambient temperature [51]. The fibers recovered using this procedure can achieve tensile strength of 50–65% of those of new carbon fibers [48]. The primary health and safety risk of this procedure is the dust from the recycling system, which may be reduced by using correct engineering controls, personal protective equipment, and good ventilation with particle and dust extraction.

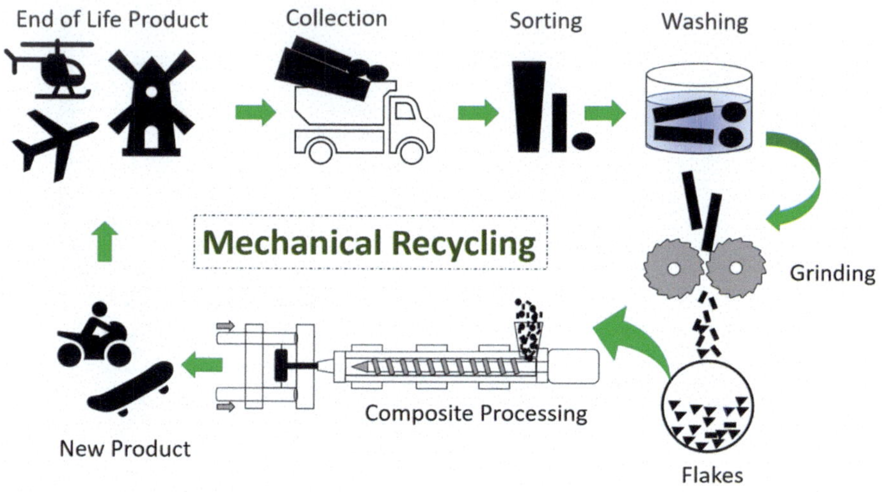

Fig. 4.3 Mechanical recycling of composites waste

2.1.2 Hydrocyclones

Hydrocyclones are cyclonic separators that use aqueous solutions to separate product phases by differences in gravity. The waste is then combined with water to move the solution into the chamber. A hydrocyclone has a cylindrical feed section with tangential feed, an overflow section with a vortex finder, and a conical section with an apex. There are no moving parts in a cyclone. The water and trash mixture is pushed through a hydrocyclone chamber, which is increasingly utilised in modern recycling and separation plants and can be an efficient approach for material separation. At the chamber's inflow, a mixture of waste and water turns into a vortex, producing centrifugal acceleration several times greater than gravity. As a consequence, lighter materials migrate toward the centre, where an air interface and vortex finder remove this material from the top. As seen in Fig. 4.4, heavier materials flow toward and along the outer surface and run down a tapering sidewall; these particles then go downward and exit from the bottom. Thus, the hydrocyclone may be utilised to separate the different composite wastes (like CFRP and GFRP) coming from various goods that have reached the end of their useful life.

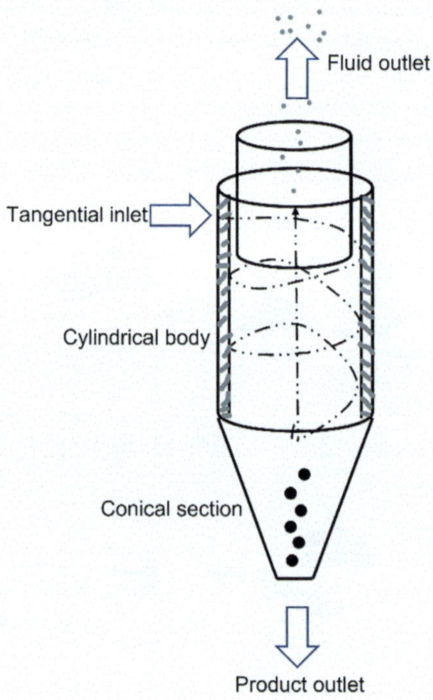

Fig. 4.4 Hydrocyclone recycling process

2.2 *Thermal Recycling*

Thermal recycling employs high temperature and fluids in recycling of CFRCs and GFRCs to disintegrate the matrix and recover the fibers. The various thermal methods used for recycling is explained in the following section.

2.2.1 Pyrolysis Process

The primary objective of the method known as pyrolysis is to convert the organic components of waste composites into their molecular forms (gas or liquid). The remaining fraction will be composed of various fillers and fibers. The waste composite and reinforcing materials are processed in an environment with very little to no oxygen at temperatures ranging from 400 to 1000 °C, depending on the kind of waste composite and reinforcing material. In this temperature range, GFs and CFs are equally suitable options. This process can result in the production of tar and other heavy liquids, which can then be utilised to manufacture oil and other products. Even if the calorific values (CVs) of the gaseous by-products such as CH_4, H_2, CO_2, and other hydrocarbons are low, these gaseous by-products can still be exploited as sources of energy. This method is also referred to as a self-sustaining recycling process because of the by-products that it produces. In order to accomplish the most effective and comprehensive recovery of solid materials, fuels, and chemical compounds, it is necessary to optimise the process parameters of the pyrolysis process [30]. The term pyrolysis is often used in place of thermolysis when organic compounds are decomposed at high temperature. The difference between pyrolysis and thermolysis is that pyrolysis is the decomposition of a material or compound caused by heat in the absence of oxygen or any other reagents, whereas thermolysis is the decomposition or dissociation of a material as a result of being heated.

2.2.2 Fluidized Bed Recycling

Since 2000, the University of Nottingham, United Kingdom has been developing fluidized bed recycling (Fig. 4.5), which is already operating now [48]. After shredding to 6–20 mm [26], the composite waste is placed in a bed of 1 mm silica sand fluidized by hot air at 450–550 °C and 10–25 kPa [51]. The primary disadvantage of this process includes pollutant gas (health risk), and the use of a lot of energy, based on the recycling conditions. The reaction temperature ensures the thermoset matrix decomposes without fiber damage. Due to attrition, thermal degradation, and matrix damage during quick heating, fibers are removed and separated. The recycled fiber length ranges from 5 to 10 mm, with 10–75% of the tensile strength as compared to virgin fibers [48, 51]. This process of composite recycling of the waste at its EOL is used widely in industries though the process has few disadvantages. Thus, there is a need to develop a sustainable recycling method to recycle the composite waste.

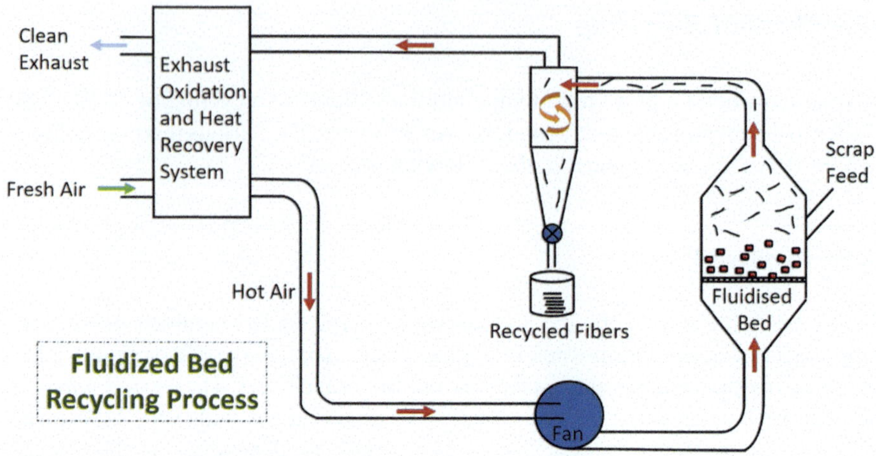

Fig. 4.5 Fluidized bed recycling of composites waste

2.3 *Chemical Recycling*

Recycling methods for FRP composites that are chemical or allied electrochemical often include the use of a broad variety of solvents, including subcritical and supercritical fluids, and are typically carried out in the presence of pressure and temperature.

2.3.1 Solvolysis Using Subcritical/Supercritical Solvent

Solvolysis separates fibers from thermoset matrix by decomposing it with solvents or heated solvents. Solvolysis provides many methods of recycling due to the availability of a broad variety of solvents paired with diverse processing parameters such as pressure and temperature, and often in combination with a catalyst. Supercritical, subcritical, and near-critical chemical processes are the most common [27]. Water is preferred over other solvents due to its environmental friendliness. Other solvents, including methanol, acetone, and ethanol as well as their mixtures with water at low critical temperature and pressure values, have also been investigated [48]. Because the chemical recycling technique may generate longer rFs and without any residue on the surface, high-quality carbon/glass fibers with minimal mechanical properties can be achieved. The chemical treatment employs a lower processing temperature than pyrolysis and provides cleaner and longer rFs; nevertheless, solvolysis necessitates expensive equipment that can withstand rusting at higher pressure, temperature, and the solvents may have negative effects on the human physiology and environment. This fluidized bed process features frequently restrict its industrial-scale applicability [22].

Fig. 4.6 Electrochemical recycling of composites waste

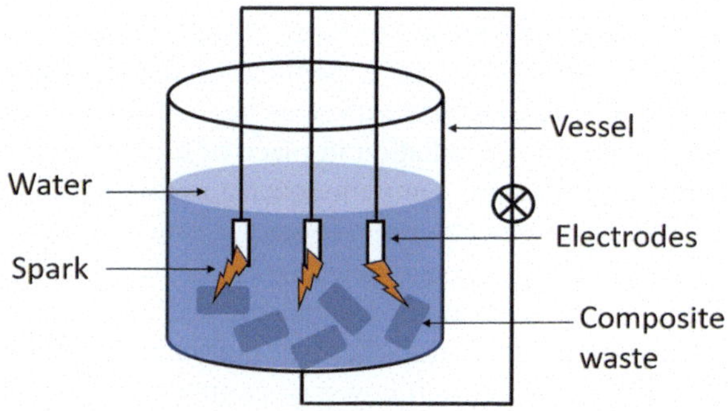

2.3.2 Solvolysis Using Solvent at Ambient Pressure

Solvolysis has distinct advantages in terms of reducing damage of recycled carbon fibers (rCFs) at ambient pressure. In order to recycle carbon fiber reinforced plastics, Liu et al. used a moderate chemical recycling technique using $ZnCl_2$/ethanol catalyst [21]. The rCFs were not damaged by the low treatment temperature (200 °C), and the recovered polymer matrix was proven to retain good strength and modulus. Hou's group created an effective chemical recycling approach that uses $AlCl_3/CH_3COOH$ as a catalyst to achieve recycling of the composite waste [42]. About 97% of the cured epoxy was recovered under optimal recycling conditions. Tensile strength was maintained at 98% of that of virgin carbon fibers (vCFs) in the recovered fibers.

2.3.3 Electrochemical Recycling

Pulsed electrical discharges are employed in the high voltage fragmentation (HVF) process of electrochemical recycling to break down solid material. Previously, rocks were fragmented to recover precious components in mining. The process of HVF is depicted in Fig. 4.6. Between the electrodes, a high-voltage electric pulse is produced that travels through the vessel's materials. Overconsumption of energy has been the primary limitation since 75–80% of the energy is used by the vessel itself rather than for processing waste materials [48]. High voltage fragmentation techniques and mechanical recycling have been compared in research [25]. The energy required in HVF is 2.6 times greater than that of mechanical processing, however this technology yields longer and cleaner recovered fibers than mechanical recycling.

2.4 Microwave Irradiation Recycling

The idea of using microwave radiation instead of traditional heating to provide the necessary heat to boost reaction efficiency has been studied over the past several decades and is verified to be more energy-efficient and ecologically sound [11, 38]. According to the literature, chemical recycling method exhibits the greatest breakdown rate of resin matrix with clean fibers. Numerous studies have demonstrated that using microwave irradiation in place of traditional heat sources can result in faster chemical processes. Thus microwave-assisted chemical method (MACM) would result in faster reaction time, better performance, faster and eco-friendly processing [2, 9, 10, 50]. The general close-loop steps of the MACM method used to recover fibers in glass fiber reinforced plastic (GFRP), and carbon fiber reinforced plastic (CFRP) are shown in the schematic diagram (Fig. 4.7). The various steps involved in MACM method are as follows; (i) cutting of the composite waste sample into small pieces, (ii) immersing the cut composite waste into chemical solution for certain time at room temperature, (iii) microwave exposure to the solution at the optimized setting of microwave power and time, (iv) cooling of the solution at room temperature, (v) cleaning of the recovered fibers. Table 4.1 shows the various green solvents used for the recycling of CFs/GFs from the waste composites keeping in mind the sustainability aspect. Table 4.2 highlights the previous research work carried out in the reclamation of fibers using various approaches.

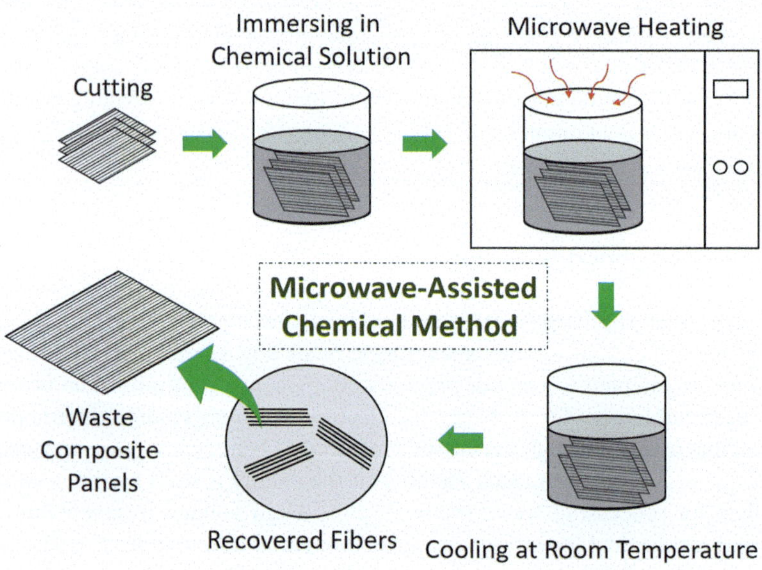

Fig. 4.7 A novel microwave-assisted green chemical recycling of composites waste

Table 4.1 Recycling of composites waste using green chemical solvents (a sustainable approach)

References	Year	Matrix	Fiber	Recycling method	Chemical solvent	Parametric conditions*	Output	Recycled fibers strength compared to virgin fibers
Lester et al. [20]	2004	Epoxy	CF	Microwave heating	—	P: 3 kW t: 8 s	rCF	—
Xu et al. [43]	2013	Epoxy	CF	Chemical	Acetic acid (CH_3COOH), Hydrogen peroxide (H_2O_2) and N, N-dimethylformamide (DMF)	T: 90 °C t: 30 min	rCF	>95%
Das et al. [6]	2018	Epoxy	CF	Pyrex glass bottle in an oil bath	CH_3COOH and H_2O_2	T: 65 °C t: 4 —> 10 h	rCF, Epoxy, and CH_3COOH	—
Zhao et al. [49]	2019	Epoxy	—	Microwave assisted catalysis	Dichloromethane, and Diethylenetriamine (DETA)	P: 300 W t: 50 min	Recycled epoxy	—
Zabihi et al. [46]	2020	Epoxy	CF	MACM	H_2O_2 and tartaric acid (TA)	P: 800 W t: 1–5 min	rCF	~92%
Zabihi et al. [47]	2020	Epoxy	GF	MACM	H_2O_2 and TA	P: 800 W t: 1.5–6 min	recycled glass fibers (rGF)	~92%
Rani et al. [39]	2021	Epoxy	GF	MACM	H_2O_2 and CH_3COOH	P: 700 W t: 3 min	rGF	97.2%

*Where, P: Power, T: Temperature, and t: time

Table 4.2 Recycling techniques used by researchers for processing a wide variety of composites waste into usable fibers, and energy

References	Matrix	Fiber	Recycling method	Parameters	Characterizations	Inference
García et al. [13]	Thermoset	GF	Mechanical	–	Compression test, Bending test, Shrinkage stability, Durability test, SEM, Infrared spectroscopy	• Recovered fibers containing dust or large size splinters affects the mechanical properties • The mechanical properties of the micro concrete were found to increase for optimized mechanical recycling
Yazdanbakhsh et al. [44]	Resin	GF	Mechanical	–	Tensile, Flexural, Compressive, Toughness, Polymer burn-off test	• GFRP needles used as concrete coarse aggregate • Needles replaced 5 to 10% of concrete's coarse aggregate by volume • Needles didn't influence fresh concrete's stability or workability • Although adding needles didn't affect tensile, compressive, and flexural strength, it increased toughness • Transversely oriented needles increase concrete's mechanical performance

(continued)

Table 4.2 (continued)

References	Matrix	Fiber	Recycling method	Parameters	Characterizations	Inference
López et al. [24]	Polyester	GF	Thermolysis	T: 550 °C t: 3 h	Differential thermal analyser (DTA), Thermogravimetric analyser (TGA), Bomb calorimeter (BC), Gross calorific value (GCV), Gas chromatography (GC), Mass spectrometry (MS), Rheology, Scanning electron microscope (SEM), X-ray diffraction (XRD)	• Thermolysis is an efficient technique for extracting useful materials and energy • Output in the form of solid (E-glass fiber and char), liquid (oil) and gas • Gross calorific value (GCV) of the oil was found to be high as compared to gas • It's possible that recycled fiber might be used to create tiles with a variety of aesthetic attributes
Cunliffe et al. [3]	Polyesters, Phenolic, Epoxy, and Polypropylene	GF/CF	Pyrolysis	T: 350–800 °C	Size exclusion chromatography (SEC), TGA, Difference thermo gravimetry (DTG), Gross calorific value (GCV)	• Both the mass balance of products and the composition of the pyrolysis gas were significantly affected by the polymer matrix and the pyrolysis temperature

(continued)

Table 4.2 (continued)

References	Matrix	Fiber	Recycling method	Parameters	Characterizations	Inference
Kim et al. [17]	Epoxy	CF	Pyrolysis (super-heated-steam method)	T: 550 °C	Single fiber test, IFSS, SEM, XPS, TGA, Single fiber contact angle test, FTIR	• No matrix char residue on the surface of fiber • Reaction time, pyrolysis temperature, and super-heated steam feeding rate strongly affects the recycling efficiency • The tensile strength of rCF as compared to vCF was 90%
Pickering et al. [33]	Epoxy	CF	Fluidised bed	T: 500 °C	Tensile testing	• The method is especially well suited for treating mixed and contaminated CFRP trash • There was no decline in modulus of tensile and the tensile strength decreased by 18% • Typically, 5–10% of the energy needed to make vCF is used in the carbon fiber recovery process

(continued)

Table 4.2 (continued)

References	Matrix	Fiber	Recycling method	Parameters	Characterizations	Inference
Xu et al. [43]	Epoxy	CF	Chemical	T: 90 °C t: 30 min	SEM, XRD, Raman spectroscopy, Single fiber test, TGA	• Two-step chemical method for recycling of high-quality fibers • The strength of rCF was > 95% and decomposition ratio of epoxy was > 90% • The degree of graphitization decreased slightly compared to the vCF • Promising process to be scaled up in future as the fibers obtained were clean and smooth

(continued)

Table 4.2 (continued)

References	Matrix	Fiber	Recycling method	Parameters	Characterizations	Inference
Das et al. [6]	Epoxy	CF	Chemical	T: 65 °C t: 4 —> 10 h	UV – vis spectroscopy, TGA, SEM, EDS, XPS, Single fiber test, FTIR, NMR, GC, MS, Rheology, Tensile, Optical microscope	• Single-step chemical recycling process • Long and clean fibers were recovered • No significant loss in tensile strength of recycled fiber was reported • Distillation process was performed for solvent recovery • Almost complete recovery of the solvent was reported (>90%) • No gaseous emission from the process thus an eco-friendly recycling process

(continued)

Table 4.2 (continued)

References	Matrix	Fiber	Recycling method	Parameters	Characterizations	Inference
Zhu et al. [51]	Epoxy	CF	Electrochemical	T: Room temperature (RT) T: 72 h	TGA, Single fiber test, IFSS, SEM, AFM, XPS	• The process uses nontoxic electrolyte at ambient temperature and atmospheric pressure for recycling of fibers • Process is simple and green compared to conventional processes • Even at RT and atmospheric pressure the depolymerization of epoxy and recovery of CF was found to be effective • 100% removal of epoxy • Potential process which can be industrially scaled

(continued)

Table 4.2 (continued)

References	Matrix	Fiber	Recycling method	Parameters	Characterizations	Inference
Leißner et al. [19]	Galvanized plastics, CFRP, and Electrode foils	—	Electrochemical (High voltage fragmentation)	Total pulse: 500 Pulse rate: 1–4 Hz	Inductively coupled plasma optical emission spectrometry (ICP-OES), Size distribution, Density, Degree of fiber liberation	• Results of this technology are not superior to conventional mechanical recycling process • Degree of filling, Number of pulses, and Pulse rate were found to influence the performance of the process • Larger percentage of energy consumed in the process was transferred into heat • The process is not efficient as it demands higher energy and investment cost
Kim et al. [18]	Epoxy	CF	Solvolysis using supercritical water	T: 405 ± 2 °C p: 280 ± 10 bar t: 10, 30, 60, and 120 min	Raman microscope, FTIR, XPS, SEM, TGA, Single fiber test, X-ray micro computed tomography (micro-CT), thermal conductivity meter, GC–MS, Steady-state heat flow measurement apparatus	• Supercritical water without catalyst or oxidant was used for recycling • Green process as only supercritical water is used About 99.5% of the epoxy was removed from the fiber surface

(continued)

Table 4.2 (continued)

References	Matrix	Fiber	Recycling method	Parameters	Characterizations	Inference
Liu et al. [23]	Epoxy	CF	Solvolysis using subcritical water	T: 315 °C T: 30 min	GC- MS, Single fiber test, SEM, XPS, TGA,	• Phenol and potassium hydroxide were employed with subcritical water for chemical recycling • The phenoxy radical generated from phenol anion was shown to have a significant effect on the cleavage of chemical bonds in the crosslinked network of cured epoxy resin, which was largely responsible for the exceptional breakdown efficiency observed

(continued)

Table 4.2 (continued)

References	Matrix	Fiber	Recycling method	Parameters	Characterizations	Inference
Lester et al. [20]	Epoxy	CF	Microwave heating	P: 3 kW t: 8 s	Tensile test, SEM, TGA, Atomic force microscopy (AFM)	• Initial experiments showed that it is possible to separate polymer from fiber • Long fibers can be obtained using this process unlike other known recycling processes • Quartz sand and glass wool were used inside the microwave to carry out the experiments • Nitrogen gas was used to prevent combustion of fibers • Comparable properties were obtained for the rCF More viable technology compared to fluidised bed

(continued)

Table 4.2 (continued)

References	Matrix	Fiber	Recycling method	Parameters	Characterizations	Inference
Zabihi et al. [47]	Epoxy	GF	MACM	P: 800 W t: 1.5–6 min	TGA, Single fiber test, FTIR, SEM	• Green chemical solvent used for the recycling of the composite waste • MACM is a rapid and low-cost process • Microwaves accelerate the reaction and effectively decomposes the matrix • Low energy consumption process • Compared to virgin glass fiber (vGF) the percentage reduction in mechanical properties was less

(continued)

Table 4.2 (continued)

References	Matrix	Fiber	Recycling method	Parameters	Characterizations	Inference
Zabihi et al. [46]	Epoxy	CF	MACM	P: 800 W t: 1–5 min	TGA, Single fiber test, FTIR, SEM, Optical microscope, Raman spectra, X-ray photoelectron spectroscopy (XPS), Interfacial shear strength (IFSS), Interlaminar shear strength (ILSS), Tensile and Flexural tests, Single fiber fragmentation tests (SFFT)	• Integration of green chemistry with microwaves for recycling • Sustainable approach for recycling of FRP waste • Compared to vCF, the percentage reduction in mechanical properties was less • Compatibility of the rCF with epoxy is more compared to virgin fiber • IFSS and ILSS increased by ~ 44% and ~ 17%, respectively • Flexural properties of the remanufactured composites were highly comparable with the virgin composites
Rani et al. [39]	Epoxy	GF	MACM	P: 700 W t: 180 s	TGA, Single fiber test, Fourier transform infrared (FTIR), Tensile properties, SEM	• Output was rGF • As the volume ratio of the chemical solvent grew, so did the rate of breakdown of the GFRP composite • SEM micrographs revealed that the surface of rGF was clean and smooth • Compared to virgin GFs, the maximum tensile strength of rGFs was 99.8%

* where, p: Pressure, P: Power, T: Temperature, t: time, V: fluidising velocity

3 Recent Developments in Recycling of Composites

Currently industries are focussing on many applications which incorporates the recycled synthetic fibers obtained from recycling of the composites from high-end applications. The following are the various fields where composites industries are getting attracted these days.

- Wind energy sector
- Solar panels
- Automotive
- Aerospace industry
- Marine (boat hull)
- Energy storage tanks (hydrogen tank)
- Electronics (mobile)
- Furniture
- Construction (bridges)
- Bicycles

The various composites industries worldwide have now started contributing towards recycling compositse waste after its end of life. The latest developments are shown in Table 4.3.

4 Conclusions and Future Scope

The sustainable use of FRPCs in various industries is critically dependent on their ease of recycling. Mechanical recycling, pyrolysis, fluidized beds, chemical and microwave-assisted chemical recycling are some of the currently employed recycling techniques. It is necessary to develop new, affordable recycling methods to efficiently collect and recycle valuable carbon and glass fibers. Several prospective approaches to dissolve thermoset matrices under moderate settings that cause less environmental damage and preserve the mechanical properties of the fibers have been demonstrated to be superior and more cost-effective recycling alternatives. These techniques also seek to maximize performance of the remanufactured composites fabricated utilizing reclaimed fibers (CF/GF) with/without recycled polymer matrices.

Recyclable fiber length, laboratory and industrial scale feasibility, surface defects, process suitability for different fibers, and set-up cost, are all factors that can affect the success of thermal, chemical, and mechanical recycling methods. Researchers took into account the drawbacks of traditional recycling techniques and turned their attention to hybrid approaches that mitigate these issues. Microwave-assisted chemical recycling is an area of active investigation among the hybrid approaches. It is the most sustainable, environment friendly and energy-efficient way to recycle the composites waste. It has the potential to be the most effective technology for recovering fibers or

Table 4.3 Recent developments in the field of recyclability of composites

References	Company	Inference	Objective(s)
Composites World [53]	Siemens Gamesa	• World's first 81-m wind turbine equipped with a recyclable blade • It was installed at Kaskasi offshore wind power project in Essen, Germany • This project is a turning point for offshore wind power farms • Recycled material: Fiberglass • Electricity is being generated with this turbine	Recycling technology
Composites World [54]	Teijin and Fujitsu	• The collaboration between the two companies will develop a platform for increasing the environmental value of recycled materials • Applying information on the environmental impact of recycled materials and components	Information technology-based solution
Composites World [55]	Igus Inc	• Single-use reused plastics (old fishing nets) have been used to fabricate the first prototype of the bicycle • The idea of fabricating the igus-bike came from the problems faced by beach bikes • Other versions, such as an e-bike, have also been planned • The first model of a bicycle is to be available by the end of the year 2022	Product development
Composites World [56]	Levante and Advanced Composites Solution	• Developed portable and origami-inspired solar panel frame via scrap carbon fibers from the automotive industry • The developed panel won the 2022 Best Practices Award for Innovation	Product development

(continued)

Table 4.3 (continued)

References	Company	Inference	Objective(s)
Composites World [57]	Swancor	• Launched recyclable and reusable thermosetting epoxy resin • CleaVER technology was used to recycle polymer • The process does not generate pollution and waste solvents • Recycled fibers show equivalent mechanical performance to virgin fibers	Product development and Fiber recycling process
Composites World [58]	National Composites Centre with partners B&M Longworth and Cygnet Texkimp	• Successfully reclaimed continuous carbon fibers from the pressure tank • Recycled fibers were used to fabricate new sustainable pressure vessel filament winding • The pressurized steam-based composites recycling process, also known as DEECOM process, was used for the reclamation of fibers	Product development and Hydrogen energy
Composites World [59]	CompPair Technologies and Decathlon	• Completed joint fiber-reinforced plastic (FRP) recycling project • Produced bicycle shoe sole with glass fiber prepregs	Product development
Composites World [60]	Voith Composites, V-carbon	• Aims to develop a closed-loop solution for CFRP used in hydrogen tanks	Case study and Hydrogen energy
Composites World [61]	Decathlon	• Developed athletic shoe (Traxium Compressor) with repurposed thermoplastic waste • The shoe (soccer) was announced for the first time recently in October 2021 • Single-step molding process was used to fabricate the shoe	Case study and Product development
Composites World [62]	C^{e}cence, National Composites Centre, and Gen^{2} Carbon	• Developed sustainable airplane seatback • Substantial weight savings in the cabin interior and a further reduction in CO_2 per aircraft	Case study

(continued)

Table 4.3 (continued)

References	Company	Inference	Objective(s)
Composites World [63]	Boeing	• Created ecoDemonstrator program to reduce its environmental footprint • The focus of the program was to improve the entire aviation ecosystem by using sustainable materials, etc. • Tested cabin interior sidewalls made up of recycled carbon fiber composite • These sidewalls reduce the noise and the waste going to landfills	Product development
Composites World [64]	Anmet and GP Renewables Group	• Installed the first recycled pedestrian and bicycle footbridge made from wind turbine blade • Solution to decommissioned turbine blades	Product development
Composites World [65]	Carbon Rivers	• Developed glass fiber recycling technology • The U.S. Department of Energy's (DOE) Wind Energy Technologies Office funded the project • The company achieved 99.9% purity of recycled glass fiber	Recycling technology
Composite World [66]	Aimplas	• Developed carbon fiber reinforced thermoplastic car door panel prototype using fresh fibers • Product was developed by the integration of microwaves into the resin transfer molding process • The use of microwaves has reduced energy consumption and produced better parts • The material's recyclability was validated by manufacturing a car door handle comprising 50% recycled material	Product development

(continued)

Table 4.3 (continued)

References	Company	Inference	Objective(s)
Composites World [67]	Longworth	• Developed pressurized steam-based DEECOM process to separate resin and fiber • The process uses high-temperature steam and pressure • Process claims reclamation of high-quality fibers and potential to reuse resins	Recycling technology
Composites World [68]	Teijin and Fuji Design	• Developed precision pyrolysis recycling technique • The process involves cyclical heating and cooling • The process claims intact fibers without surface oxidation damage of fiber • Recycled fiber strength was 93% of the virgin fiber • The company is also working on reclaiming resin matrices, including thermoplastics	Recycling technology

transforming waste composite into a valuable supply of fibers for composite products, both of which have environmental advantages. Further investigation is needed to make the microwave-assisted chemical method (MACM) approach a more sustainable recycling process to deal with the end-of-life waste before it can be used on an industrial scale. The following are the areas where research efforts are required, as well as potential avenues for future development, in order to enhance and refine this method of fiber recovery.

1. Determination of dielectric properties of the fibers is important as quality of microwave recycled fibers (rFs) can vary according to their dielectric properties.
2. To check the effectiveness of the MACM process in recycling of the composite scraps reinforced with hybridized fibers.
3. Investigations on alternative potential green solvents for recycling of the composites waste.
4. Complete understanding of the interplay between microwave irradiation and chemical solvents is necessary for better quality fibers after recovery.
5. Optimization of the various microwave parameters (power, exposure time, and temperature) for achieving the best properties of the rFs.
6. To determine the influence of the microwave power on the mechanical properties of the rFs.
7. Entanglement limits rFs reuse in remanufactured composites. Re-aligning rFs just like virgin fibers can improve their utilization.

8. Performing the tecno-economic analysis of the MACM process and comparing it with the existing and well-established recycling processes.

Acknowledgements The author Tejas Pramod Naik who is the Commonwealth Split-site PhD Scholar would like to acknowledge Commonwealth Scholarship Commission (CSC) in the United Kingdom (provides the main UK government scholarship), for the financial support provided through its CSC reference No.: **INCN - 2023 - 402**.

References

1. Albers H, Greiner S, Seifert H, Kuehne U (2009) Recycling of wind turbine rotor blades. Fact or fiction? Recycling von Rotorblaettern aus Windenergieanlagen. Fakt oder Fiktion? DEWI-Magazin
2. Bassyouni FA, Abu-Bakr SM, Rehim MA (2012) Evolution of microwave irradiation and its application in green chemistry and biosciences. Res Chem Intermed 38:283–322. https://doi.org/10.1007/s11164-011-0348-1
3. Cunliffe AM, Jones N, Williams PT (2003) Recycling of fibre-reinforced polymeric waste by pyrolysis: thermo-gravimetric and bench-scale investigations. J Anal Appl Pyrolysis 70:315–338. https://doi.org/10.1016/S0165-2370(02)00161-4
4. Dang W, Kubouchi M, Sembokuya H, Tsuda K (2005) Chemical recycling of glass fiber reinforced epoxy resin cured with amine using nitric acid. Polymer (Guildf) 46:1905–1912. https://doi.org/10.1016/j.polymer.2004.12.035
5. Dang W, Kubouchi M, Yamamoto S et al (2002) An approach to chemical recycling of epoxy resin cured with amine using nitric acid. Polymer (Guildf) 43:2953–2958. https://doi.org/10.1016/S0032-3861(02)00100-3
6. Das M, Chacko R, Varughese S (2018) An efficient method of recycling of CFRP waste using peracetic acid. ACS Sustain Chem Eng 6:1564–1571. https://doi.org/10.1021/acssuschemeng.7b01456
7. Das M, Varughese S (2016) A novel sonochemical approach for enhanced recovery of carbon fiber from CFRP waste using mild acid–peroxide mixture. ACS Sustain Chem Eng 4:2080–2087. https://doi.org/10.1021/acssuschemeng.5b01497
8. Deng J, Xu L, Zhang L, et al. (2019) Recycling of carbon fibers from CFRP waste by microwave thermolysis. Processes 7:207. https://doi.org/10.3390/pr7040207
9. Ding A, Wang J, Ni A, Li S (2019) A new analytical solution for cure-induced spring-in of L-shaped composite parts. Compos Sci Technol 171:1–12. https://doi.org/10.1016/j.compscitech.2018.12.004
10. Dudley GB, Richert R, Stiegman AE (2015) On the existence of and mechanism for microwave-specific reaction rate enhancement. Chem Sci 6:2144–2152. https://doi.org/10.1039/C4SC03372H
11. Gairola S, Naik TP, Sinha S, Singh I (2022) Corncob waste as a potential filler in biocomposites: A decision towards sustainability. Compos Part C Open Access 9:100317. https://doi.org/10.1016/j.jcomc.2022.100317
12. Gairola S, Sinha S, Singh I (2022) Novel millet husk crop-residue based thermoplastic composites: waste to value creation. Ind Crops Prod 182:114891. https://doi.org/10.1016/J.INDCROP.2022.114891
13. García D, Vegas I, Cacho I (2014) Mechanical recycling of GFRP waste as short-fiber reinforcements in microconcrete. Constr Build Mater 64:293–300. https://doi.org/10.1016/j.conbuildmat.2014.02.068
14. Geyer R, Jambeck JR, Law KL (2017) Production, use, and fate of all plastics ever made. Sci Adv 3:25–29. https://doi.org/10.1126/sciadv.1700782

15. Hyde JR, Lester E, Kingman S et al (2006) Supercritical propanol, a possible route to composite carbon fibre recovery: A viability study. Compos Part A Appl Sci Manuf 37:2171–2175. https://doi.org/10.1016/j.compositesa.2005.12.006
16. Jagadeesh P, Mavinkere Rangappa S, Siengchin S, et al. (2022) Sustainable recycling technologies for thermoplastic polymers and their composites: a review of the state of the art. Polym Compos 1–32. https://doi.org/10.1002/pc.27000
17. Kim K-W, Lee H-M, An J-H et al (2017) Recycling and characterization of carbon fibers from carbon fiber reinforced epoxy matrix composites by a novel super-heated-steam method. J Environ Manage 203:872–879. https://doi.org/10.1016/j.jenvman.2017.05.015
18. Kim YN, Kim Y-O, Kim SY et al (2019) Application of supercritical water for green recycling of epoxy-based carbon fiber reinforced plastic. Compos Sci Technol 173:66–72. https://doi.org/10.1016/j.compscitech.2019.01.026
19. Leißner T, Hamann D, Wuschke L et al (2018) High voltage fragmentation of composites from secondary raw materials – Potential and limitations. Waste Manag 74:123–134. https://doi.org/10.1016/j.wasman.2017.12.031
20. Lester E, Kingman S, Wong KH et al (2004) Microwave heating as a means for carbon fibre recovery from polymer composites: a technical feasibility study. Mater Res Bull 39:1549–1556. https://doi.org/10.1016/j.materresbull.2004.04.031
21. Liu T, Zhang M, Guo X et al (2017) Mild chemical recycling of aerospace fiber/epoxy composite wastes and utilization of the decomposed resin. Polym Degrad Stab 139:20–27. https://doi.org/10.1016/j.polymdegradstab.2017.03.017
22. Liu Y, Farnsworth M, Tiwari A (2017) A review of optimisation techniques used in the composite recycling area: State-of-the-art and steps towards a research agenda. J Clean Prod 140:1775–1781. https://doi.org/10.1016/j.jclepro.2016.08.038
23. Liu Y, Liu J, Jiang Z, Tang T (2012) Chemical recycling of carbon fibre reinforced epoxy resin composites in subcritical water: Synergistic effect of phenol and KOH on the decomposition efficiency. Polym Degrad Stab 97:214–220. https://doi.org/10.1016/j.polymdegradstab.2011.12.028
24. López FA, Martín MI, Alguacil FJ et al (2012) Thermolysis of fibreglass polyester composite and reutilisation of the glass fibre residue to obtain a glass–ceramic material. J Anal Appl Pyrolysis 93:104–112. https://doi.org/10.1016/j.jaap.2011.10.003
25. Mativenga PT, Shuaib NA, Howarth J et al (2016) High voltage fragmentation and mechanical recycling of glass fibre thermoset composite. CIRP Ann 65:45–48. https://doi.org/10.1016/j.cirp.2016.04.107
26. Meng F, McKechnie J, Turner TA, Pickering SJ (2017) Energy and environmental assessment and reuse of fluidised bed recycled carbon fibres. Compos Part A Appl Sci Manuf 100:206–214. https://doi.org/10.1016/j.compositesa.2017.05.008
27. Meng F, Olivetti EA, Zhao Y et al (2018) Comparing life cycle energy and global warming potential of carbon fiber composite recycling technologies and waste management options. ACS Sustain Chem Eng 6:9854–9865. https://doi.org/10.1021/acssuschemeng.8b01026
28. Naik TP, Gairola S, Singh I, Sharma AK (2022) Microwave hybrid heating for molding of Sisal/Jute/HDPE composites. J Nat Fibers 00:1–15. https://doi.org/10.1080/15440478.2022.2100553
29. Naik TP, Singh I, Sharma AK (2022) Processing of polymer matrix composites using microwave energy: A review. Compos Part A Appl Sci Manuf 156:106870. https://doi.org/10.1016/j.compositesa.2022.106870
30. Onwudili JA, Miskolczi N, Nagy T, Lipóczi G (2016) Recovery of glass fibre and carbon fibres from reinforced thermosets by batch pyrolysis and investigation of fibre re-using as reinforcement in LDPE matrix. Compos Part B Eng 91:154–161. https://doi.org/10.1016/j.compositesb.2016.01.055
31. Osborne M (2022) At least 85 percent of U.S. plastic waste went to landfills in 2021 | Smart News| Smithsonian Magazine. In: Smithson. Mag. https://www.smithsonianmag.com/smart-news/the-us-recycled-just-5-percent-of-its-plastic-in-2021-180980052/. Accessed 26 Sept 2022

32. Pickering SJ (2006) Recycling technologies for thermoset composite materials—current status. Compos Part A Appl Sci Manuf 37:1206–1215. https://doi.org/10.1016/j.compositesa.2005.05.030
33. Pickering SJ, Turner TA, Meng F, et al (2015) Developments in the fluidised bed process for fibre recovery from thermoset composites. CAMX 2015 - Compos Adv Mater Expo 2384–2394
34. Pimenta S, Pinho ST (2011) Recycling carbon fibre reinforced polymers for structural applications: technology review and market outlook. Waste Manag 31:378–392. https://doi.org/10.1016/j.wasman.2010.09.019
35. Pimenta S, Pinho ST (2012) The effect of recycling on the mechanical response of carbon fibres and their composites. Compos Struct 94:3669–3684. https://doi.org/10.1016/j.compstruct.2012.05.024
36. Piñero-Hernanz R, García-Serna J, Dodds C et al (2008) Chemical recycling of carbon fibre composites using alcohols under subcritical and supercritical conditions. J Supercrit Fluids 46:83–92. https://doi.org/10.1016/j.supflu.2008.02.008
37. Rana RS, Kumar J, Singh Y et al (2022) Ultrasonic welding of banana/ bagasse based polypropylene composites. J Nat Fibers 19:15808–15823. https://doi.org/10.1080/15440478.2022.2133054
38. Rani M, Choudhary P, Krishnan V, Zafar S (2021) A review on recycling and reuse methods for carbon fiber/glass fiber composites waste from wind turbine blades. Compos Part B Eng 215:108768. https://doi.org/10.1016/j.compositesb.2021.108768
39. Rani M, Choudhary P, Krishnan V, Zafar S (2022) Development of sustainable microwave-based approach to recover glass fibers for wind turbine blades composite waste. Resour Conserv Recycl 179:106107. https://doi.org/10.1016/j.resconrec.2021.106107
40. Sanjay MR, Madhu P, Jawaid M et al (2018) Characterization and properties of natural fiber polymer composites: a comprehensive review. J Clean Prod 172:566–581. https://doi.org/10.1016/j.jclepro.2017.10.101
41. Tapper RJ, Longana ML, Yu H et al (2018) Development of a closed-loop recycling process for discontinuous carbon fibre polypropylene composites. Compos Part B Eng 146:222–231. https://doi.org/10.1016/j.compositesb.2018.03.048
42. Wang Y, Cui X, Ge H et al (2015) Chemical recycling of carbon fiber reinforced epoxy resin composites via selective cleavage of the carbon–nitrogen bond. ACS Sustain Chem Eng 3:3332–3337. https://doi.org/10.1021/acssuschemeng.5b00949
43. Xu P, Li J, Ding J (2013) Chemical recycling of carbon fibre/epoxy composites in a mixed solution of peroxide hydrogen and N, N-dimethylformamide. Compos Sci Technol 82:54–59. https://doi.org/10.1016/j.compscitech.2013.04.002
44. Yazdanbakhsh A, Bank LC, Rieder K-A et al (2018) Concrete with discrete slender elements from mechanically recycled wind turbine blades. Resour Conserv Recycl 128:11–21. https://doi.org/10.1016/j.resconrec.2017.08.005
45. Yip HLH, Pickering SJ, Rudd CD (2002) Characterisation of carbon fibres recycled from scrap composites using fluidised bed process. Plast Rubber Compos 31:278–282. https://doi.org/10.1179/146580102225003047
46. Zabihi O, Ahmadi M, Liu C et al (2020) Development of a low cost and green microwave assisted approach towards the circular carbon fibre composites. Compos Part B Eng 184:107750. https://doi.org/10.1016/j.compositesb.2020.107750
47. Zabihi O, Ahmadi M, Liu C et al (2020) A sustainable approach to the low-cost recycling of waste glass fibres composites towards circular economy. Sustainability 12:641. https://doi.org/10.3390/su12020641
48. Zhang J, Chevali VS, Wang H, Wang C-H (2020) Current status of carbon fibre and carbon fibre composites recycling. Compos Part B Eng 193:108053. https://doi.org/10.1016/j.compositesb.2020.108053
49. Zhao X, Wang X-L, Tian F et al (2019) A fast and mild closed-loop recycling of anhydride-cured epoxy through microwave-assisted catalytic degradation by trifunctional amine and subsequent reuse without separation. Green Chem 21:2487–2493. https://doi.org/10.1039/C9GC00685K

50. Zhou J, Xu W, You Z et al (2016) A new type of power energy for accelerating chemical reactions: the nature of a microwave-driving force for accelerating chemical reactions. Sci Rep 6:25149. https://doi.org/10.1038/srep25149
51. Zhu J-H, Chen P, Su M et al (2019) Recycling of carbon fibre reinforced plastics by electrically driven heterogeneous catalytic degradation of epoxy resin. Green Chem 21:1635–1647. https://doi.org/10.1039/C8GC03672A
52. JEC World - The Leading International Composites Show. https://www.jec-world.events/. Accessed 26 Sept 2022a
53. Siemens Gamesa RecyclableBlades installed at RWE offshore wind farm | CompositesWorld. https://www.compositesworld.com/news/siemens-gamesa-recyclableblades-installed-at-rwe-offshore-wind-farm. Accessed 8 Oct 2022b
54. Teijin, Fujitsu to develop blockchain-based commercial platform for recycled materials manufacturing | CompositesWorld. https://www.compositesworld.com/news/teijin-fujitsu-to-develop-blockchain-based-commercial-platform-for-recycled-materials-manufacturing. Accessed 8 Oct 2022c
55. Igus Inc. rolls out urban bicycle made entirely from recycled plastic | CompositesWorld. https://www.compositesworld.com/news/igus-inc-rolls-out-urban-bicycle-made-entirely-from-recycled-plastic. Accessed 8 Oct 2022d
56. Italian startup, Advanced Composites Solutions develop portable solar panel via recovered carbon fiber | CompositesWorld. https://www.compositesworld.com/news/italian-startup-advanced-composites-solutions-develop-portable-solar-panel-via-recovered-carbon-fiber. Accessed 8 Oct 2022e
57. Swancor launches recyclable thermosetting epoxy resin | CompositesWorld. https://www.compositesworld.com/products/swancor-launches-recyclable-thermosetting-epoxy-resin. Accessed 8 Oct 2022f
58. UK team successfully recycles reclaimed continuous carbon fibres from composite pressure tanks | National Composites Centre. https://www.nccuk.com/news/uk-team-successfully-recycles-reclaimed-continuous-carbon-fibres-from-composite-pressure-tanks/. Accessed 8 Oct 2022g
59. CompPair and Decathlon successfully complete composite recycling project | CompositesWorld. https://www.compositesworld.com/news/comppair-and-decathlon-successfully-complete-composite-recycling-project. Accessed 8 Oct 2022h
60. Voith Composites, V-Carbon to develop closed-loop solution for CFRP used in hydrogen tanks | CompositesWorld. https://www.compositesworld.com/news/voith-composites-v-carbon-to-develop-closed-loop-solution-for-CFRP-used-in-hydrogen-tanks. Accessed 8 Oct 2022i
61. Athletic shoe developed with repurposed thermoplastic waste | CompositesWorld. https://www.compositesworld.com/articles/athletic-shoe-developed-with-repurposed-thermoplastic-waste. Accessed 8 Oct 2022j
62. Cecence, NCC and Gen 2 Carbon develop sustainable airplane seatback | CompositesWorld. https://www.compositesworld.com/news/cecence-ncc-and-gen-2-carbon-develop-sustainable-airplane-seatback-. Accessed 8 Oct 2022k
63. Boeing ecoDemonstrator tests recycled carbon fiber sidewalls, acoustic upgrades for aeroengines and more | CompositesWorld. https://www.compositesworld.com/news/boeing-ecodemonstrator-tests-recycled-carbon-fiber-sidewalls-acoustic-upgrades-for-aeroengines-and-more. Accessed 8 Oct 2022l
64. Anmet installs first recycled wind turbine blade-based pedestrian bridge | CompositesWorld. https://www.compositesworld.com/news/anmet-installs-first-recycled-wind-turbine-blade-based-pedestrian-bridge. Accessed 8 Oct 2022m
65. Carbon Rivers Makes Wind Turbine Blade Recycling and Upcycling a Reality With Support from DOE | Department of Energy. https://www.energy.gov/eere/wind/articles/carbon-rivers-makes-wind-turbine-blade-recycling-and-upcycling-reality-support. Accessed 8 Oct 2022n
66. RECONTRAS project validates use of microwaves and laser welding to obtain recyclable composites | CompositesWorld. https://www.compositesworld.com/news/recontras-project-validates-use-of-microwaves-and-laser-welding-to-obtain-recyclable-composites. Accessed 8 Oct 2022o

67. Pressurized steam-based composites recycling for full fiber reclamation | CompositesWorld. https://www.compositesworld.com/articles/pressurized-steam-based-composites-recycling-for-full-fiber-reclamation. Accessed 8 Oct 2022p
68. Precision pyrolysis: Controlled, heat-based composites recycling for higher material properties |CompositesWorld. https://www.compositesworld.com/articles/precision-pyrolysis-controlled-heat-based-composites-recycling-for-higher-material-properties. Accessed 8 Oct 2022q

Drilling of Ceramics Using Microwave Energy at 2.45 GHz

Gaurav Kumar

Abstract The current chapter discusses the development in microwave drilling over the years. Microwave drilling is a non-conventional machining process that uses a metallic tool to convert microwave energy into thermal energy, which is then used for the machining of the workpiece. Material is removed from the workpiece during microwave drilling due to the combined action of microwave heating and plasma heating. Microwave drilling has been successfully used for drilling holes in various materials like concrete, glass, ceramics, bones, metals, etc. A 6 mm diameter and 13 mm deep hole in alumina was drilled in 2 min at a microwave input power of 0.9 kW. The smallest hole of 0.5 mm diameter was drilled in borosilicate glass and soda-lime glass; however, defects were present around hole. Use of dielectric in dynamic mode helps in reducing the defects like heat affected zone, overcut; but machining time gets increased slightly.

Keywords Microwave drilling · Alumina · Glass · Power · Heat affected zone

1 Introduction

Global demand for micro-machined components has significantly increased due to Micro Electromechanical Systems (MEMS) and other devices used in micro-electronics (semiconductor devices, Integrated Circuit (IC) technology), aerospace, medical, and other industries [1, 4]. This requires that the manufacturing technologies are robust and capable of producing precise micro components with complex features in less processing time. Micro-machining techniques like ultrasonic machining, micro-electrochemical machining, electrochemical discharge machining, laser beam machining, electric discharge machining, etc. can produce micro-machined components with reasonable precision. However, there are inherent limitations like low

G. Kumar (✉)
Department of Mechanical Engineering, National Institute of Technology Rourkela, Rourkela, Odisha 769008, India
e-mail: kumarg@nitrkl.ac.in

M. K. Singh et al. (eds.), *Composite Materials Processing Using Microwave Heating Technology*, Composites Science and Technology,
https://doi.org/10.1007/978-981-97-2772-8_5

material removal rate, conditional processing (for example, dependence on electrical conductivity and ductility of the workpiece), etc. The microwave drilling has shown great potential to overcome the above limitations. The evolution of microwave drilling over the years has been summarized in subsequent sections.

2 Microwave Drilling

Microwave drilling is a thermal energy-based machining process that uses a concentrator to convert electromagnetic energy into thermal energy, which is then used for the machining of the workpiece. Jerby et al. developed the microwave drill for the first time whose schematic is shown in Fig. 5.1. The drilling machine consisted of a magnetron, an impedance-matching tuner, a directional coupler, a coaxial guide, and an extendable monopole antenna (Drill bit). The localized concentration of the microwave caused a hot spot to form at the surface of the material directly below the drill bit, which was pierced by the drill bit made up of Tungsten. The concentrator was pulled out mechanically to create the hole. Pressurized airflow was used for cooling the coaxial structure to sustain 1 kW power transmission. The developed machine drilled holes in various materials like glass, concrete, ceramics, basalt, and silicon. The newly invented process demonstrated its potential by drilling a 2 mm diameter hole up to 2 cm depth in concrete within a minute with precision and accuracy comparable to a mechanical drill [8]. Holes of diameters ranging from 0.5 mm to 13 mm were drilled on concrete, silicon, glass, plastic, ceramics, bones, rocks, and wood using microwave drilling [2, 5, 7–9]. A maximum drilling speed of 1.2 mm/s was achieved while drilling concrete with a 2.4 mm diameter drill at an effective microwave power of 0.7 kW [6].

Later, Meir and Jerby designed a new experimental set-up and reported that materials like glass, nylon, etc., can be drilled even at a lower input power of 10 W than the previously reported input power of 600 W. The transistor-based microwave drill consisted of a coaxial structure, a 1 mm diameter movable electrode, an amplifier, an isolator, and a reflectometer. The solid-state amplifier, which employed a single laterally diffused metal oxide semiconductor field-effect transistor (LDMOS-FET) evaluation board, generated microwave power up to 140-W with positive feedback. The improved experimental set-up drilled materials like glass, silicon, ceramics, and basalt at lower power than the previous experimental set-up [19].

A domestic microwave oven, a spring with one end fixed to the top beam of the resonant chamber and the other to a metallic tool were used in the experiment by George et al. to drill various metallic materials. The tool, under the application of the spring force, penetrated the workpiece softened due to microwave hybrid heating to drill holes in aluminium, copper, and stainless-steel sheet of 1 mm thickness [3]. Lautre et al. reported drilling of materials using a domestic microwave oven. The jigs and fixtures made of microwave transparent materials were used to hold the concentrator just above the surface of the workpiece. A low deadweight of 15.7 g was used to gravity feed the tool. The entire set-up was kept inside a multi-mode

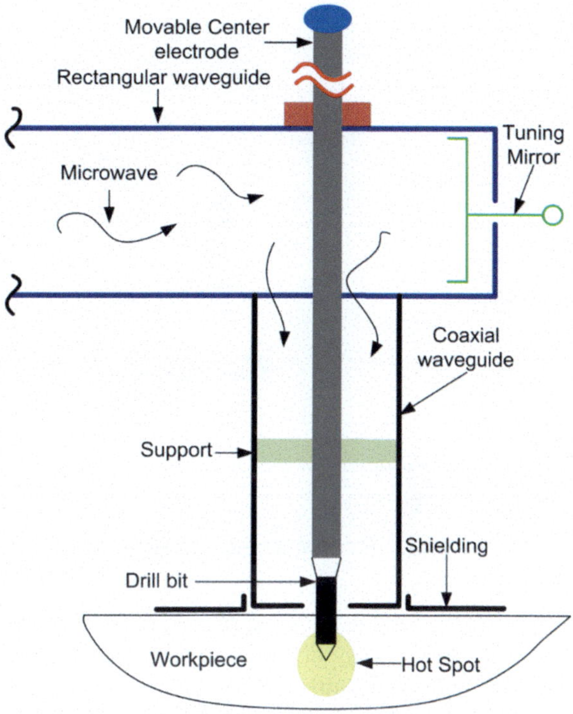

Fig. 5.1 Microwave drilling process [8]

domestic microwave applicator's resonating cavity (440 × 420 × 210 mm). The input powers of 90 W, 180W, 360W, 600 W, and 900 W considering the duty cycle (DC) as 20%, 25%, 50%, 75%, and 100%, respectively, were applied during the drilling of materials like soda-lime glass and perspex. It was reported that a tool made of materials like copper, stainless steel, etc., helped concentrate the microwave at its tip and generated plasma, which was used for machining of the workpiece [14–17]. Drilling of holes in borosilicate glass using microwave energy at input power of 700 W (Fig. 5.2) was reported by Kumar and Sharma. Use of transformer oil dielectric and concentrators such as graphite and WC produced better results and significantly reduced the heat affected zone (HAZ) [12]. Use of controlled feed rate, constant machining gap and preheating improved the quality of hole significantly. Immersion of tool and workpiece contact zone inside a dielectric (Transformer oil) up to a depth of 45 mm reduced HAZ, overcut and improved the circularity of holes significantly [11]. Later, Singh and Sharma proposed another method which consisted of a domestic microwave oven, a pin vice fixture, a pin vice, and a cylindrical pointed tool of ϕ1 mm to drill a hole in metallic materials without applying mechanical force. To ensure the formation of plasma channels, the tool made of thoriated tungsten was kept at a specific distance from the workpiece's surface. The plasma channels melted and/or ablated the workpiece locally. The arrangement could drill holes in 0.6 mm thick stainless steel (SS) [21].

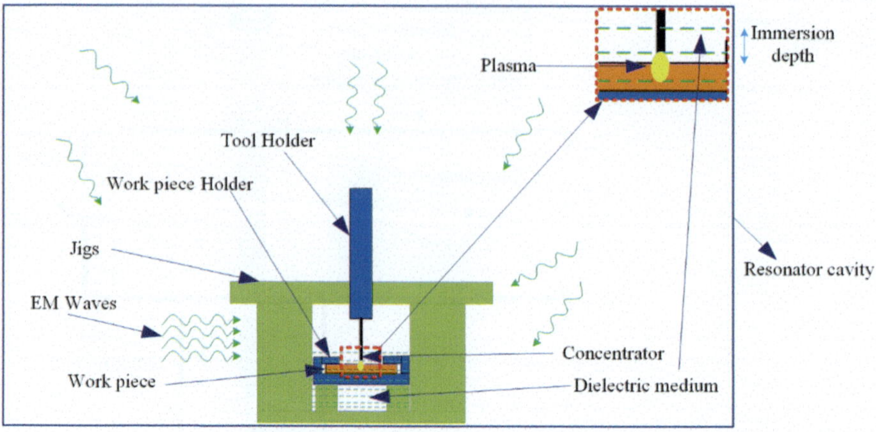

Fig. 5.2 Microwave drilling of glass using dielectric [12] (Adapted from Kumar and Sharma)

3 Theory of Microwave Drilling

Jerby et al. reported that the microwave drilling is governed by the concentration of electromagnetic energy in an area smaller than the microwave wavelength using a drill (Fig. 5.3). The localized electromagnetic energy dissipated in the workpiece due to its dielectric loss factor generated intense thermal energy that softened/melted the material. The drill bit antenna was inserted manually to shape the hole's boundary [8]. Lautre et al. reported that plasma formed at the tip of the concentrator generated high thermal energy which in turn, melted/ablated the target area in the workpiece [16]. Kumar and Sharma reported that the plasma heating as well as microwave heating caused micromachining of target material in the workpiece [11].

Jerby et al. reported that Maxwell's equations and Fourier equations govern the microwave drilling process. The Electromagnetic waves resonate in the resonating chamber of the microwave applicator or coaxial structure of the microwave drill [6]. The electric field (E) at the tip of the antenna or concentrator is given by (Pozar

Fig. 5.3 Mechanism of Microwave drilling process [6] (Adapted from Jerby et al.)

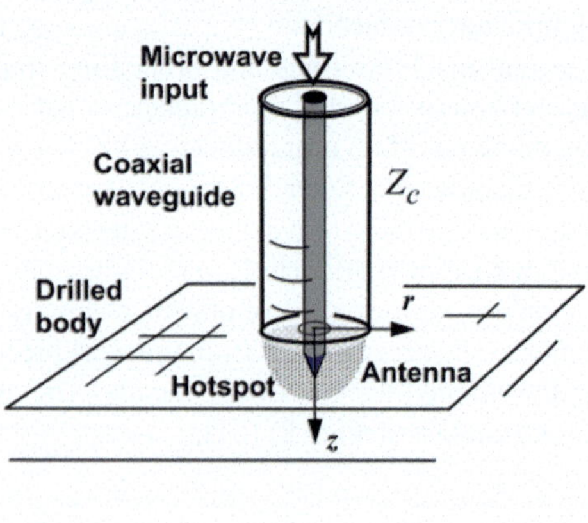

2021):

$$\nabla \times \mu_r^{-1}(\nabla \times E) - k_0^2\left(\varepsilon_r - \frac{j\sigma_e}{\omega\varepsilon_0}\right)E = 0 \tag{5.1}$$

where μ_r is the relative permeability, E is the electric field intensity, k_0 is the wavenumber in free space, ε_r is the relative permittivity, σ_e is the electrical conductivity, ω is the angular frequency, and ε_0 is the permittivity of vacuum.

The intensity of electric fields around the tool's tip is very high and ionizes the medium around it, resulting in plasma formation. The newly ionized atoms and molecules collide with others and ionize more atoms in an avalanche process, which leads to spark formation at the tool tip (Fig. 5.4). Initially, material removal occurs due to the heating action of plasma produced at the concentrator's tip. The electron's input power, $P_{i,e}$ during discharge, is given by Lieberman and Alan [18]:

$$P_{i,e} = \frac{e^2 N_e}{m_e \vartheta_{m,e}} E^2 \tag{5.2}$$

where e is an electron's charge, N_e is the electron's number, m_e is the electron's mass, $\vartheta_{m,e}$ is the electron's collision frequency.

The plasma generated at the tip of the concentrator interacts with the target material placed just underneath it (Fig. 5.4). Initially, the plasma formed at the concentrator's tip heats the workpiece surface locally. Consequently, the material just beneath the concentrator gets partially melted/ablated as shown in Fig. 5.4. This rapidly increases the temperature of the material near the melted/ablated surface. Plasma formed at the tool tip exerts recoil pressure on molten material in the machining zone, and Marangoni stress is developed due to the temperature gradient in the machining zone. Thus, molten/ablated glass gets ejected away from the machining zone due to the recoil pressure and Marangoni stress.

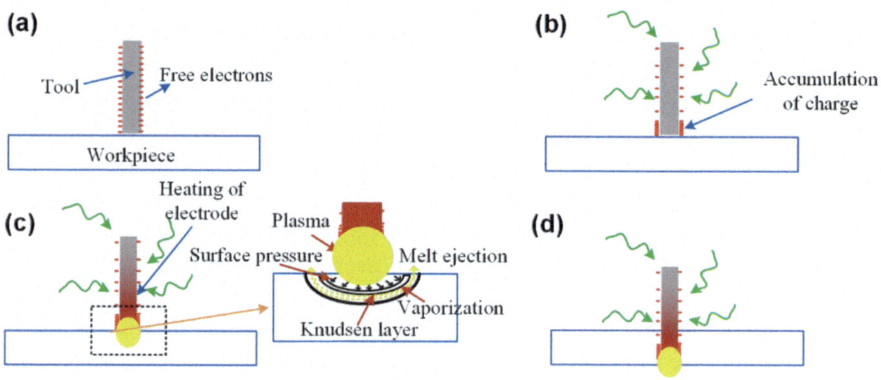

Fig. 5.4 Mechanism of the microwave drilling process [12] (Adapted from Kumar and Sharma)

Further, Xi and Tinga [22] reported that the dielectric constant and dielectric loss factor of the material increase exponentially as the temperature in the exposed zone rises [22]. When the dielectric loss factor reaches a certain value, microwave heating also aids in material removal. The power density (P_d) due to microwave heating is given by [11]:

$$Pd = \pi f \varepsilon\varepsilon{\prime\prime}|E|^2 \tag{5.3}$$

where f is frequency, ε'' is dielectric loss factor.

Thus, it can be inferred from Eqs. (5.2) and (5.3) that the change in electric field intensity and dielectric loss factor affect the power density around the tool tip which will impact the precision and accuracy of the hole drilled and defects around it. With an increase in temperature, microwave heating also assists in hole drilling. Thus, material gets removed due to microwave heating as well as heating due to plasma formed at tool's tip.

4 Drilling of Materials

There are various ways of drilling engineering materials using microwave energy, as described in the previous sections, but the core principles of microwave drilling are nearly the same in all methods. Some researchers had used magnetron to generate microwave energy [8, 12, 14], whereas some used a transistor-based system [19] to generate microwaves. The use of a concentrator to convert microwave energy into thermal energy remained the same in all the methods. A variety of engineering materials were drilled using these methods. The subsequent sections describe the characteristics of the holes drilled in these materials.

4.1 Drilling of Alumina and Thermal Barrier Coatings

Jerby et al. reported microwave drilling of alumina. The microwave drill was able to drill hole (Diameter = 6 mm; Depth = 13 mm) in alumina in 2 min at a microwave input power of 0.9 kW. The molten material on the bore perimeter was mechanically removed during or shortly after the drilling [8]. Jerby and Thompson showed the selectivity of microwave drilling by drilling holes only in ceramic coatings on metallic substrates. The developed technology may be useful in fabricating advanced gas turbine components. A microwave drilling process was used to uncover an array of holes in ceramic thermal barrier coatings. It was observed from a top view of the hole after microwave irradiation that the hole surface was smoother, and microwave drilling didn't change the microstructure of the material lying underneath the thermal barrier coating (TBC). However, ceramic at the circumference of the hole drilled by microwave was found to be dense and cracked due to the molten pool generated in the microwave drilling. The microwave drilling operation was observed to be terminated once it reached the metal surface. Further, they reported that drilling operation was found to be better in porous TBC as compared to dense TBC [9]. Jerby et al. reported that microwave drilling might add a local sintering effect and, thus, strengthen the materials shaping the wall in ceramics and aggregated materials [8].

4.2 *Drilling of Glass*

Holes of diameter 0.5 mm, 1.6 mm and 2.4 mm were drilled in borosilicate glass and soda-lime glass. Pre & post heating of glass minimized the crack formation around hole drilled in glass [5]. Later, Meir and Jerby [19] reported that a transistor-based microwave drill was capable of drilling crackless holes on the glass even in the power range of power (10–100 W), thus eliminating the need for preheating before drilling. However, a decrease in power from 100 to 80 W increased the time to melt (TTM) from 10 to 17 s. Drilling of ~ϕ1 mm hole in a soda-lime glass of 1 mm thickness was reported, but tool wear and material removal rate were not reported. Simulation results showed that a time of 17 s was needed to reach the softening temperature of glass near the tip of the electrode during microwave drilling of a 1 mm thick soda-lime glass plate at an input power of 80 W and a frequency of 2.1 GHz. The electric field distribution showed that the electric field intensity at the tool's tip was more than 3.5×10^6 V/m, and the axial component of the electric field was more dominant. The origin of heat from a point 0.1 mm below the tip of the tool, as revealed by the normalized heat flux vectors, confirmed that the heat is induced locally. It was reported that the efficiency of the drilling process could be enhanced by the ejection of plasma from the surface at a high temperature in the workpiece towards the monopole antenna. The electrode was lifted for a few seconds after the localized microwave heating of the target material to generate plasma, which assisted the local preheating and enhanced the efficiency of the microwave drilling process. However, mechanical action was needed to remove the porous glass from the machining zone, which deteriorated the quality of the hole in case of drilling of the thicker glass plate.

Lautre et al. reported that low power (90 W) was more effective as compared to 900 W in microwave drilling of glass as high power produces a larger diameter plasma sphere which gets spread over the surface of the workpiece and produces an undesired effect like heat-affected zone. Thermal imbalance was the prime reason for the circular crack and linear radial crack. The tool was found to be stuck in the partially melted glass in the machining zone after drilling up to a certain depth due to its ineffective flushing. This resulted in deformation and burning of the tip of the tool. Further, the presence of cracks around the hole was significantly affecting the hole accuracy. The hole overcut at the entrance and exit was ranging from 0.31 mm to 0.37 mm while drilling soda-lime glass with the tool of 0.25 mm diameter. The taper was approximately 1.2° while drilling a 1.2 mm thick workpiece; however, circularity of the hole was very poor at the entrance [16]. Photoelasticity and simulation were used to study the induced stress during microwave drilling of glass [15]. Simulation study revealed that thermal loading (90–900 W) generated sufficient stress to initiate a crack in the surface of the workpiece. Interaction of different heat-affected region, variation of stress from tensile to compressive at the extreme boundary of the HAZ region in radially outward direction, relatively lower temperature of the tool as compared to molten glass surface temperature were some of the cited reasons behind the initiation of crack. Surface treatment of glass top surface helped in reducing the crack dimension up to a certain extent but didn't eliminate it completely. The precursors aided in reducing thermal shock in glass during machining by allowing energy to be transferred gradually from the plasma sphere at the machining zone to the glass surface [15]. Precursors reduced the thermal intensity of the plasma sphere, which, in turn, reduced thermal stress. In terms of crack control with minimal HAZ, glycerin and perspex precursors produced the best results because the liquid precursor attenuates the heat energy before it reaches the glass specimen. As a microwave absorber, the precursor reduces the induced thermal stress in the drilling zone. However, the inefficient dissipation of extra thermal energy from the machining zone due to the low volume of precursors used was the main reason behind the significant heat affected zone around the hole [14]. Kumar and Sharma used different oil dielectrics and concentrators while drilling holes in borosilicate glass plate using microwave energy at input power of 700 W. The dielectric's dielectric constant and the concentrator's electric conductivity was found to influence the shape and intensity of the plasma. A graphite concentrator with a diameter of 0.6 mm at an immersion depth of 35 mm in transformer oil produced the hole with minimum defects. Higher immersion depth helped in better heat dissipation from the machining zone and helped in minimizing the defect to a larger extent [12]. Further, microwave drilling with a conical tip tool produced a higher quality hole than drilling with a cylindrical tip tool. Further, optimum feed rate and machining gap helped in further decreasing the defects in the drilled workpiece by synchronizing the energy channelization from tool to workpiece and better flushing of the glass residue from the machining zone [13]. It can be observed from Fig. 5.5 that an optimum machining gap of 0.3 mm provides the hole with minimum heat affected zone and minimum roundness error as it facilitates better circulation of dielectric. It was also observed that preheating helps in minimizing thermal stress by minimizing temperature gradient which in turn minimizes the crack formation around the hole. The best quality hole in stagnant dielectric was drilled at a machining gap of 0.3 mm, feed rate of 1.2 mm/s, immersion depth of 45 mm, and preheating time of 10 s).

Further, Kumar et al. observed that combination of flowing dielectric and lower input power produced hole in borosilicate glass with minimum defect due to efficient removal of unwanted thermal energy and glass residue from the machining zone. However, the time required to drill hole was relatively higher due to low input power. Efficient heat flushing from the machining zone minimized the thermal stress around the hole drilled to a large extent. The parametric combination of input power of 70 W and flow rate of 204 cm^3s^{-1} produced hole with minimum HAZ, overcut, and thermal stresses during microwave drilling [10].

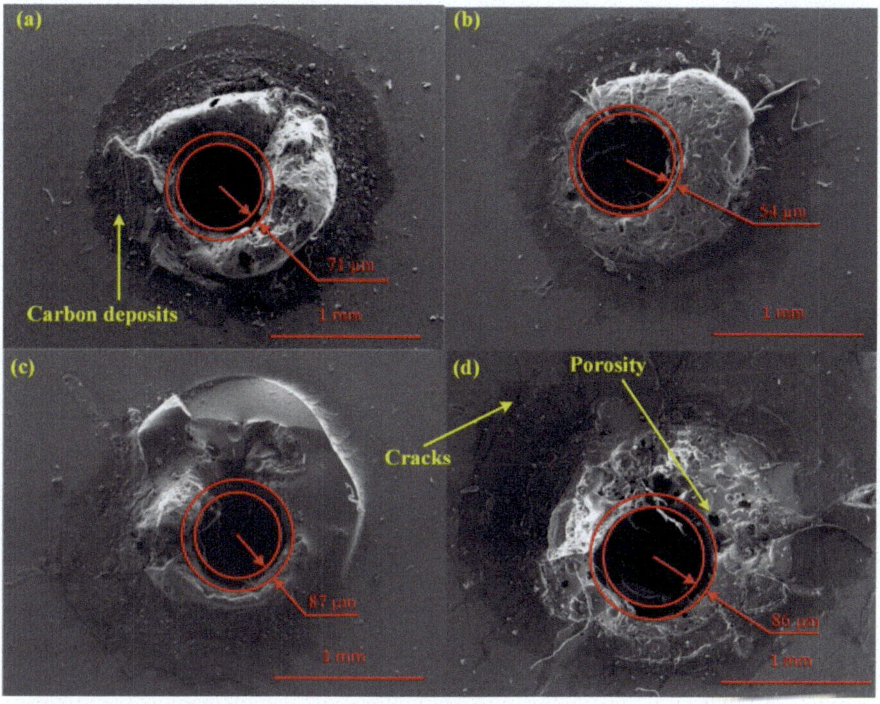

Fig. 5.5 Image of hole drilled in borosilicate glass [13] (Adapted from Kumar and Sharma)

Therefore, it is concluded from the above literature that there is a need to have better control on the process to have better machining performance and drill hole having better surface quality as the process has a huge potential in the field of rapid manufacturing.

5 Challenges in Microwave Drilling

Many advanced machining processes were developed to fabricate micro components with a high degree of precision at low cost and time. Microwave drilling of materials has been explored in the last two decades. However, the process is yet to be matured for industrial applications. This is primarily due to less studied physics of the processes and limited research on microwave drilling. The challenges associated with microwave drilling are briefly discussed in the following sections:

5.1 Limited Information

The physics of microwave drilling is largely unexplored; a few reports about the mathematical modeling are available [6]. The mathematical model used to describe the microwave drilling process neglected the plasma effect during the microwave drilling process and considered only the heating effect due to microwave. Further, there are no mathematical models available that explain the mechanism of material removal from the workpiece during the microwave drilling process. The physics of spark formation during microwave drilling is still unexplored. Further, the inability to access real-time data of temperature, electron density, etc. of plasma during the machining process makes it difficult to develop its mathematical model and explain the physics.

5.2 Power

The input power in the microwave drilling process plays an important role as it controls the amount of thermal energy during machining. It is challenging to calculate the variation and exact value of temporal power during the drilling process due to the inability to access the machining zone and the complexity of the mathematical formulation. Therefore, it becomes difficult to precisely calculate the value of power required to machine a specific workpiece during the microwave drilling process.

5.3 Special Tooling Requirement

The materials which are required for developing zigs and fixture during the drilling process should be transparent to microwave. The jigs and fixtures made of metallic material will reflect the microwave and may decrease the process efficiency; the magnetron may also get damaged. The jigs and fixtures made of a material having higher dielectric loss will absorb microwave and will produce a heating effect which may damage the fixture itself and may affect machining accuracy.

6 Scope for Future Work

Any investigation opens many other possibilities for further studies. A few of such possibilities which offer scopes for future research include:

(1) A system to facilitate axial rotation of concentrator tool can be developed which may further help in reducing defect like roundness error, HAZ etc. in the workpiece.

(2) The effect of tool surface roughness, cone angle, etc. on process performance can be explored in future.
(3) Effect of applying magnetic field, ultrasonic vibration etc. on process performance can also be explored in future.
(4) The microwave heating effect can further be incorporated in the model used in the present simulation study to obtain more realistic results.
(5) The physics of discharge formation at tool tip can also be explored through simulation study.

7 Conclusions

Based on above study, following conclusions can be drawn:

1. Microwave drilling is an advanced machining process which is used for machining of materials irrespective of its conductivity.
2. Material gets removed from the workpiece due to microwave heating as well as plasma heating in microwave drilling.
3. Microwave drilling can be used for controlled material removal from ceramic coatings on metallic substrates which may be useful in fabricating advanced gas turbine components.
4. The smallest hole of 0.5 mm has been drilled in glass using microwave energy at 2.45 GHz.
5. Dielectric in dynamic mode has proven to be useful in removing excess thermal energy and residue from machining zone, which in turn improves the hole quality significantly.

References

1. Aramcharoen A, Mativenga PT, Yang S et al (2008) Evaluation and selection of hard coatings for micro milling of hardened tool steel. Int J Mach Tools Manuf 48:1578–1584
2. Eshet Y, Mann RR, Anaton A et al (2006) Microwave drilling of bones. IEEE Trans Biomed Eng 53:1174–1182
3. George TJ, Sharma AK, Kumar P (2012) A Feasibility study on drilling of metals through microwave heating. i-manager's J Mech Eng 2:1–6
4. Huo D (2013) Micro-cutting: fundamentals and applications. John Wiley and Sons
5. Jerby E, Aktushev O, Dikhtyar V, et al. (2004) Microwave drill applications for concrete, glass and silicon. In: Proc 4th World Congr Microw Radio-Frequency Appl, pp 156–165
6. Jerby E, Aktushev O, Dikhtyar V (2005) Theoretical analysis of the microwave-drill near-field localized heating effect. J Appl Phys 97
7. Jerby E, Dikhtyar V (2006) Drilling into hard non-conductive materials by localized microwave radiation. In: Microw Radio Freq Process–Rep from 8th Int Conf Microw High Freq Heat, pp 687–694
8. Jerby E, Dikhtyar V, Aktushev O, Grosglick U (2002) The microwave drill. Science 298:587–589

9. Jerby E, Thompson AM (2004) Microwave drilling of ceramic thermal-barrier coatings. J Am Ceram Soc 10:308–310
10. Kumar G, Mishra RR, Sharma AK (2021) On finite element analysis of material removal rate in microwave drilling of borosilicate glass. Mater Today Proc 41:759–764
11. Kumar G, Mishra RR, Sharma AK (2020) On defect minimization during microwave drilling of borosilicate glass at 2.45 GHz using flowing dielectric and optimized input power. J Therm Sci Eng Appl 13(3):031021
12. Kumar G, Sharma AK (2018) Role of dielectric fluid and concentrator material in microwave drilling of borosilicate glass. J Manuf Process 33:184–193
13. Kumar G, Sharma AK (2020) On processing strategy to minimize defects while drilling borosilicate glass with microwave energy. Int J Adv Manuf Technol 108:3517–3536
14. Lautre NK, Sharma AK, Das S, Kumar P (2015) On crack control strategy in near-field microwave drilling of soda lime glass using precursors. J Therm Sci Eng Appl 7:1–15
15. Lautre NK, Sharma AK, Kumar P, Das S (2015) A photoelasticity approach for characterization of defects in microwave drilling of soda lime glass. J Mater Process Technol 225:151–161
16. Lautre NK, Sharma AK, Pradeep K, Das S (2015) A simulation approach to material removal in microwave drilling of soda lime glass at 2.45 GHz. Appl Phys A Mater Sci Process 120:1261–1274
17. Lautre NK, Sharma AK, Pradeep K, Das S (2020) Experimental evaluation of a microwave drilling process in perspex. J Test Eval 48:20180103
18. Lieberman MA, Alan JL (2005) Principles of plasma discharges and materials processing. John Wiley & Sons
19. Meir Y, Jerby E (2012) Localized rapid heating by low-power solid-state microwave drill. IEEE Trans Microw Theory Tech 60:2665–2672
20. Pozar DM (2011) Microwave Engineering. 2nd edn. John Wiley and Sons
21. Singh A, Sharma AK (2019) Effect of Drilling Conditions on Microwave-Metal Discharge during Microwave Drilling of Stainless Steel. Appl Mech Mater 895:253–258
22. Xi W, Tinga W (1993) Error analysis and permittivity measurements with re-entrant high temperature dielectrometer. J Microw Power Electromagn Energy 28:104–112

Casting of Materials Using Microwave Energy

T. Lachana Dora, Shruti C. Bhatt, Nilesh D. Ghetiya, and Radha Raman Mishra

Abstract Microwave energy has been widely used in various material processing methods such as drying, sintering, joining, and cladding processes. However, its use has been very limited in the field of casting of materials. Microwave casting has gained significant attention in recent years due to its potential to reduced processing time and energy consumption, improved quality of castings, and reduced porosity over conventional casting methods. This chapter provides a brief idea about the fundamentals associated with the microwave casting process. Major research reported in this area have been highlighted. The role of different process parameters as well as the challenges and opportunities associated with the microwave casting process have been briefly discussed.

Keywords Casting · Microwave energy · Metallic materials · Melting · Solidification

1 Introduction

Microwave energy-based processing of engineering material has emerged as a novel technique for many applications due to its eco-friendly nature [1–3]. Relevant literature on microwave energy-based technologies indicate that these processes improve material properties and offer significant advantages in terms of reduced cost, time saving, and high energy efficiency as compared to the conventional processes [4, 5]. It is noted that the use of microwave energy in the field of material processing has significantly grown (Fig. 6.1) over the years; however, due to their distinctly

T. L. Dora · R. R. Mishra (✉)
Department of Mechanical Engineering, Birla Institute of Technology and Science Pilani, Pilani 333031, India
e-mail: rraman.mishra@pilani.bits-pilani.ac.in

S. C. Bhatt · N. D. Ghetiya
Department of Mechanical Engineering, Institute of Technology, Nirma University, Ahmedabad 382481, India

M. K. Singh et al. (eds.), *Composite Materials Processing Using Microwave Heating Technology*, Composites Science and Technology,
https://doi.org/10.1007/978-981-97-2772-8_6

Fig. 6.1 A timeline of microwave processing of metallic materials

unique characteristics of material interaction, it is difficult to process all engineering materials with microwaves. However, it is worth noting that various researchers have attempted processing of all the major categories of engineering materials, including metals, ceramics, polymers, and their composites using microwave energy [6, 7]. Based on the processing temperature, microwave processing of materials can be classified into three major categories—(i) Low temperature (<500 °C) which is suitable for synthesis, polymer processing, and drying applications), (ii) Medium temperature (500 °C to 1000 °C) which is suitable for sintering applications of metals, ceramics, and glass. 3) High temperature (>1000 °C) which is suitable for advanced material processing such as high melting point metals and alloys, ceramics, and composites [2]. Initially, it was considered that metals reflect microwave energy and microwave processing of metals in any form is impossible. However, in year 1999, successful microwave sintering of metals/alloys provided new possibilities for microwave energy-based metal processing. In the last two decades, microwave energy has been used in various metal processing methods like sintering, cladding, joining, drilling, melting, and additive manufacturing. Figure 6.1 indicates a timeline for different materials and processing techniques reported by various researchers.

Among all reported microwave processing techniques, microwave sintering of metallic materials is the most established process for metal powders, alloy powders, and composite powders. Microwave sintering offers significant advantages in terms of energy saving and improved material properties compared to conventional sintering [8–10]. In microwave cladding of composite and ceramic powders having varying sizes (micro/nano) on the bulk metallic substrates have been reported with complete melting and diffusion of the clad layer with substrate [11–13]. Microwave joining of similar and dissimilar metallic materials using pure/alloy powders have been reported with improved mechanical properties [14–16]. Microwave energy has also been used in drilling of thin metal sheets to develop the proof of concept [17]. Microwave energy-based melting and solidification of materials of low temperature metals and alloys was first reported in the year 2005. A plenty of literature have been reported on experimental and simulation studies on microwave melting/casting of metallic materials [18–20]. Incremental solidification of the metal and alloys have been reported using localized microwave heating. The present chapter focuses on microwave casting techniques and a detailed discussion on fundamentals, casting approaches, experimental and simulation studies, and casting outcomes have been presented.

2 Fundamentals

2.1 Heating Mechanism

The most crucial step in casting process is heating of the charge material which can be in the form of powder or bulk metallic materials. In conventional heating processes, energy gets transferred from surface of the charge to core of the charge and the rate of heat flow is constrained by temperature, thermal conductivity, heat capacity, and thermal diffusivity of the charge. It leads to non-uniform heating, super heating above melting point, defects in cast, and energy wastage. Microwave energy offers volumetric heating that gets evolved from core of the charge to surface of the charge. It is attributed to dissipation of energy at atomic level through interaction of waves with material at atomic level [21]. In microwave casting, the charge material can be exposed to microwaves using two different techniques: (1) Direct heating, and (2) Indirect heating. In case of direct heating, the material is directly exposed to microwaves (Fig. 6.2a); whereas, indirect heating, which is also known as hybrid heating, utilizes a susceptor material (highly microwave absorber) to prevent reflection of microwaves from the surface of material (Fig. 6.2b). Microwave processing of metallic powders ($d < 100\ \mu m$), ceramics and plastics are possible by both the techniques. On the other hand, processing of the bulk metallic materials can only be possible by indirect heating [22]. In case of direct heating, microwaves easily interact with materials and generate heat inside the treated substance but overheating of the materials may cause hot spot formations and creates thermal runaway [23]. Thermal instabilities generate non-uniformity and cracking in materials with direct heating. Indirect heating approach is more significant to address the issues with direct microwave heating. In indirect heating or microwave hybrid heating different material components are required for casting applications in a controlled atmosphere.

Due to their excellent conductivity and low penetration depth, metallic-based materials have shown distinct microwave heating properties than ceramic-based materials. However, the skin depth of metal surges with increases in temperature. The electromagnetic and thermophysical properties of metals varies with temperature during microwave irradiation. The heating and melting of charge material can be obtained by frequency-transient response with heat conduction Eq. (1). However,

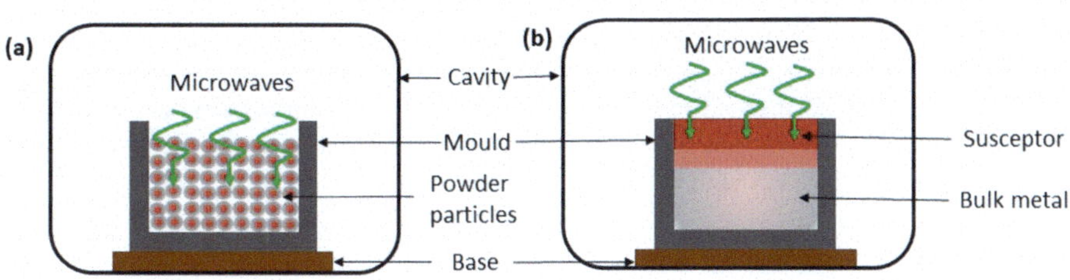

Fig. 6.2 Microwave energy-based heating **a** direct heating **b** indirect heating

the convective and radiative losses are dominant at high temperature compared to conduction loss. The estimated heat losses from the surface of material can be expresses as Eq. (2). The microwave heating of bulk metals with active heat loss mechanism has been depicted in Fig. 6.2. The microwave absorption rate varies in case of powder metal and bulk metal due to their size, shape and electromagnetic properties. Microwave casting can be applied to polymers and metallic materials including powders, composite powders, and bulk metals. The direct heating is not possible for the bulk metals or metallic powders above 100 μm due to low penetration depth and reflection of microwaves from the surface at room temperature. Microwave Hybrid heating enables microwave heating of coarser and bulk metals. Initially, temperature of the material increases due to heat transfer; consequently, it starts absorbing microwave energy above its critical temperature. In hybrid heating, the tooling assembly design and selection of materials for tooling are an important step to process bulk metals and powder-based metals. Moreover, the electromagnetic and thermo-physical properties of the metal charge determine rate of heat dissipation inside a material due to microwave-material interaction.

$$\nabla \times \frac{1}{\mu_r}(\nabla \times E) - k_0{}^2\left(\varepsilon_r - j\frac{\sigma}{\omega\varepsilon_0}\right)E = 0 \tag{1}$$

where, $\mu_r = \mu_r{}' - j\mu_r{}''$, $\varepsilon_r = \varepsilon_r{}' - j\varepsilon_r{}''$

E = Electric field, ε_0 = Vaccum permittivity, k_0 = Wavenumber of free space, ω = Angular frequency, μ_r = Relative permeability, ε_r = Relative permittivity, σ = Electrical conductivity

$$\rho C_p \frac{dT}{dt} = \nabla(\mathrm{k}\nabla\mathrm{T}) + Q_e \tag{2}$$

where, T = Temperature, ρ = density, k = Thermal conductivity and C_p = Heat transfer coefficient.

2.2 Microwave Casting Approaches

Microwave energy-based casting can be classified based on the solidification techniques used for developing the cast. A general classification of the microwave casting is shown in Fig. 6.3.

Different elements of in-situ and ex-situ microwave casting process are shown in Fig. 6.4. Multi tooling assembly setup is required for in-situ casting process. This assembly setup is made up of (1) Pouring basin (2) Sprue (3) Mold (4) Susceptor and, (5) Casket. Pouring basin, sprue, mold, and susceptor work as microwave absorber due to high electric loss factor. The casket works as a microwave transparent material whereas charge works as a microwave reflective material. Microwave casting of materials can be achieved with two different methods: (i) In-situ microwave casting, (ii)

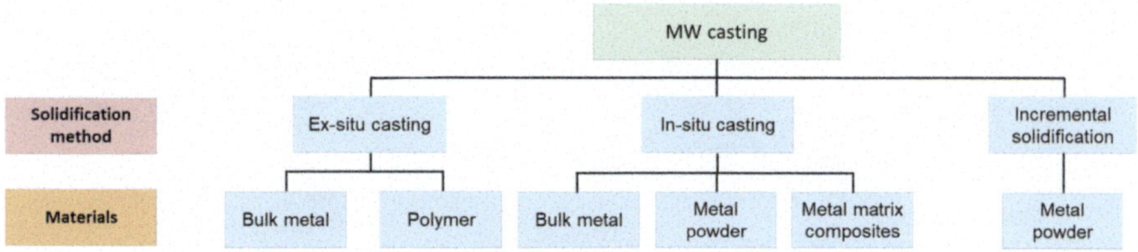

Fig. 6.3 Classification of microwave casting process

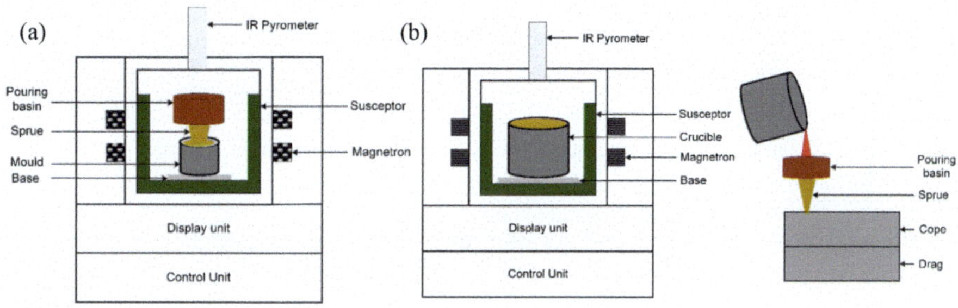

Fig. 6.4 Schematic illustration of **a** in-situ microwave casting and **b** ex-situ microwave casting

Ex-situ microwave casting [24]. The required steps to melt and cast charge material has been depicted in Fig. 6.4a, b. In the first method, melting, pouring and solidification of charge gets accomplished inside the microwave applicator. Whereas, in the second approach the charge is melted inside the applicator cavity and transferred to a mould that is positioned outside the cavity. The first approach offers control over the mould preheating temperature, flow of the molten metal, and grain structure refinement by putting the mould assembly with the charge within the cavity. On the other hand, the second approach utilises microwave radiation for melting of the charge. Pouring and solidification steps are similar to conventional casting procedure.

2.3 Process Parameters

Figure 6.5 illustrates the various parameters involved in microwave casting. The parameters can be categorised in three different groups based on their association with charge, microwave exposure and solidification. The mass, shape, size and composition of the charge governs microwave absorption in it and time required for melting the charge. The parameters that need to be controlled while microwave exposure are power, frequency, time, surrounding medium and design of mold/fixture. These parameters control the charge heating and subsequent melting of the charge, preheating temperature of mold, meting time of the charge. The parameters associated with solidification step are mold composition, mold temperature, gating design,

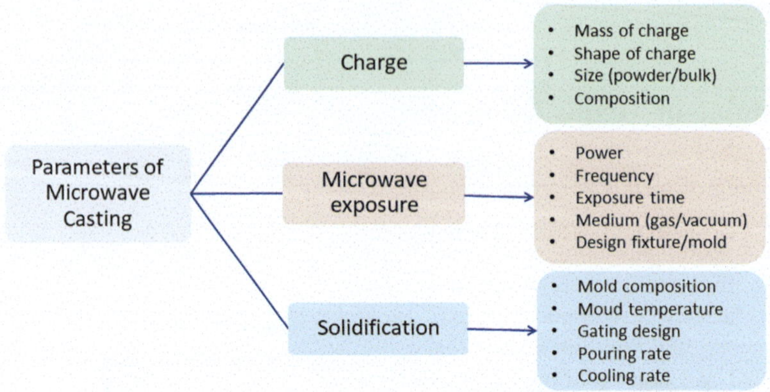

Fig. 6.5 Process parameters involved in microwave-based casting metallic material

pouring rate of molten metal, and cooling rate of melt. These parameters affect cast properties such as microstructure, and mechanical properties.

3 Experimental and Simulation Studies

Based on the available literature, most of the theoretical and experimental research on microwave casting of metallic based material were focused on either bulk metal or powder metal. The experimental and simulated analysis has been reported for pure metals (tin, lead, zinc), metals alloy powders (Al1050, AA7039, Al6061), and metal matrix composites considering powder and bulk form of charge material. Microwave heating and melting performance has been investigated by Xu et al. [25] for Tin powder. The improved heating efficiency considering volumetric heating rate were reported with less energy consumption. An investigation to characterize the microwave melting of metallic copper powder was conducted, and a quantitative description of the copper powders' microwave heating curve showcase quick melting of Cu powder with high densification [26]. In another study, microwave heating of different metallic powder materials (like Fe, Cu, Ti, Sn and Au) was carried out [27]. Due to eddy current loss and magnetic reversal loss, Fe-based metallic powder melted completely, whereas, Cu and Ni-based powders melted moderately.

Researchers at Oak Ridge Y-12 National Laboratory (TN, USA) melted and moulded different ferrous and non-ferrous materials with varying volume and mass employing microwave energy [8]. A patent application has been filed in the USA for a process setup to melt metal using microwave energy as the primary heat source [9]. Further, Agrawal [28] reported melting of bulk metals like Al, Cu, and SS inside the applicator with special setup arrangement. Microwave melting phases and role of oxide formation were analysed with numerical simulation [18]. The bulk metal melting and time–temperature characteristics were reported by Chandrasekharan et al. [29]. In this study, melting of Pb, Sn, Al and, Cu was reported by considering power variation of microwave energy and comparing it with conventional energy. Lingappa et al. [30] have examined the impact of conventional and microwave energy on the rate of energy consumption and melting time for non-ferrous bulk metals. In overall comparison the microwave-based melting was found to be fast and efficient as compared to traditional approach. Mishra and Sharma [31] reported on the in-situ casting of bulk Al-alloy using microwave energy. Further, the effect of various parameters like susceptor, mold, power level, location, and applicator design propertied on melting time and grain structure refinement have been reported [32, 33]. These heating phases enhances microstructure properties of material due to uniform heating and cooling. The microwave applicator and mold assembly/fixture parameters effect properties of the cast [34]. For example, thermal and electromagnetic properties of crucible and mold significantly impact the solidification process and hence microstructural and mechanical properties of the cast. The major outcomes of the recently reported literature have been presented in Table 6.1.

The majority of the literature reported on microwave casting has been centred around metals and alloys. An India patent was filed in 2016 on the advancement of

Table 6.1 Major process outcome on microwave melting/casting of metals

Material	Input parameters	Results	Ref
Pb, Sn, Al, and Cu	Power: 1.3 kW Charge wt.: Pb and Sn (up to 150 gm), Al and Cu (up to 25 gm)	Microwave and conventional energies were compared. Power: Reduced by 2 times Melting time: Decreased by 2 times	[29]
60Sn40Pb	Power: 0.9 kW Crucible: Silica Heating time:1.5 min	UTS: 50.2867 MPa Hardness: 10.33 HV	[35]
Al1050	Power: 0.9 kW Frequency:2.45 GHz Crucible: Graphite Heating time: 20 min	UTS: 83.9054 MPa Ductility:12.93% Hardness: Increase by 11%	[36]
Tin, Al, Zn, and Brass	Power: 0.9 kW Crucible: Graphite Charge wt.: 50, 100, and 150 gms	Compared to conventional melting: Melting time: Decreased, Power consumption: Low	[30]
AA7039	Power: 1.4 kW Charge wt.: 100 gm Heating time: 15.5 min	Porosity: <2% Hardness: 112 HV	[18, 31]
AZ91	Power: 1.4 kW Crucible: Graphite Charge wt.: 50, 100, and 150 gms	Due to graphite mould Melting time: Reduced (127 s)	[37]
AA6061	Power: 1 kW Frequency:2.45 GHz Mold: Graphite	Difference in experimental and computational property is 5%	[38, 39]

metal matrix composites (MMCs) using the microwave casting method. Microwave casting of Ni powder based MMCs was reported by Singh et al. [40–42]. The metal goes through different heating phases during microwave irradiation and SiC inclusions as a reinforcing material within the powder mixture substantially improved the microwave absorption capability, leading to the rapid attainment of high temperatures due to bi-directional heating. In another study, microwave casting of Ni-based matrix reinforced with WC–Co was carried out using microwave hybrid heating (MHH) approach [43]. Compared to pure Nickel, the microwave cast MMC shows a 2.6 folds increase in hardness. Kaushal et al. [44] have demonstrated the successful fabrication of EWAC-Cr_3C_2 MMCs utilizing microwave casting process. Moreover, the tribological examinations of the cast specimens revealed that the significant wear loss in cast sample were mainly attributed the formation of craters and plastic deformation. Recently, Pal et al. carried out cast SS316 based MMCs with different reinforcements like EWAC + WC–Co [45, 46], EWAC + Cr_7C_3 [47] using microwave irradiation. Both, WC–Co and Cr_7C_3 reinforced MMC shows 2 times higher hardness than the SS316 matrix material. The major outcomes of the recently reported literature on microwave casting composites have been presented in Table 6.2.

Table 6.2 Major process outcome on microwave casting of composites

Material	Input parameters	Results	Ref
EWAC + SiC (10 wt.%)	Power: 0.9 kW Heating time: 18–24 min Susceptor: Pulverized coal	Hardness: 980 HV Porosity: 1.68%	[40]
EWAC + SiC (5 and 10 vol.%)	Power: 0.9 kW Heating time: 18–24 min Preheat temp.: 200 ^{0}C	Hardness: 1085 ± 210 HV Porosity: 1.68 to 1.75%	[41]
EWAC + SiC (5 and 10 vol.%)	Power: 0.9 kW Preheat temp.: 200 ^{0}C	Hardness: 980 ± 208 HV (10 vol.%) UTS: 442 ± 20 MPa (10 vol.%)	[42]
EWAC + WC-8Co (5 and 10 vol.%)	Power: 0.9 kW Heating time: 10–24 min	Hardness: 788 ± 52 HV Microstructure: Equiaxed non-uniform grain	[43]
EWAC + Cr_3C_2	Power: 0.9 kW	High sliding velocity and high load: high wear loss	[44]
SS316 + 15EWAC + WC-12Co (WC—5, 10, 15 wt.%)	Power: 0.9 kW	Maximum microhardness at carbide phase	[45]
SS316 + 10EWAC + WC-12Co (3 wt.%)	Power: 0.9 kW	Hardness: 648.16 HV Grain size: 23 μm	[46]
SS316 + 10EWAC + Cr_7C_3 (5 wt.%)	Power: 0.9 kW	Density: 97% Porosity: 1.98% Grain size: 17 ± 2 μm	[47]

It has been reported by various researchers that materials processed by microwave energy have shown better microstructural (i.e., fine and homogeneous) and mechanical properties. However, the uniform and volumetric heating nature of microwave have put a significant effect on the grain boundary, phases generation and distribution of precipitates of melted metals. Due to the continuous stirring of molten metal throughout the microwave melting process by bubbling and wavy action with volumetric heating improves the homogeneity and density of microwave processed cast sample. The dense cast structure can obtain with less porosity due to equiaxial growth by uniform heating compared to conventional one. The environment during solidification has an impact on the material's intermetallic precipitates and grain structure, to produce a casting with a sound form, controlled solidification is required. Due to a faster cooling rate and formation of more nucleation site, the microstructure characteristics in in- situ microwave casting reveals a finer and denser structure when appropriate environmental control is applied.

4 Challenges and Opportunities

Microwave casting is at very nascent stage at present and possess various challenges as compared to conventional casting process. However, the available literature indicates its potential in terms of some unique characteristics such as time and energy saving, and improved properties in the casts. Some challenges in microwave casting are highlighted in Fig. 6.6. The electromagnetic field of microwaves transfer the energy, hence any irregularities in field strength and distribution in side microwave cavity will produce uneven heating. The electromagnetic field is governed by the processing applicator design; hence the design of microwave applicator cavity and magnetron position are very important aspects. Materials frequently experience physical and structural changes during microwave processing and that have an impact on their dielectric/magnetic characteristics. Therefore, microwave heating efficacy depends on the process conditions. Process modelling and control may be affected by sudden changes in the heat-generating efficiency of microwaves. It is crucial to comprehend the interaction efficacy of microwaves with materials. Microwave processing of casting process is crucial considering the effect of electromagnetic field's characteristics, material properties, structural changes of the material during microwave processing, the size and shape of the object being heated, and the physics of the microwave/materials interactions. New studies in microwave device design, microwave/material interactions, and material processing are contributing to the continued growth in interest of researchers globally and in microwave casting technology as well. The microwave casting process offers a huge opportunity in future in terms of control over microstructure and cast properties for different applications. A few techniques for controlled solidification can be used in the microwave casting process to enhance the cast characteristics by regulating microwave exposure time, power, preheating temperature of mould, and flow of molten metal into the mould.

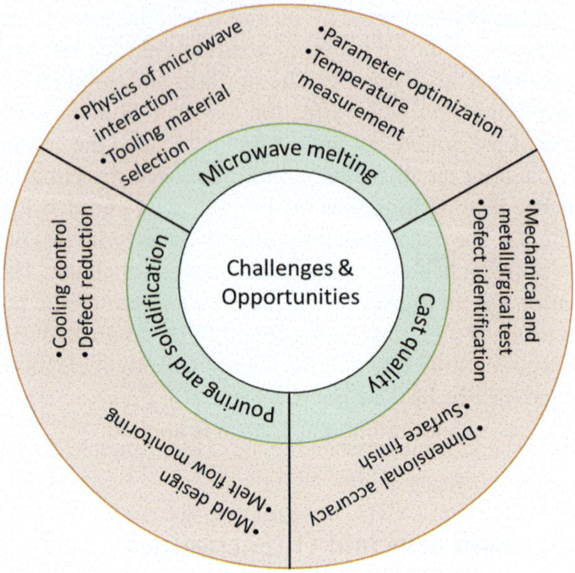

Fig. 6.6 Challenges and opportunities in Microwave based casting/melting

References

1. Gupta M, Leong EWW (2008) Microwaves and metals. John Wiley & Sons, Singapore
2. Mishra RR, Sharma AK (2016) A Review of research trends in microwave processing of metal-based materials and opportunities in microwave metal casting. Crit Rev Solid State Mater Sci 41:217–255. https://doi.org/10.1080/10408436.2016.1142421
3. Mishra RR, Sharma AK (2016) Microwave-material interaction phenomena: heating mechanisms, challenges and opportunities in material processing. Compos Part A Appl Sci Manuf 81:78–97. https://doi.org/10.1016/j.compositesa.2015.10.035
4. Padmavathi C, Upadhyaya A, Agrawal D (2011) Effect of microwave and conventional heating on sintering behavior and properties of Al-Mg-Si-Cu alloy. Mater Chem Phys 130:449–457. https://doi.org/10.1016/j.matchemphys.2011.07.008
5. Oghbaei M, Mirzaee O (2010) Microwave versus conventional sintering: a review of fundamentals, advantages and applications. J Alloys Compd 494:175–189. https://doi.org/10.1016/j.jallcom.2010.01.068
6. Agrawal D (2010) Latest global developments in microwave materials processing. Mater Res Innov 14:3–8. https://doi.org/10.1179/143307510X12599329342926
7. Chandrasekaran S, Ramanathan S, Basak T (n.d.) Microwave material processing—a review, pp 1–124
8. Padmavathi C, Upadhyaya A, Agrawal D (2012) Microwave assisted sintering of Al-Cu-Mg-Si-Sn alloy. J Microw Power Electromagn Energy 46:115–127. https://doi.org/10.1080/08327823.2012.11689830
9. Xu F, Li Y, Hu X, Niu Y, Zhao J, Zhang Z (2012) In situ investigation of metal's microwave sintering. Mater Lett 67:162–164. https://doi.org/10.1016/j.matlet.2011.09.052
10. Gupta M, Wong WLE (2005) Enhancing overall mechanical performance of metallic materials using two-directional microwave assisted rapid sintering. Scr Mater 52:479–483. https://doi.org/10.1016/j.scriptamat.2004.11.006
11. Gupta D, Sharma AK (2014) Microwave cladding: a new approach in surface engineering. J Manuf Process 16:176–182. https://doi.org/10.1016/j.jmapro.2014.01.001

12. Zafar S, Sharma AK (2015) Dry sliding wear performance of nanostructured WC-12Co deposited through microwave cladding, Tribol. Int. 91 (2015). https://doi.org/10.1016/j.triboint.2015.06.023
13. Gupta D, Bhovi PM, Sharma AK, Dutta S (2012) Development and characterization of microwave composite cladding. J Manuf Process 14:243–249. https://doi.org/10.1016/j.jmapro.2012.05.007
14. Srinath MS, Sharma AK, Kumar P (2011) Investigation on microstructural and mechanical properties of microwave processed dissimilar joints. J Manuf Process 13:141–146. https://doi.org/10.1016/j.jmapro.2011.03.001
15. Bansal A, Sharma AK, Kumar P, Das S (2014) Characterization of bulk stainless steel joints developed through microwave hybrid heating. Mater Charact 91:34–41. https://doi.org/10.1016/j.matchar.2014.02.005
16. Tamang S, Aravindan S (2022) Joining of dissimilar metals by microwave hybrid heating: 3D numerical simulation and experiment. Int J Therm Sci 172:107281. https://doi.org/10.1016/j.ijthermalsci.2021.107281
17. das S, Sharma AAK (2012) Microwave drilling of materials, Barc Newsl. 15–21
18. Mishra RR, Sharma AK (2017) On melting characteristics of bulk Al-7039 alloy during in-situ microwave casting. Appl Therm Eng 111:660–675. https://doi.org/10.1016/j.applthermaleng.2016.09.122
19. Wiedenmann O, Ramakrishnan R, Kili E, Saal P, Siart U, Eibert TF, Kili E, Saal P, Siart U, Eibert TF () A multi-physics model for microwave heating in metal casting applications embedding a mode stirrer. In: 2012 7th Ger. Microw. Conf., pp 1–4
20. Marahadige SL, Sridharmurthy SM, Jayraj AH, Mahabaleshwar US, Lorenzini G, Lorenzini E (2018) Development of copper alloy by microwave hybrid heating technique and its characterization. https://doi.org/10.18280/ijht.360425
21. El Khaled D, Novas N, Gazquez JA, Manzano-Agugliaro F (2018) Microwave dielectric heating; applications on metals processing. Renew Sustain Energy Rev 82:2880–2892. https://doi.org/10.1016/j.rser.2017.10.043
22. Warren B, Awida MH, Fathy AE (2012) Microwave heating of metals. IET Microwaves, Antennas Propag 6:196. https://doi.org/10.1049/iet-map.2010.0494
23. Santos T, Valente MA, Monteiro J, Sousa J, Costa LC (2011) Electromagnetic and thermal history during microwave heating. Appl Therm Eng 31:3255–3261. https://doi.org/10.1016/j.applthermaleng.2011.06.006
24. Mishra RR, Sharma AK (2019) Microstructural characteristics and tensile properties of in-situ and ex-situ microwave casts of Al-7039 alloy. Mater. Res. Express. 6. https://doi.org/10.1088/2053-1591/ab5ab1
25. Xu L, Peng J, Bai H, Srinivasakannan C, Zhang L, Wu Q, Han Z, Guo S, Ju S, Yang L (2017). Application of microwave melting for the recovery of tin powder. https://doi.org/10.1016/J.ENG.2017.03.006
26. Xu L, Srinivasakannan C, Peng J, Guo S, Xia H (2017) Study on characteristics of microwave melting of copper powder. J Alloys Compd 701:236–243. https://doi.org/10.1016/j.jallcom.2017.01.097
27. Buchelnikov VD, Louzguine-Luzgin DV, Xie G, Li S, Yoshikawa N, Sato M, Anzulevich AP, Bychkov IV, Inoue A (2008) Heating of metallic powders by microwaves: experiment and theory. J Appl Phys 104:113505. https://doi.org/10.1063/1.3009677
28. Agrawal D (2006) Microwave sintering , brazing and melting of metallic. In: Sohn Int. Symp. Adv. Process. Met. Mater. 183–192
29. Chandrasekaran S, Basak T, Ramanathan S (2011) Experimental and theoretical investigation on microwave melting of metals. J Mater Process Technol 211:482–487. https://doi.org/10.1016/j.jmatprotec.2010.11.001
30. Lingappa SM, Srinath MS, Amarendra HJ (2018) Melting of bulk non-ferrous metallic materials by microwave hybrid heating (MHH) and conventional heating: a comparative study on energy consumption. J Braz Soc Mech Sci Eng 40. https://doi.org/10.1007/s40430-017-0921-7

31. Mishra RR, Sharma AK (2016) On mechanism of in-situ microwave casting of aluminium alloy 7039 and cast microstructure. Mater Des 112:97–106. https://doi.org/10.1016/j.matdes.2016.09.041
32. Mishra RR, Sharma AK (2017) Effect of susceptor and mold material on microstructure of in-situ microwave casts of Al-Zn-Mg alloy. Mater Des 131:428–440. https://doi.org/10.1016/j.matdes.2017.06.038
33. Mishra RR, Sharma AK (2020) Effect of input microwave power and insulation on microstructure and tensile properties of cast Al 7039 alloy produced at 2.45 GHz. J Microw Power Electromagn Energy 54:312–329. https://doi.org/10.1080/08327823.2020.1838049
34. Bhatt SC, Ghetiya ND (2022) Review on effect of tooling parameters on microwave processing of metallic materials with special emphasis on melting/casting application. Int J Met. https://doi.org/10.1007/s40962-021-00743-z
35. Gouthama TR, Harisha G, Manjunatha YR, Kumara SMM, Srinath MS, Lingappa MS (2016) Melting of tin using muffle furnace and microwave energy and its characterization. IOP Conf Ser Mater Sci Eng 149. https://doi.org/10.1088/1757-899X/149/1/012100
36. Shashank Lingappa M, Srinath MS, Amarendra HJ (2017) Microstructural and mechanical investigation of aluminium alloy (Al 1050) melted by microwave hybrid heating. IOP Publishing. https://doi.org/10.1088/2053-1591/aa7aaf
37. Sharma AK, Mishra RR, Yadav J (n.d.) Resource-efficient melting of bulk magnesium alloy AZ91 using microwave energy at 2 . 45 GHz, pp 1–6
38. Bhatt SC, Ghetiya ND (2023) 3D Multiphysics simulation of microwave heating of bulk metals with parametric variations. Chem Eng Process Process Intensif 184:109271. https://doi.org/10.1016/j.cep.2023.109271
39. Bhatt SC, Ghetiya ND, Joshi M (2022) Multiphysics simulation and validation of microwave melting characteristics of AA6061 by finite element analysis. Adv Mater Process Technol 8:1557–1568. https://doi.org/10.1080/2374068X.2021.1948708
40. Singh S, Gupta D, Jain V (2018) Microwave melting and processing of metal–ceramic composite castings. In: Proc Inst Mech Eng Part B J Eng Manuf 232:1235–1243. https://doi.org/10.1177/0954405416666900
41. Singh S, Gupta D, Jain V (2016) Novel microwave composite casting process: theory, feasibility and characterization. Mater Des 111:51–59. https://doi.org/10.1016/j.matdes.2016.08.071
42. Singh S, Gupta D, Jain V (2019) Microwave processing and characterization of nickel powder based metal matrix composite castings. Mater Res Express 6 (2019). https://doi.org/10.1088/2053-1591/ab2138
43. Singh S, Gupta D, Jain V (2018) Processing of Ni-WC-8Co MMC casting through microwave melting. Mater Manuf Process 33:26–34. https://doi.org/10.1080/10426914.2017.1291954
44. Kaushal S (2022) Microstructure and tribological characterization of composite castings developed through in-situ microwave hybrid heating. Int J Met 16:2150–2161. https://doi.org/10.1007/S40962-022-00757-1/METRICS
45. Pal J, Gupta D, Singh TP (2023) Influence of tungsten carbide-cobalt reinforcement on flexural behavior of iron-based composite developed through microwave. https://doi.org/10.1177/09544062231181826. https://doi.org/10.1177/09544062231181826
46. Pal J, Gupta D, Singh TP (2022) Processing and characterization of SS316 based metal matrix composite casting through microwave hybrid heating, https://doi.org/10.1177/09544062221104443. https://doi.org/10.1177/09544062221104443
47. Pal J, Gupta D, Singh TP (2023) Development of iron-based composite materials through electromagnetic radiations in a domestic microwave oven. https://doi.org/10.1177/09544062231181832. https://doi.org/10.1177/09544062231181832

Application of Microwave-Metal Discharge for Metallic Material Removal

Pranjal Gupta, Apurbba Kumar Sharma, and Inderdeep Singh

Abstract Microwave irradiation at 2.45 GHz is considered a clean and economical energy source. Material processing using microwave energy has emerged as a potential solution for various applications. The discharge effect has gained immense popularity while exploring material processing in the past few decades. Microwave-induced discharges are currently utilized in several material processing applications such as pyrolysis, pollutant removal, material synthesis, sintering, etc. The present work investigates the physical aspects of microwave-metal discharge and emphasizes its ability to machine metal-based materials. This chapter briefly reviews the different theories proposed regarding the mechanism of microwave-metal discharge, its effects, potential applications, and the related challenges. The article also presents some results emerging from a few experimental investigations performed to understand the impact of microwave-metal discharges on the ablation characteristics of AISI 304 stainless steel in atmospheric air medium. The study compares the effect of microwave-metal discharge exposed to AISI 304 for the time durations of 30 s, 40 s, and 50 s. The irradiated and ablated areas formed by the discharge were evaluated at different time intervals. The article also suggests possible plasma confinement strategies to further improve the machining performance.

Keywords Microwave-metal discharge · Plasma · Material removal · Thermal damage

1 Introduction

Microwaves are electromagnetic radiations ranging from 300 MHz to 300 GHz in the electromagnetic spectrum. The radiation has two components – electrical and magnetic, which are orthogonal to each other. Microwaves were predominately

---

P. Gupta · A. K. Sharma (✉) · I. Singh
Department of Mechanical and Industrial Engineering, IIT Roorkee, Roorkee, Uttarakhand 247667, India
e-mail: aks@me.iitr.ac.in

M. K. Singh et al. (eds.), *Composite Materials Processing Using Microwave Heating Technology*, Composites Science and Technology,
https://doi.org/10.1007/978-981-97-2772-8_7

utilized in communication systems before the heating ability of microwave radiation was realized. In 1945, Percy Spenser accidentally found the heating effect of microwaves. Following the discovery, significant developments have occurred in the technique of microwave heating to process various materials for numerous industrial applications. Today, microwaves are used to process materials that involve zero heating to melting and ablating materials with melting points in excess of 2000 °C. Microwave heating offers unique advantages in material processing compared to traditional processing routes, such as uniform and rapid heating, reduced processing time, and superior mechanical properties. It is also considered clean and environment friendly. The dielectric and magnetic losses incurred within a material due to the realignment of charges are primarily responsible for causing the heating effect. High conductivity materials such as metals are subjected to conduction losses due to the abundance of free electrons, which causes the heating effect when exposed to microwaves. However, the heating is limited to the surface of metals since most of the radiation is reflected. Moreover, any presence of sharp discontinuities in conductive materials causes the generation of abrupt electric discharges due to an increase in the electron density and subsequent ionization of the surrounding medium [1].

Electrical discharges are generated when the intensity of electric field within a material surpasses the breakdown strength of the media surrounding the material. In gas medium, electric discharges depending on gas pressure are distinguished based on the current–voltage characteristics [2]. These discharges are classified into four broad categories based on the magnitude of current and voltage and are shown in Fig. 7.1. Townsend discharge occurs at very low current magnitudes of the order of $\sim 10^{-8}$ A, whereas a low value of current ($\sim 10^{-2}$ A) and high value of voltage (~1 kV) at low pressure (~mbar) gives rise to glow discharge which is utilized mostly in diverse industrial applications. Corona discharge also occurs at low values of current (10^{-6} A) and exists at atmospheric pressure conditions. Lastly, Arc discharges, on the other hand, develop initially at higher voltages, followed by lower voltage and higher current magnitudes.

Unlike gas, discharges in liquid medium are more challenging to establish due to their higher density, thus requiring a higher electric field to cause the breakdown. Also, the hydrostatic pressure exerted by the surrounding liquid medium greatly influences the discharge characteristics. Considerable studies have been conducted to study the discharge characteristics in various liquid media such as water, hydrocarbon-based liquids such as benzene, toluene, cyclohexane, etc., and oils with aromatic compounds. In liquid media, the breakdown often initiates with the formation of streamers (a streamer is an ionized plasma column that originates in a high electric field region). Gradually the streamer propagates and bridges the gap between the two electrodes, eventually generating a discharge.

In recent times, microwave-induced metal discharges have gained much attention owing to their high thermal capacity and ease of generation compared to typical electric discharges. Microwave-metal discharges are currently being utilized in numerous material processing applications such as pyrolysis, pollutant removal, sintering, material synthesis, etc. Despite such extensive involvement in various

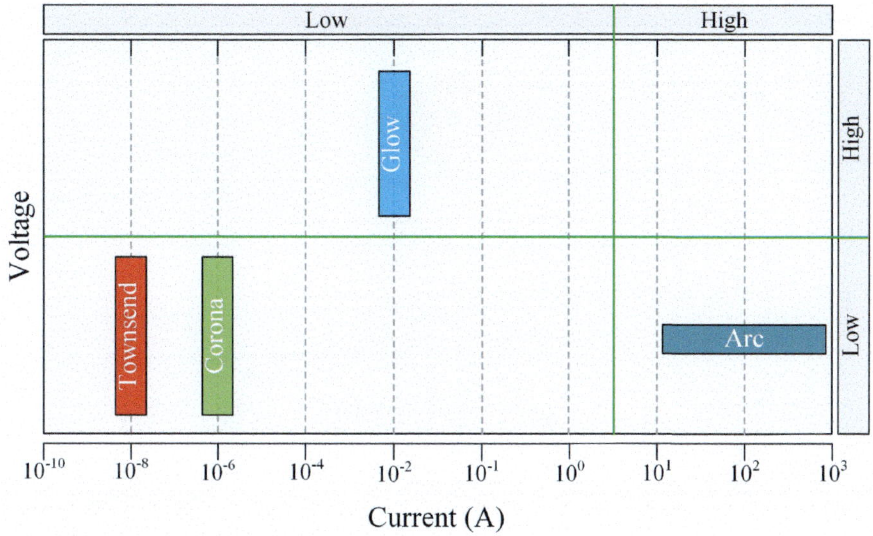

Fig. 7.1 Current–voltage characteristics of discharges in gaseous media [2]

research domains, a sufficiently clear and exhaustive understanding of the nature, origination, and mechanism of discharge is far from maturity.

2 Microwave-Metal Interaction and Associated Special Effects

The interaction of microwaves with a material depends on its absorption characteristics of the incident electromagnetic energy. Materials are categorized as transparent, absorbing, or opaque based on their absorption characteristics [3]. The nature of the bulk metallic materials is opaque towards microwave energy due to their high reflectivity and low absorption capability. Direct coupling of microwaves to bulk metallic materials is challenging due to their high conductivity. The frequent movement of the available free electrons between the conduction and valence band inculcates high conductivity in metal-based materials such as alloys and composites, this high conductivity restrains the bulk absorption of microwave energy. Thus, while most radiations get reflected, the penetration is significant only up to a few microns. This micron-sized depth measured from the surface of a bulk metallic material is known as surface penetration depth or skin depth [4]. However, as a special case, bulk heating of metallic materials is feasible provided, they are present in sufficiently small-sized particles or powders [5]. Bulk heating of select metallic materials are being attempted by several researchers within limited success.

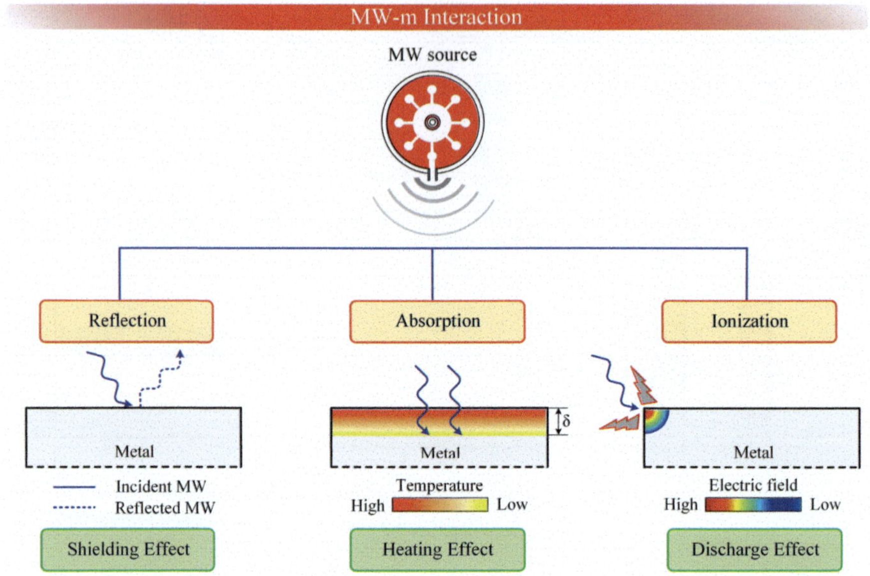

Fig. 7.2 Effects associated with MW-m interaction

Discharges take place when bulk metallic materials with sharp edges, tips, or sub-microscopic irregularities are exposed to microwave irradiation. These electrical discharges on metallic objects or films are known as microwave-metal discharges or microwave-induced metal discharges. The discharges are produced as a result of localized electric field intensification around the sharp features, which leads to ionization of the medium surrounding the material. The discharges induce heating, plasma, and photo-catalysis within the exposed metallic materials [1].

The interaction of microwaves with metallic materials can be broadly classified as the reflection, absorption/partial absorption, and ionization based on the size, amount, geometry, and dielectric properties of the medium surrounding the metal [6]. These phenomena result in opacity or shielding effect, heating effect and discharge effects. The mechanistic representation of these effects is demonstrated in Fig. 7.2. Each effect may exist independently or in combinations based on the metal properties.

2.1 Shielding Effect

The metal-based materials get poorly coupled with microwave radiation since the penetration is insignificant. This poor coupling causes formation of constructive and destructive interference at the surface of the metallic object. It is due to the overlap of the incident and reflected waves. As a result, the distribution of electric field intensity gets altered, and causes changes in power absorption characteristics. Constructive interference increases power absorption due to resonance, whereas the destructive interference causes a decrease in power absorption characteristics.

2.2 Heating Effect

The induced eddy current on the surface of metallic materials is responsible for causing surficial joule heating. However, in the case of ferromagnetic materials, magnetic loss is the reason for the joule heating. Skin depth is a function of electrical and magnetic properties of a material. Since metallic alloys and composites have high conductivity and high permeability, skin depth is of the order of a few microns for a given frequency. The metals could effectively be heated if present in the form of particles having a size comparable to the skin depth of metals due to complete penetration inside the material.

2.3 Discharge Effect

Besides reflection and heating effects, microwave radiation, when imposed on metal-based material, causes charges to accumulate on its surface due to low skin depth. The redistribution in the arrangement of charges depends on the shape of the conductor. A unique phenomenon gets triggered when metallic conductors having sharp tips, edges, or uneven discontinuities are subjected to microwaves, known as the discharge effect or microwave-induced metal discharge. This is because sharp entities cause the charges to align closer, leading to localized charge densification. It results in electric field intensification, followed by the avalanche breakdown of the medium surrounding the metallic conductor. A few charges gain sufficient kinetic energy during the avalanche effect, and escape out of the metallic conductor, causing an electrical discharge.

3 Microwave Metal Discharge and Metallic Material Removal

Fundamentally, a discharge is localized breakdown of surrounding medium in the presence of electromagnetic field (here, microwaves). The breakdown mechanism is different in gases, liquid and solid medium. A sufficiently clear understanding regarding the mechanism is not present to date; however, several researchers have characterized the nature and occurrence of these discharges while working in different technical domains.

Whittaker and Mingos [7] studied the interaction of microwaves with metallic powders while performing solid-state reactions. They reported that microwave radiations get attenuated in a conductive medium due to the displacement of charge carriers by the electric field. This causes induced current on the surface of the conductors. The authors further reported that ohmic heat was produced when the traveling charge conductors collided with the lattice photons, leading to discharge formation.

Hu et al. [8] investigated the discharge mechanism by imparting microwave irradiation over a constantan wire of diameter 3 mm; they reported that electrons tended to escape from the wire terminal. An arc discharge was created when the induced potential exceeded the coulomb potential.

Recently, Cai et al. [9] investigated the dynamic characteristics of microwave-induced metal discharge using the photo-voltage conversion method. Iron, copper and aluminium metal rods with varying lengths (3–5 cm) and diameters (0.5–3 mm) were used to investigate the discharge effect. The authors reported that the skin effect of metallic materials enabled the current generation, which eventually resulted in joule heating of the metallic tips. This joule heating exited the valence-band electrons to escape from the metallic tips, and the breakdown potential of the gaseous medium around the metal created microwave discharge. Further, collisions of the excited metallic electrons with free electrons present in the gaseous phase exhibited a noticeable thermal and plasma effect.

Despite a shallow understanding regarding the physical nature of microwave-induced discharges, researchers have utilized these discharges for processing wide range of materials in numerous technical domains. Material processing techniques such as, sintering, pyrolysis, pollutant removal, material synthesis, etc., processed using microwave-induced discharges are gaining immense popularity owing to the superior thermal capacity of these discharges, which fulfills their processing requirements [1]. Characterization of the thermal aspects of microwave-induced discharges have also been reported.

Hussain et al. [10] performed pyrolysis of polystyrene utilizing the microwave-metal interaction. They employed a wire-gauge mesh of commercially available iron to focus microwaves and generate high temperatures, temperatures close to 1200 °C were reported.

In one of the studies on the heating effect of microwave discharge caused by metal, Qiao et al. [11] reported that when stainless steel strips embedded in quartz sand were exposed to microwave irradiation, an intense discharge phenomenon was observed. The partial melting and bonding of strips with sand revealed a high-temperature rise during discharge. Experimental and simulation studies on the coupled effect of wave absorption and metal discharge generation due to microwaves revealed that the presence of microwave-absorbing material, such as activated carbon, could increase the intensity of discharge. Localized hotspots with temperatures of 1000–2000 °C were produced, which may be exploited in potential heating applications.

Singh and Sharma [12] showed the feasibility of microwave-metal discharge for machining sub-millimeter-sized holes on thin steel sheets and proposed an hypothesis for the process. The authors suggested that the interaction of microwaves with metallic (tungsten) tool tip induces a surface current owing to the skin effect (Fig. 7.3).

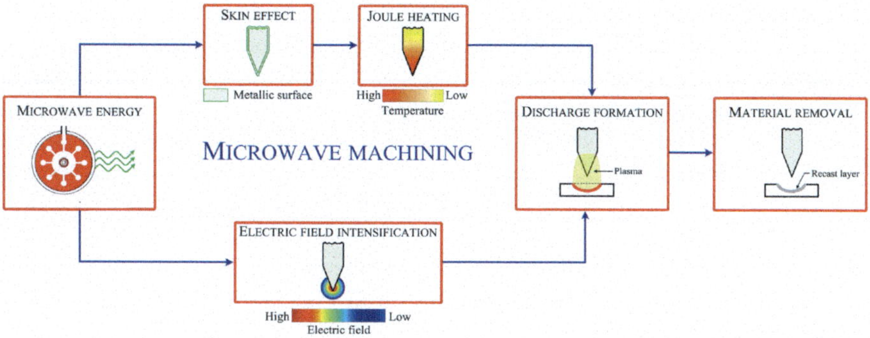

Fig. 7.3 Flowchart representing steps involved in material removal using microwave energy

As a result, the temperature of the metallic tool increases due to joule heating. The surface current also causes electric field intensification at the pointed tip of the metallic tool due to radius of curvature effect. The intensification of electric field, along with joule heating, favours the formation of discharge originating from the tip of the metallic tool. As the discharge grows and becomes stable, it imparts thermal energy to the metallic workpiece (say, steel) kept beneath the tool. The thermal energy of discharge was sufficient enough to cause the ablation of the workpiece. As a result, through holes were formed on the metallic workpiece. A flowchart describing the steps involved in the microwave machining process is depicted in Fig. 7.3. Based on the hypotheisis proposed, the authors reported drilling of steel steels (thickness: 0.5 mm) using a tungsten electrode (or tool) with a diameter of 1.0 mm. The workpiece and tool were separated by a fixed gap of about 1–2 mm. The plasma column formed in the study claimed to have reached a temperature close to 4100 °C. The methodology was economical and simpler to implement compared to similar thermal energy-based machining processes such as electric discharge machining, laser beam machining, and electron beam machining. However, the unconfined plasma formed due to microwave-meal discharge posted several defects on the workpiece, such as wide heat affected zone, recast bead, micro-craters, and micro-cracks. Moreover, the oxidation of tungsten tool material left oxides deposits near the hole boundary, which degraded the integrity of the machined holes.

4 Experimental Details

The experimental trials were performed inside a domestic microwave applicator (Make: LG; Model: MS2021CW) having a maximum rated power of 700 W. The volume of the microwave applicator cavity was 20 L. AISI 304 stainless steel (thickness: 0.5 mm) and tungsten (diameter: 1.0 mm, tool tip angle: 12°–15°) were chosen as the work and tool materials respectively. The relevant properties of the materials are mentioned in Table 7.1. The experimental setup comprises of a workpiece held in a microwave transparent fixture made up of Teflon. The assembly was placed over a refractory base as shown in Fig. 7.4. The refractory base acts as a dead load. It inhibits the reflected microwave from reaching the magnetron and cause damage. The tool was gripped in a holder made of mild steel and kept at a fixed gap of 1.0 mm over the workpiece. A stepper motor coupled with MACH 3 (open source) software was utilized to precisely maintain the gap. The tool-work gap is also known as discharge gap since the discharge is formed between the tool and the workpiece. The experimental conditions adapted for the present study are summarized in Table 7.2. Microwave metal discharge trials were conducted for durations of 30 s, 40 s, and 50 s, respectively.

The images of the samples after being exposed to microwave metal discharge were captured using a stereoscope (Make: Nikon: Model: SMZ 745 T). The irradiated and ablated areas were measured and analyzed using Image J software.

Table 7.1 Physical properties of AISI 304 stainless steel and tungsten [13]

Properties	AISI 304 (Work)	Tungsten (Tool)
Thermal conductivity (W/m–K)	16.2	174
Melting temperature (°C)	1450	3410
Vaporization temperature (°C)	3200	5660
Density (kg/m^3)	8000	19,300
Specific heat capacity (J/kgK)	130	500
Thermal expansion coefficient (1/K)	17.2×10^{-6}	4.40×10^{-6}

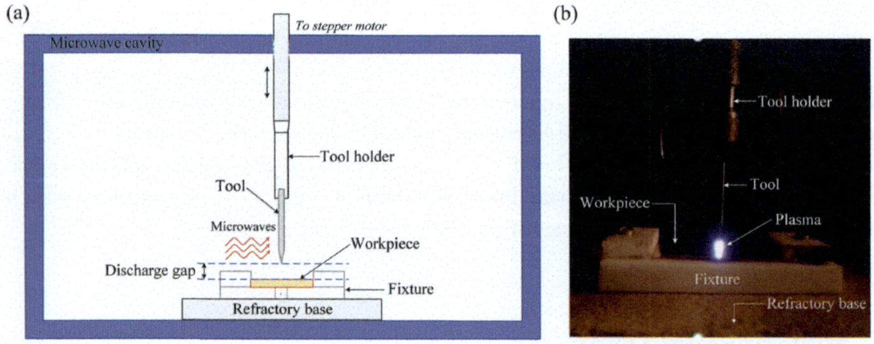

Fig. 7.4 **a** Schematic representation of the experimental setup **b** actual view of the machining in progress

Table 7.2 Experimental parameters and machining conditions

Parameters/conditions	Description
Microwave power (W)	700
Frequency (GHz)	2.45 GHz
Discharge gap (mm)	1.5
Drilling medium	Atmospheric air
Drilling time (s)	30, 40, and 50

5 Results and Discussion

AISI 304 stainless steel samples having a thickness of 0.5 mm were exposed to microwave metal discharge generated from the pointed end of tungsten tool as shown in Fig. 7.4b. The interaction of the discharge with the metallic workpiece leads to rise in temperature of the workpiece, as a result of continuous transfer of thermal energy from the plasma column. Gradually, the temperature of the material rises and reaches above the melting point and later above its vaporization temperature to finally ablate the material. It was identified that a minimum duration of 30 s was necessary to completely remove the work material. Figure 7.5 shows the stereoscopic images of the machined samples at different time durations. The macrographs reveal the presence of a wide thermally damaged region surrounding the ablated area on the workpiece. This region identified as the irradiated area represents the overall area on the workpiece which gets exposed to the discharge. The irradiated area, apart from the ablated area, was characterized by the presence of recast bead, partially ablated region, and surficial oxides. The irradiated and the ablated areas evident on the workpiece are schematically shown in Fig. 7.6a.

Visual inspection of the machined samples suggests the non-uniform temperature distribution of plasma column throughout its geometry. The maximum temperature of the plasma existed at the core since the material was removed entirely and the temperature diminishes gradually outward in the radial direction as observed by the presence of partially melted zone and surficial damage. Olsen [14], while studying the thermal and electrical properties of argon plasma, observed a similar trend in

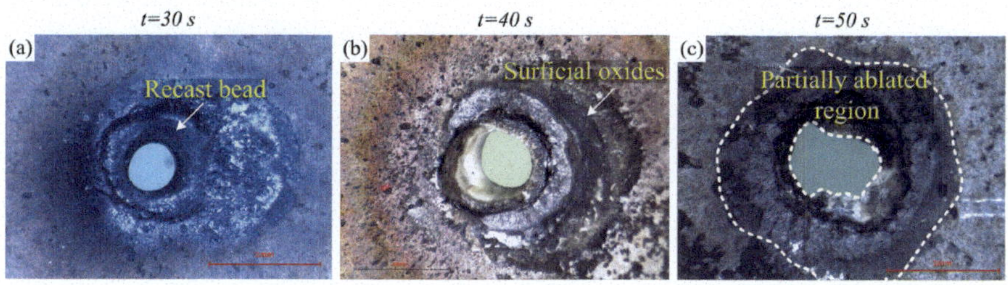

Fig. 7.5 Metallic samples exposed to microwave metal discharge for different time durations

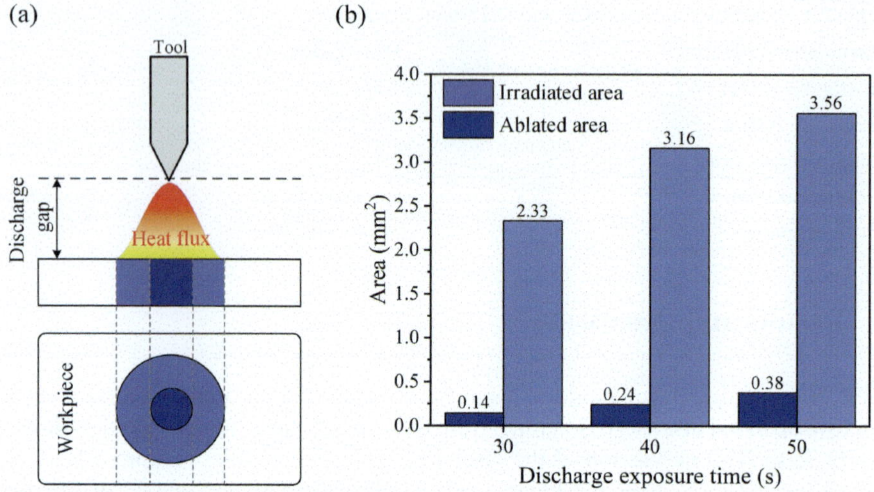

Fig. 7.6 **a** Schematic representation of irradiated area and ablated area, **b** Graphical representation of ablated and irradiated areas at different exposure duration

temperature distribution. The author applied spritzer's theory to measure the radial temperature distributions in the plasma column.

An increase in both irradiated and ablated areas were observed as the time duration was increased. This was due to continuous absorption of microwave energy into the plasma channel. Figure 7.6 (b) graphically represents the irradiated and ablated areas obtained at the discharge exposure intervals of 30 s, 40 s and 50 s. The ratio of abated area to irradiated area was 6.18%, 7.62%, and 10.64% for discharge exposure durations of 30 s, 40 s and 50 s, respectively. This indicates that a typical microwave metal discharge-induced plasma column consists of several isotherms having maximum temperature in the core to the least in the periphery.

Figure 7.7 shows typical stereoscopic images of the tool in unused condition and after being exposed to discharge for a time duration of 50 s. It was observed that due to oxidation of tungsten in atmospheric air condition, most of the tool material gets oxidized and eventually vaporized. Also, the prolonged exposure to the high-temperature discharge caused melting of the tool since tiny blisters were evident on the surface of the tool. As a result of melting and vaporization, the tool wear causes degradation of the sharp tip, which diminishes the uniformity of discharge at later stages.

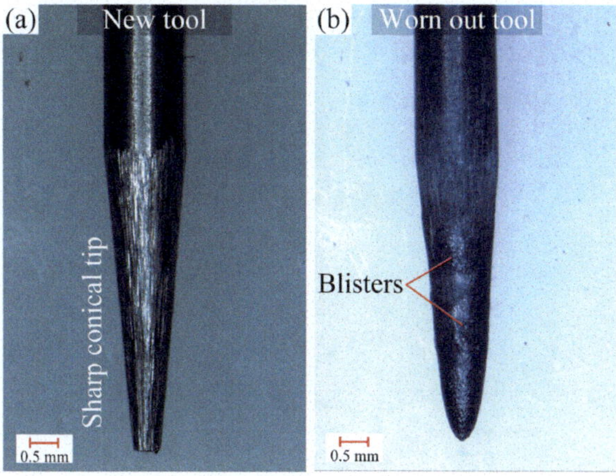

Fig. 7.7 Tool condition **a** before trial **b** after exposure to discharge for 50 s

6 Possible Strategies for Plasma Confinement

The results confirmed that, microwave metal discharge is effective in a removal of metallic material and thus could be identified as a feasible machining process for metal-based materials. However, a significant difference between irradiated and ablated areas due to unconstrained plasma forms a considerable challenge. It results in a wide heat-affected zone and deteriorates the integrity of the machined feature. Hence, the plasma must be constrained in order to make this process commercially viable. There lies a broad scope to improve the effectiveness and efficiency of this process by confining the plasma, thus minimizing the difference between irradiated and ablated areas. The following possible strategies/ techniques are suggested to confine the plasma in order to yield better machining performance.

6.1 Magnetic Field Assistance

Application of external magnetic field in plasma energy based machining processes such as electric discharge machining and laser induced plasma micro machining (LIPMM) has been reportedly effective in confining the plasma column. Zhang et al. (2021) conducted LIPMM to machine single crystal silicon in the presence of external magnetic field as shown in Fig. 7.8. The authors reported that the existence of 0.1 T magnetic field regulated and confined the geometrical features of plasma by controlling the motion of charged particles; a maximum reduction of 12.64% in heat affected zone was obtained with better surface morphologies.

In another research, Govindan et al. [15] employed an external magnetic field while performing electric discharge machining (EDM) of stainless steel 304 using copper as the tool material in the presence of oxygen medium. They reported that the magnetic field induces Lorentz force within the plasma column, thus making it

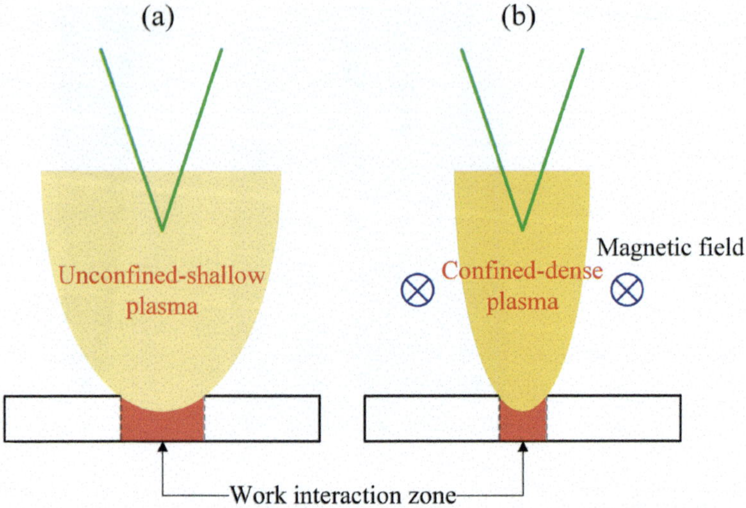

Fig. 7.8 **a** Schematic representation of LIPMM, **b** Influence of external magnetic field on plasma in LIPMM

confined. As a result, crater dimensions were reduced, and material erosion became more uniform.

The external magnetic field could effectively be employed in microwave-metal discharge plasma to help confine its geometry, which would yield better machining characteristics with lesser heat-affected zone and better surface integrity. However, special arrangements must be incorporated to shield the electromagnet (typical source for magnetic field) from microwave radiation inside the cavity.

6.2 Optimizing Tool Tip Angle

The radius of curvature of the tool tip controls the electric field intensity during dielectric breakdown of a medium. Sharp tips and edges in a conductor allow the free charges to align more closely along the edge of metallic conductor. Figure 7.9 schematically explains the effect of tool tip angle on charge distribution. The higher tool tip angle (θ) in case-2 as compared to case-1, results in a greater inter-electron distance (Δ) in case-2 as compared to case-1. This is attributed to the higher surface area, which decreases the electron density. Owing to a sharp electron density in case-1, a more intense discharge is created, which inculcates undesirable thermal damage while machining. Ma et al. [16] performed an electrostatic study of the curvature effects in high-voltage differential application. The authors concluded that the electric field intensity is practically infinite along a sharp edge and decreases significantly as the value of the radius of curvature increases. The extremely high

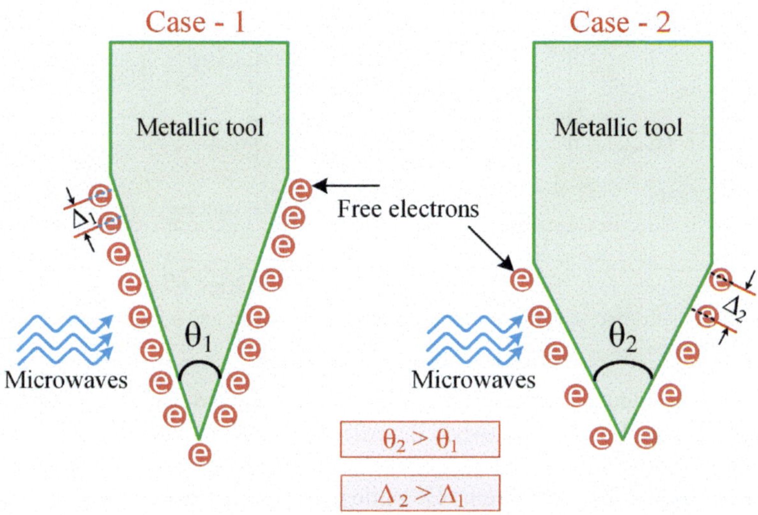

Fig. 7.9 Effect of tip angle on charge distribution in metallic tool during microwave exposure

electric field amplitude potentially generates discharge of extremely high intensity and may be detrimental for high voltage differential devices. At the same time, a much larger radius of curvature may induce a deficit electric field enhancement required for the breakdown of the surrounding medium. Hence an optimum value of tool electrode tip angle must be selected in order to control the discharge intensity. A reduced discharge intensity will facilitate the confinement of geometrical boundaries of plasma and will ensure controlled erosion of material from the workpiece.

6.3 Effect of Liquid Dielectric Medium

Researchers have incorporated low microwave-absorbing liquids such as hydrocarbon oils to generate microwave-induced discharges for various industrial applications [17]. The prime advantage offered is the attenuation of microwave energy within the liquid dielectric medium which diminishes the intensity of discharge. Also, the hydrostatic pressure applied by a much denser liquid medium helps to confine the boundaries of plasma column. A comparative overview of discharge generated in atmospheric and liquid dielectric medium is schematically represented in Fig. 7.10.

Kumar and Sharma [18] performed the drilling of borosilicate glass using graphite, tungsten carbide, and stainless steel as the tool materials. They employed liquid-based dielectric media such as transformer oil, palm oil and coconut oil as the drilling media. The authors concluded that, as compared to air medium, liquid dielectrics significantly confined the plasma column and minimized the thermal gradient around the edges of the hole. As a result, thermally-induced defects such as heat-affected zone

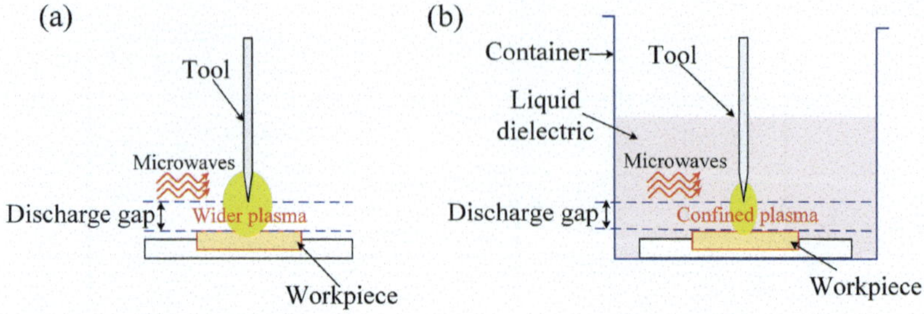

Fig. 7.10 Schematic representation of discharge generated in **a** air and **b** liquid dielectric medium

and cracks were reduced considerably. However, the breakdown of liquid dielectric into burnt hydrocarbons led to carbon deposits near the hole boundaries.

Therefore, liquid dielectric media could be utilized to generate microwave-metal discharge for metallic material removal while maintaining a discharge gap, as shown in Fig. 7.10. The use of liquid dielectric medium may prove beneficial in confining the plasma column and help improve the integrity of the machined surfaces.

7 Conclusions and Future Perspective

Microwave heating is commonly considered dielectric-based heating; however, in the present text, discharge effect associated with microwave energy has been explored. Microwave-metal discharge is generated when metallic objects having sharp tips or discontinuities are exposed to microwave energy. The immense heat produced by the discharge has proved its potential for rapid processing of material such as machining. The discharge was triggered from sharp tip of tungsten electrode and was subjected to AISI 304 stainless steel. It was observed that a minimum duration of 30 s was essential to completely ablate the AISI 304 stainless steel material. On increasing the discharge duration, the machined area also increases. The wide spatial distribution of plasma column contributed to significant difference between irradiated area and the ablated area. It was also identified that the electrode material experienced excessive wear and thermal degradation due to oxidation of tungsten in atmospheric air medium. Further, possible plasma confinement strategies such as magnetic field assistance, optimizing tool tip angle, and the use of liquid dielectric medium have been suggested to improve the process in the near future.

Acknowledgements The authors acknowledge the financial support by Science and Engineering Board (SERB), Department of Science and Technology (DST), Government of India, under the Grant no. CRG/2018/004305.

References

1. Sun J, Wang W, Yue Q et al (2016) Review on microwave-metal discharges and their applications in energy and industrial processes. Appl Energy 175:141–157. https://doi.org/10.1016/j.apenergy.2016.04.091
2. Descoeudres A (2006) characterization of electrical discharge machining plasmas 3542
3. Gupta M, Eugene WWL (2011) Microwaves and Metals
4. Mishra RR, Sharma AK (2016) Microwave-material interaction phenomena: heating mechanisms, challenges and opportunities in material processing. Compos Part A Appl Sci Manuf 81:78–97. https://doi.org/10.1016/j.compositesa.2015.10.035
5. Padmavathi C, Upadhyaya A, Agrawal D (2011) Effect of microwave and conventional heating on sintering behavior and properties of Al-Mg-Si-Cu alloy. Mater Chem Phys 130:449–457. https://doi.org/10.1016/j.matchemphys.2011.07.008
6. Sun J, Wang W, Yue Q (2016c) Review on microwave-matter interaction fundamentals and efficient microwave-associated heating strategies. Materials 9. https://doi.org/10.3390/ma9040231
7. Gavin Whittaker A, Mingos P (2000) Arcing and other microwave characteristics of metal powders in liquid systems. Journal of the Chemical Society, Dalton Transactions 1521–1526. https://doi.org/10.1039/b000462f
8. Hu C, Xi X, Huang Z, Zhan Z (2008) Simple analysis of mechanism of microwave sintering of metal powder. Mater Rev 22:329–332
9. Cai D, Tan Y, Zhang L et al (2021) Dynamic characteristics and energy analysis of microwave-induced metal discharge. J Energy Inst 100:277–284. https://doi.org/10.1016/j.joei.2021.12.002
10. Hussain Z, Khan KM, Hussain K (2010) Microwave-metal interaction pyrolysis of polystyrene. J Anal Appl Pyrolysis 89:39–43. https://doi.org/10.1016/j.jaap.2010.05.003
11. Qiao Z, Wang Z, Zhang C et al (2012) PVAm–PIP/PS composite membrane with high performance for CO_2/N_2 separation. AIChE J 59:215–228. https://doi.org/10.1002/aic
12. Singh A, Sharma AK (2020) On microwave drilling of metal-based materials at 2.45 GHz. Appl Phys A Mater Sci Process 126:1–11. https://doi.org/10.1007/s00339-020-03994-5
13. Gupta P, Singh A, Sharma AK, Singh I (2023) Influence of liquid dielectric medium on microwave-metal discharge-based drilling of AISI 304 stainless steel. Appl Phys A 129(2):150
14. Olsen HN (1959) Thermal and electrical properties of an argon plasma. Phys Fluids 2:614–623. https://doi.org/10.1063/1.1705962
15. Govindan P, Gupta A, Joshi SS et al (2013) Single-spark analysis of removal phenomenon in magnetic field assisted dry EDM. J Mater Process Technol 213:1048–1058. https://doi.org/10.1016/j.jmatprotec.2013.01.016
16. Ma N, Jones VF, Miller EJ, Kozaczek KJ (2019) An electrostatic study of curvature effects on electric field stress in high voltage differentials. J Electrostat 101:103370. https://doi.org/10.1016/j.elstat.2019.103370
17. Lebedev YA (2021) Microwave discharges in liquid hydrocarbons: Physical and chemical characterization. Polymers 13.https://doi.org/10.3390/polym13111678
18. Kumar G, Sharma AK (2018) Role of dielectric fluid and concentrator material in microwave drilling of borosilicate glass. J Manuf Process 33:184–193. https://doi.org/10.1016/j.jmapro.2018.05.010
19. Zhang Y, Zhang Z, Zhang Y et al (2021) Study on machining characteristics of magnetically controlled laser induced plasma micro-machining single-crystal silicon. J Adv Res 30:39–51. https://doi.org/10.1016/j.jare.2020.12.005

Joining of Metals Through Microwave Hybrid Heating

Rahul Samyal, Ankush Thakur, Raman Bedi, Ashok Kumar Bagha, and Akhil Khajuria

Abstract The source of heat is one of the crucial parameters that has varied significantly over the years for the fabrication of metal joints. A requirement for a new source of heat is justified by the fact that a single source of heat is not suitable for joining all the metals. Knowing the disadvantages of conventional welding processes, a novel heat source was developed for fabricating metal joints through microwave radiations. Microwaves are electromagnetic waves with a wavelength ranging between 1 to 1 mm and a frequency ranging between 0.3 to 300 GHz, respectively. Microwave heating of the materials depends upon the value of skin depth. Since metals have a low value of skin depth, metals are considered a reflector of microwaves at ambient temperature. However, at elevated temperatures, the metal starts absorbing microwaves. Thus, susceptor material is used to provide an initial rise in the temperature of metals. This heating of metals through susceptor material is termed microwave hybrid heating (MHH). Defining a new source of heat that provides various advantages like uniform heating, volumetric heating, eco-friendly, and energy efficiency has grabbed the attention of researchers. The MHH technique has been extensively used for joining different ceramics, composites, and metals. However, the metallurgical aspect of the microwave-processed joint is still not explored in detail. The present study describes the metallurgical aspects of the joints and base metals reported earlier and the scope of future study for particular metal joining.

Keywords Microwave welding · Exposure time · Interface material · Selective microwave hybrid heating

---

R. Samyal (✉) · A. Thakur · R. Bedi · A. K. Bagha · A. Khajuria
Department of Mechanical Engineering, Dr. B. R. Ambedkar National Institute of Technology, Jalandhar 144008, India
e-mail: rahulsamyal52@gmail.com

M. K. Singh et al. (eds.), *Composite Materials Processing Using Microwave Heating Technology*, Composites Science and Technology,
https://doi.org/10.1007/978-981-97-2772-8_8

Nomenclature

E	Electric field
μ_r	Relative permeability
μ_o	Permeability of free space
σ	Electrical conductivity
ε_o	Permittivity of free space
ε_r	Relative permittivity
K_o	Wave number of free space
ε'	Dielectric constant or permittivity
ε''	Energy factor
μ'	Penetration of magnetic field on material
μ''	Magnetic loss factor
δ	Skin depth
μ	Magnetic permeability
ρ	Resistivity
c	Speed of light
T	Temperature of the specimen
h	Heat transfer coefficient
f_c	Cut off frequency
T_{ext}	External temperature
K	Thermal conductivity
Q	Heat transfer
C_p	Specific heat at constant pressure
ρ'	Density
$\acute{\omega}$	Angular frequency
$\sigma E^2{}_{rms}$	Resistive loss
$\acute{\omega}\varepsilon_r'' E^2{}_{rms}$	Magnetic loss
$\acute{\omega}\mu r'' H^2{}_{rms}$	Dielectric loss
Q_{rms}	Power dissipated

1 Introduction

Over the years, metal joining is an essential process for manufacturing every type of structural assembly. Thus, welding is a well-recognized process for joining similar or dissimilar parts through heat and pressure. From the originated year, the process has developed dramatically due to the demand for sound joint, energy efficiency, reliability, and other factors. From time to time, the variation and improvisation in the processing parameters enable the engineers to join almost all the metals and alloys. The source of heat is one of the parameters that has varied significantly over the years, which directly impacts the quality of the joint [24, 32, 49, 60, 64]. With the change in the heat source, the heating rate varied; thus metallurgy of the adjoining parts also eventually varied [3, 35, 39, 50]. Thus, the welding process classification is based on a heat source (arc, flames, laser, etc.) and type of interaction (fusion and solid state). There are almost 35 types of welding processes [28, 30, 59]. Fusion welding processes involve the fusion of the adjoining metal with or without filler metal; however, mostly filler metal is used. Among all the fusion welding processes, electric arc welding is extensively used in industries where heat is generated through an arc [42]. The temperature generated in electric arc welding is 3500 ºC which is more than enough to melt all the metals. However, heat generation is so rapid and uncontrollable that it causes microstructural transformation in adjoining metals. Owing to its rapid heating ability, it is not possible to use for the joining of thin metals. Thus, the heat source was upgraded to laser and electron beam technology. But these processes require high maintenance costs and low energy conversion efficiency [51].

Conventional welding processes have a few limitations, like rapid heating, extensive power consumption, low energy conversion, and emitting harmful gases into the environment [29, 41, 55, 57, 58, 65, 83, 85]. It is a justified act that, eventually, there is a need for a new heat source that can provide uniform heating and is eco-friendly [27]. Defining a new source of heat that provides various advantages like uniform heating, volumetric heating, eco-friendly, and energy efficient [21, 43]. Microwave processing is a novel route of heating materials through electromagnetic radiation absorption and conversion [36]. Microwaves are electromagnetic waves with a wavelength ranging between 1 to 1 mm and a frequency ranging between 0.3 to 300 GHz, respectively. For the application of heating, 2.45 GHz frequency has been used for material processing [22, 23, 25, 26, 38]. Based on the different temperature ranges, microwave processing is categorized as follows:

1. Low-temperature processing (T < 500 ºC) uses synthesis and drying processes, for example, in polymer matrix, wood textile, rubber, and food industries.
2. Moderate temperature processing (500 < T < 1000 ºC), such as glass melting, brazing, drilling of non-metals, sintering of metallic powder, etc., comes under this domain.
3. Microwave processing at high temperatures (T > 1000 ºC) involves processing ceramics, bulk metals, Nano-materials, and MMCs. This domain also includes cladding, casting, welding, etc., of high-conducting materials.

1.1 Dielectric Heating

Dielectric heating is the process of heating material at a high frequency caused by molecular dipole rotation. These dipoles have considerable mass, polarizing at frequencies such as radio waves and microwaves. Initially, the dipoles are randomly aligned because of no electric field, resulting in a zero-dipole moment applicator. However, when an external alternating electric field is generated in the microwave, the polarization of the dipole begins. This makes the randomly aligned dipoles in an orientation parallel to that electric field. In the next half-cycle, the direction of the electric field changes, because of which the molecules rotate at 180°. This alignment changes every cycle, which agitates the molecules leading to dielectric heat generation [48]. The heating of materials via microwaves occurs at the molecular level in which microwaves transfer ionic moments. Therefore, the heating is strongly dependent on the dissipation factor (or dielectric loss or loss tangent factor) [63, 72]. This dissipation factor describes the amount of energy that is converted into heat, which is presented by the equation below:

$$\tan\delta = \frac{\varepsilon''}{\varepsilon'} \tag{1.1}$$

When the microwave passes through the material, the microwave absorption capabilities decrease. As a result, the electric field intensity decreases rapidly. The reduction of power dissipation depends on the relative permeability of materials and the loss tangent factor.

1.2 Role of the Magnetic Field in Microwave Heating

When microwaves interact with the material, heat is generated by energy conversion. As microwaves are electromagnetic waves, energy conversion depends on electric and magnetic fields. The mathematical formulation for this energy conversion has been mentioned below:

$$P = 2\pi[\varepsilon_0\varepsilon' \tan\delta E^2 + \mu_0\mu' \tan\Phi H^2] \tag{1.2}$$

The above equation describes the power converted from electromagnetic waves to heat. It also presents electric (E) and magnetic fields (H). Different materials have different heating effects for E and H. Electric heating occurs in a pure electric field for low-conducting materials such as ceramics and alumina. On the other hand, high-conducting materials, like metallic powders, heat efficiently in magnetic fields [56, 78].

1.3 Skin Depth

Skin depth or penetration depth (δ) is defined as the depth of material at which the intensity of the field reduces to 1/e (36.8%). The greater the skin depth, the higher the microwave radiation absorption rate, as it allows microwaves to pass through. Conducting materials such as metals have low skin depth at room temperature. However, the skin depth increases with the temperature [40, 70]. Therefore, the increase in temperature increases the resistivity of the metals according to the following equations:

$$\rho = \rho_0(1 + \alpha \Delta T) \tag{1.3}$$

An increase in the resistivity of material accounts for an increase in skin depth:

$$\delta = \sqrt{\frac{\rho}{\pi \mu \mu_0}} \tag{1.4}$$

1.4 Classification of Materials Based on Microwave Interaction

Different materials behave differently when exposed to the microwave depending on skin depth value [52]. And based on the interaction of microwaves with different materials, they are classified as follows:

1. **Microwave reflectors:** When the microwave interacts with the microwave reflectors' surface, they get reflected. They have low skin depth; thus, heating microwave reflectors, e.g., bulk metals, are always challenging.
2. **Microwave transmitters:** Microwave radiations are transmitted through it, causing minimal heating to the material. Microwave transmitters have a loss tangent factor (tan δ) less than 0.1. Ceramics, zircon, and fused quartz are examples of microwave transmitters.
3. **Microwave absorbers:** The microwave absorber material generates heat rapidly inside the materials during the absorbed energy from the microwave field applicator. Microwave transmitters have a loss tangent factor, "tan δ" more significant than 0.1. Charcoal, graphite, and Silicon powder come into the category of microwave absorbers. The interaction of microwaves with different materials is also illustrated in Fig. 8.1.

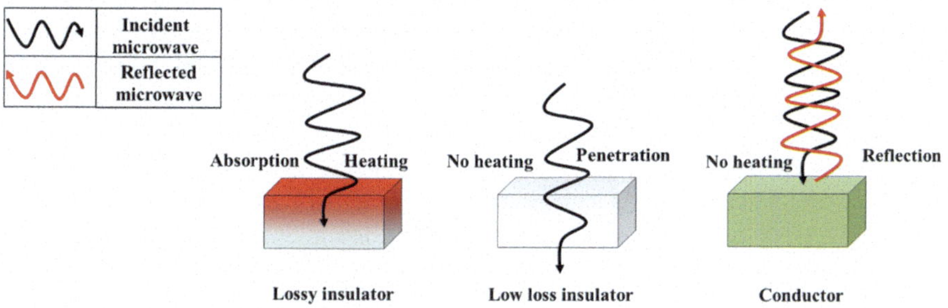

Fig. 8.1 Material classifications based on microwave interaction

1.5 Microwave Hybrid Heating (MHH)

Earlier microwave heating was only used for the application of microwave-absorbing materials due to the myth that bulk metals reflect microwaves. Skin depth is responsible for any material's absorption of microwave radiation, and skin depth is the function of temperature. At room temperature, the electrons get accumulate at the sharp edges of the bulk metals forming electron clouds and thus generating sparks inside the microwave applicator. Eventually, microwaves are reflected by bulk metal surfaces, ensuring no heating. Thus, metals are called microwave reflector materials [69]. However, with the increase in the temperature of the material, skin depth also increases. Applying similar phenomena to the processing of metals, metals were forced to absorb microwave radiation. This preliminary increase in the metal's temperature is caused by susceptor material which allows microwaves to penetrate to a minor penetration depth. At a temperature rise of above 400–600 °C, bulk metal initials the coupling with the microwave for heating, also called critical temperature. A susceptor can absorb electromagnetic energy at a higher rate, which is later converted into heat. These susceptors are carbon-based materials containing π electrons, which cause them to heat up rapidly [80]. This two-way heating is known as Microwave Hybrid Heating (MHH). Mostly graphite, silicon carbide, and charcoal powder are used as they are quite effective for this MHH. When susceptor-assisted processing is associated with direct microwave processing, it comprises some benefits like better microwave heating accuracy, lower thermal runaway, uniform processing. Characteristics of some susceptors are tabulated in Table 8.1.

Different heating types are involved in the microwave applicator depending on the interaction with the type of material. Figure 8.2 illustrates the types of heating involved in the microwave applicator. For microwave absorber materials, simple microwave heating often occurred presented in Fig. 8.2a. However, for the heating of absorber material at a specific region, the remaining part is masked with the masking material, causing microwave radiation to be absorbed within the particular region, naming it selective microwave heating (SMH). For the processing of metals under the microwave, a preliminary increase in temperature is mandatory, provided by the susceptor material, which is termed microwave hybrid heating (MHH). Further, to increase the process's efficiency, MHH was modified to the SMHH, which targets the specific region (Fig. 8.2d).

Table 8.1 Skin depth and loss tangent factor for different susceptors [20]

Susceptor material	Skin (penetration depth) (cm)	Loss tangent factor
Graphite powder (20–80 microns)	1.43	0.37
Charcoal powder	6–11	0.14–0.28
Silicon carbide	1.43	0.37
Activated carbon	0.7–3.43	0.31–0.9

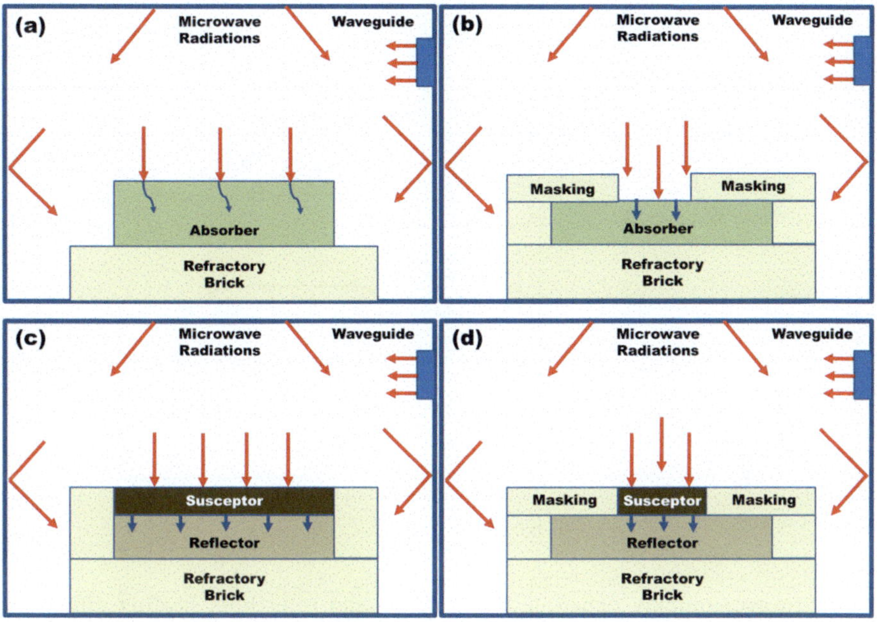

Fig. 8.2 **a** microwave heating **b** selective microwave heating **c** microwave hybrid heating **d** selective microwave hybrid heating

1.6 Advantages of Microwave Heating Over Conventional Heating

For conventional heating, the temperature gradient between the source and the target defines the mode of heat transfer as conduction, convection, and radiation. However, for microwave heating, energy is converted rather than transferred. The microwave energy penetrates through the specimen and stores energy. This stored energy creates heat throughout the bulk of the material. Thus, microwave heating exhibits a few advantages over conventional processing, such as:

1. Microwaves are penetrated through the material by molecular interaction with the electromagnetic field. Because of this, the progression of heat is uniform, as shown in Fig. 8.3.
2. Direct radiation absorption is involved in the microwave processing of materials, converting them into heat energy. This direct heat transfer leads to eliminating losses like heating furnaces, walls, and heat transfer media developed when materials were processed conventionally. So, microwave offers a faster processing heating rate than conventional heating techniques.
3. Use of susceptor material while processing materials in a microwave provides selective heating.

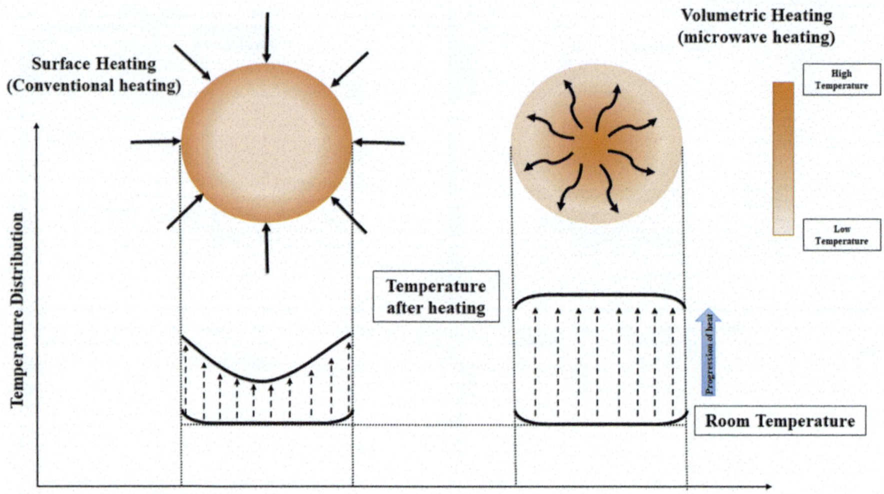

Fig. 8.3 Conventional heating versus microwave heating

4. Microwaves quickly penetrate the material and offer high heating efficiency because of minimum losses.
5. The methodology for operating materials in the microwave is simple. Also, it offers time and money savings.
6. No residue remains when a material is heated in the microwave, making it an eco-friendly way to process the material.

1.7 Evaluation of Microwave Heating of Metals

Microwave heating of the different materials started with the ease of processing materials. Initially, microwaves were used for food applications since it is easy to process. In the late 1970s, microwaves were utilized for the heating application of electrical conducting powders. Later, it was implanted for the microwave sintering of metal powders in 1999. To date, microwave sintering is the most established process among all the other microwave heating applications. Further, followed by the cladding and coating process of the metals in 2005 and 2008. MHH of metals for joining application was reported in 2009. Year-by-year evaluation of microwave heating for different applications is presented in Fig. 8.4.

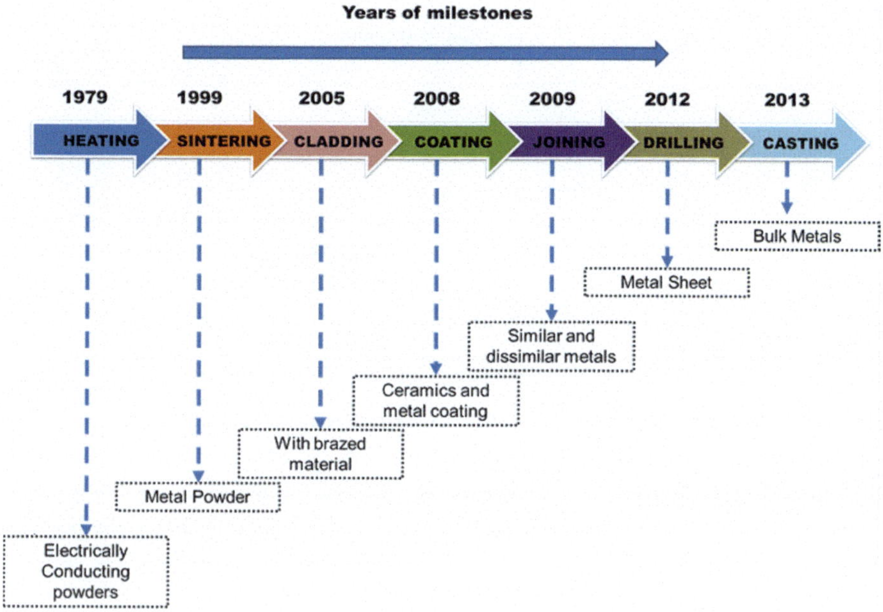

Fig. 8.4 Evaluation of microwave heating for different applications

2 Joining of Metals Using Microwave Radiations

The joining of metals is a manufacturing technique of joining two or more metals to form a strong metallurgical bond assisted by either heat or pressure. There are many conventional joining processes like welding, brazing, soldering, and adhesive bonding. In addition, there are advanced technologies such as laser beam welding, electron beam welding, and ultrasonic welding. However, microwave joining is a non-conventional joint fabrication that exhibits some unique attributes compared to traditional and non-traditional methods. The capability of localized heating using the microwave is a critical parameter for the joining process. The microwave joining depends on various factors, such as the type of metal and its physical, chemical, and metallurgical properties. Microwave joining is a practical and cost-effective method for various metals that are difficult to weld using conventional heating. Assembly used for the microwave joining application is shown in Fig. 8.5.

The Microwave joining process for the metals can be divided into the following stages:

1. Preparation of sample.
2. Selection of suitable filler powder.
3. Use of suitable susceptor medium.
4. Selection of processing parameters.

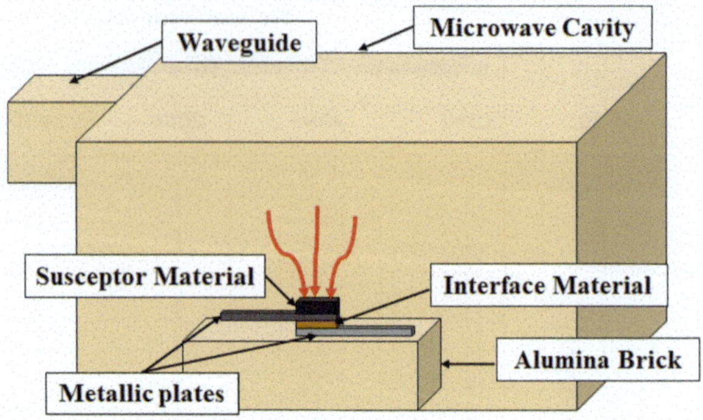

Fig. 8.5 Assembly for microwave joining of metals

2.1 *Microwave Hybrid Heating: Simulation Study*

Joint between biodegradable composites and natural fiber-reinforced composites have been produced using natural fibers from microwave heating. Later, the heating mechanism was validated using COMSOL Multiphysics software [12]. The joining of Inconel-718 in COMSOL Multiphysics software for 540 s was achieved for MHH. The maximum temperature achieved by the joint interface was observed to be 1124.8 °C [13]. Further, the microwave joining of mild steel pipe at 2.45 GHz frequency and 900 W input power through simulation in COMSOL Multiphysics. The temperature of the pipe was higher at the bottom than it was at the top because of its considerable height. The smaller size of the top pipe causes heating instability [34]. The authors successfully simulated the welding of Oxygen Free Copper and SS304 in MHH for 14 min. It has been determined that the maximum temperature of 1065 °C was attained at the susceptor material and 966 °C at the interface material. The study also reported that the volumetric diffusion and temperature would cause the reduction of the porosity of the interface material [82]. Rawat et al. [66] conducted an experimental and simulated study for the different susceptor materials. The overall arrangement was used as selective microwave hybrid heating (SMHH). Different shapes of the vertical cavity were considered for charcoal and graphite susceptor material. It was concluded that a vertical cavity with a conical shape and charcoal susceptor exhibits a better heating profile. Literature based on microwave heating of susceptor materials is tabulated in Table 8.2.

2.2 *Metallurgical Aspects of Microwave Joining of Metals*

2.2.1 Mild Steel

Mild steel is low-carbon steel with a carbon percentage of up to 0.3% and has high tensile and impact strength. Mild steel specimens with a dimension of 30 × 10 × 5 mm was joined using a microwave hybrid heating technique. Nickel powder with

Table 8.2 Literature of microwave heating of susceptor Material

References	Base metal	Specimen size (mm)	Input power/ frequency	Exposure time (s)	Results
Lin et al. [54]	Coal	Radius:25 Height:60	1000 W/ 2.45 GHz	60	The increase in the loss factor results in a reduction of thermal homogeneity. The heterogeneity of the coal increases with the exposure time. The optimum heating was achieved at an operating frequency of 2.45 GHz. Low power output is suitable for thermal homogeneity and energy saving, while high power output is suitable for differential and rapid heating. High power increases the heating rate, reducing thermal uniformity and energy saving
Tamang and Aravindan [79]	(SiC)	Radius:12 Height:38	700 W/ 2.45 GHz	300	Satisfactory validation between experimental and computational results was achieved for temperature profiles at 191 °C and 201 °C, respectively. The height of the SiC and the power output of the microwave applicator were two parameters that caused an exponential increase in the heating rate

(continued)

Table 8.2 (continued)

References	Base metal	Specimen size (mm)	Input power/ frequency	Exposure time (s)	Results
Patel et al. [62]	(SiC)	150 × 140 × 25	540 W, 700 W, 900 W/ 2.45 GHz	300	For an exposure time of 300 s and power output of 900 W, the maximum temperature for A-5 grade alumina was 576.47 °C, and for SiC was 542.05 °C. The heating capability of A5 grade alumina and SiC is equivalent
Rawat et al. [66]	Charcoal, Graphite	Upper Radius: 5 Lower Radius: 15 Height:45	900 W	80 s, 160 s, 240 s, and 300 s	Charcoal as susceptor material exhibits better temperature properties compared to Graphite. For 300 s of exposure time the maximum temperature achieved by Graphite was 215 °C However, for Charcoal susceptor, it was 655 °C. Moreover, vertical cavity of conical shape shows better heating profile for selective microwave hybrid heating compared to cylindrical shape

a 40 μm particle size was used to archive the sandwich layer. 250 MPa of tensile strength was observed for the specimen with 6% elongation [14, 17]. The obtained tensile strength for microwave welded mild steel specimen was observed to be 50% of the joint. This might be because of the poor heat treatment properties of the mild steel and microwave hybrid heating provides uniform and volumetric heating to the entire specimen. Further, the effect of the influencing parameters on the tensile strength of the microwave welded mild steel 1018 was observed. From the ANOVA analysis it was concluded that the power output of the microwave applicator and processing time are the two most influencing parameter to define the tensile strength of the joint. The study predicted that low power rated oven (800 W) gives better tensile strength (242 MPa) at an exposure time of 1000 s [31].

Utmost of the research work stated the microwave joining of metallic rods and plates; however, the microwave joining of metallic pipes has been hardly reported. The joining of metallic pipes through microwave heating is a difficult task due to the positioning of a fixture, placement of pipes inside the cavity, less wall thickness of the pipe, placement of susceptor at the joint zone, and placement of filler material at the circumference of pipes for optimal heating of the selected area. The horizontal or circumferential joining of metallic pipes was problematic due to issues related to maintaining similar susceptor thickness around the joint zone; otherwise, non-uniform heating at the joint zone occurs. The vertical joining of the pipe specimen provides ease of susceptor distribution across the joint zone. Joined mild steel pipes in the vertical position at an exposure time between the 420 s and the 510 s and observed greater joint strength at an exposure time of 480 s [33]. Carbide precipitations with micropores were observed through SEM imaging. A complete joint was formed for an exposure time of 480 s. 572 HV of Joint hardness was observed, whereas, for base metal, it was 372 HV. However, the ultimate tensile strength of the joint was reduced from base metal. The authors have attempted to circumferentially join the mild steel pipes through a selective microwave hybrid heating process using a novel cavity design, keeping all constraints in mind that arise during microwave joining. The novel cavity design was designed to overcome the challenges arising during microwave heating, like the placement of susceptor material due to the circular geometry of the specimen, proper alignment of pipes, and placement of filler material. To concentrate the microwaves directly onto the target region vertical cavity feeder was used, leading to selective heating of the filler and joint zone. Also, to explore the heating behavior inside the microwave cavity, mild steel pipe specimens have been modelled using COMSOL Multiphysics software. It was foreseen that temperature in the range of 1500–1700 °C could be attained in the joint section at an exposure time of 480 s to have an appropriate fusion of interface material with a faying surface [34]. In another investigation, the maximum temperature of 1810 °C has been observed in the susceptor, followed by 1570 °C in the joint zone at an exposure time of 900 s. It has been noticed the maximum temperature was achieved at the center of the pipe where the susceptor material has been positioned [84].

2.2.2 Inconel

Materials used for extreme environmental, heat, and pressure conditions usually are super alloys like Inconel (Ni–Cr alloy). Inconel shows oxidation resistance at a higher temperature, which are used in the aerospace industries. However, welding these super alloys is always challenging due to their tendency to crack. Bansal et al. [16] developed a joint between Inconel 718 through MHH. Ni powder was placed between slaying surfaces achieving a thickness of 0.2 mm. Joint was characterized through SEM imaging, XRD, BSE micrographs, and fractography. Further, the study was extended by using Inconel 718 as interface powder, and post-weld heat treatments were performed [16, 19]. Badiger et al. [4] used Ni-based powder for developing the microwave welded joint between Inconel 625 at an exposure time of 21 min. Refine grains were identified near the joint area through SEM imaging. Moreover, shear failure was confirmed through the fractography analysis. Badiger et al. [5] analyzed the microstructural and mechanical properties of a joint between Inconel 625 developed through microwave hybrid heating (MHH). The presence of laves phase along the grain boundaries was confirmed. Badiger et al. [8] joined Inconel-625 using MHH at two different power levels, 600 W and 900 W. Nickel-based EWAC powder is used at a 2.45 GHz frequency. The penetration depth was increased by increasing the initial temperature, which is done by using susceptor materials like coal and SiC. MHH occurs in two stages, first from the susceptor at an initial step and second by self-heating through direct microwaves. Lower diffusion was noted in the case of the specimen heated by 600 W power leading to reduced grain growth. The increase of hardness values of the specimen from the fusion zone to the interface zone was observed and maximum at the interface region. A uniaxial tensile test was conducted on UTM, and it was observed that the specimen processed at 600 W had higher yield strength than those processed at 900 W. The three-point bend test concluded that the flexural strength of the joint developed at 600 W was more than the specimen developed at 900 W. A fractography study revealed mixed mode fracture, concluding that a specimen processed at 600 W shows more ductile characteristics. Further, the study was carried out for the influencing parameters directly related to the joint properties [6, 7].

2.2.3 Dissimilar Metals

Due to the different elemental content of the joining metals, dissimilar metals are always challenging to achieve. Srinath et al. [76] used a multimode applicator with 900 W of power and 2.45 GHz of frequency. Bansal et al. [15] concluded that Inconel 718 and austenitic stainless steel (SS316L) could be joined using microwave heating, whereas Inconel 718 powder was used as an interface layer. With 600 s heating, the study revealed that the microstructure formed after joining; no interface cracking was observed. The average hardness of around 230 HV was seen at the interface.

Tamang and Aravindan [81] performed the joining of dissimilar materials, aluminum (Al) and magnesium (Mg), using Zinc (Zn) as interface powder, and the effect of different structures on the microstructures of the joint was studied. The joint formation was achieved faster when graphite susceptor was used rather than activated charcoal. However, this graphite susceptor created a random heating pattern, whereas activated charcoal produced a uniform heating pattern. It was also observed that microwave exposure time increases the Mg-Zn layer.

2.2.4 Other Materials

Srinath et al. [75] studied microwave joining for a conductive material that is possible using microwave-absorbing materials. The joining of copper was conducted using charcoal as a susceptor material. The interface material used was copper powder. Microwave heating provides uniform heating to the interface region. The microwave radiation converts copper power to copper oxide, which helps in coupling the workpiece with the microwave. The hardness of the joint was 84% of the bulk material with significant tensile strength.

Srinath et al. [77] studied that copper (high thermal conductivity material) can be joined by microwave heating in a multi-mode applicator at 900 W output power with a frequency of 2.45 GHz. To achieve MHH, charcoal powder was used as a susceptor material coupled with the microwave. The authors observed that the microwave penetrates around 88.94% volume of individual particles because of the small size of the metallic powder. It was observed that the copper powder oxidized and resulted in the formation of CuO, which improved the microwave coupling. The Scanning electron microscope microstructure of the joint is observed, and it was concluded that the metallic bond formed is of better quality with negligible pores. The presence of Carbon dioxide shows that carbon is diffused from the separator sheet. The density of microstructure in the joint is relatively more. Aravindhan and Krishnamurthy [2] performed the joining of ceramic material Alumina coating 30% zirconia using microwave heating in which sodium silicate glass was used as the filler material. The heating is conducted for 10 min. The joint strength depends on the filler material properties. The joint strength was improved by enhancing the pressure, as it would reduce the residual stress near the interlayer region.

Bajpai et al. [12] concluded that joining fibers by microwave heating provides many advantages. The specimen was joined by two techniques, first by microwave energy and second by adhesive bonding. The strength of both methods was analyzed from one other and then by parent materials. The authors used a multi-mode applicator with 900 W power and 2.45 GHz frequency for joining microwave heating. The exposure time for PLA and PP was 200 s and 250 s simultaneously. To accelerate the heating process, charcoal was used. A tensile test for parent material and composite material was conducted and it was observed that composites had a higher failure load. Further, it was observed that adherents got weaker after lap joint formation in microwave joining because of unwanted heating near the joint interface. This unwanted heating was avoided by properly varying the sample's location in the cavity and redesigning fixtures. The joint strength attained from microwave joining was better associated with adhesive joining methods. The authors concluded that for joining process power input, selection of susceptor, microwave exposure time, and location of the specimen inside the microwave cavity also play a crucial role. The literature on the microwave joining of various materials has been mentioned in Table 8.3.

2.2.5 Stainless Steel

Stainless steel is an alloy of iron having a minimum of 10.5% of Chromium in the elemental composition. This sufficient amount of Chromium forms a chromium oxide layer at the surface, preventing any further corrosion of the stainless-steel surface. Stainless steels are extensively utilized for the application of Architectural, Marine, pressure vessels, etc. due to their corrosion-resistant property. Further stainless steels are categorized as Austenitic, Ferritic, and Martensitic, with varying amounts of Chromium, Nickel, and Carbon.

Stainless steels are often difficult to weld through conventional welding due to their efficient heat-retaining properties. Thus, the MHH technique was used to apply joining SS316 using Nickel powder as an interlayer. The process was carried out for

Table 8.3 Literature of microwave joining

References	Base metal	Specimen size (mm)	Interface material	Susceptor material	Exposure time (s)	Results
Aravindan and Krishnamurthy [2]	Al_2O_3–ZrO2 composites containing 30% 23 2 zirconia	15 × 4 × 4	sodium silicate glass powder	–	300	The maximum temperature achieved at the joint was 1000 °C within 2 min. But remain constant for a further 9 min. The property of the interface material is improved by reducing the powder size of the interface material
Ahmed and Siores [1]	48%alumina-32% Zirconia-20%silica ceramics	–	–	–	–	The output power varies between 1000 to 1500 W to avoid thermal runaway and low heating. The rate of temperature rise decreases when power increases over time. So, it can control the thermal runaway, leading to improved joint quality
Bajpai et al. [12]	Polylactic acid	50 × 12 × 4	–	Charcoal	300	The composites can also be joined through microwave with lesser exposure time, and the maximum temperature was achieved of 157.2 °C
Bansal et al. [18]	SS316	25 × 15 × 4	SS316 powder	Silicon Carbide (SiC)	720	The interface has an ultimate tensile strength of 425 MPa, which is nearly 82.5% of the base material. It was also revealed that ductile fracture of the material is dominant
Gupta and Kumar [37]	Stainless Steel	–	Nickel Powder	Charcoal	900	NiSe, FeNi, Cu_2Se, and $FeSi_2$ are present in the stainless-steel joint. Stainless Steel can show good mechanical properties and microstructures with 323.16 MPa of ultimate tensile strength, 11.30% elongation, and 145.3 Hv of hardness

(continued)

Table 8.3 (continued)

References	Base metal	Specimen size (mm)	Interface material	Susceptor material	Exposure time (s)	Results
Badiger et al. [4]	Inconel 625	51 × 12 × 6	Nickel Based Powder	Charcoal	1260	The joining of Inconel-625 alloy has been successfully conducted using MHH. The tensile strength obtained was 326 MPa. The maximum hardness is found to be at the interface
Bansal et al. [13]	Inconel 718	25 × 10 × 4	Inconel 718 Powder	Silicon Carbide	540	A maximum temperature of around 1124.8 °C was observed in the joint area. The effect of the electric field is more predominant than the magnetic field
Gamit [34]	Mild Steel Pipe	–	Nickel Powder	Charcoal Powder	600	At 480 s, the left and right sides of the joint achieved 1500 °C and 1700 °C, respectively. Similarly, the temperature at the top and bottom end of the same side of the pipes was observed to be 1200 °C and 900 °C. The different temperature at the same exposure occurs due to the formation of hot spots and cold spots
Pal et al. [61]	SS304-SS316	51 × 12 × 6	Different nickel powder 70 nm,20 μm, and 50 μm	Charcoal Powder	420	The highest C content, of 5.52%, was traced at the joint region for joints developed using nickel powder of 70 nm size. On the other hand, the joint made with powder of particle size 50 μm of has the lowest carbon content. Results show that C, Ni, Fe, and Cr in the interface material show better melting of filler material and proper joint formation
Tamang and Aravindan [82]	Oxygen-free copper (OFC) /SS304	20 × 10.5 × 3	Cu powder	Graphite Powder	840	The temperature of 850 °C and 1065 °C was achieved for the interface material and graphite susceptor, respectively. The use of selective hybrid heating shows better results and lesser exposure time

390 s. SEM images observed a Fused joint layer with columnar growth and carbide precipitation at the sandwich layer. The formation of NiC_x, Cr_3C_2, and FeNi was confirmed at the joint interlayer region through XRD analysis [74]. Using 40 μm of particle size of Ni-based powder, a joint was developed for stainless steel. The Joint was mechanically characterized through Vickers hardness and tensile strength for varying proportions of Ni powder. It was observed that an interface with 95% of Ni exhibits a tensile strength of 323 MPa and hardness of 143 HV. The heating was conducted for 900 s with 700 W applied power. It was found that the joint properties have been improved due to the presence of FeNi. NiSe, $FeSi_2$ and Cu_2Se_2. For 700 s of exposure time, a maximum value of tensile strength was obtained [37]. The joining of metals through microwave hybrid heating with the same interface powder as that of parent joining metal was studied for SS316 joints with SS316 powder as interface. A sound joint was observed from SEM imaging at the interface region with epitaxial grain growth along the direction of <100> [18].

A comparative study of the two different interface powders was reported for the joining of the SS304. 99% pure Nickel powder and Ni-Based EWAC powder having equal particle size were considered. It was observed that the joint induced from 99% pure Ni powder had lesser carbon content than the EWAC-based joints. Further, the weldability of the EWAC-based joint was higher than that of Nickel powder-based joints [11]. The particle size of the interface material is the utmost influencing parameter for defining the joint strength and metallurgy of the joint. Thus, the impact of the size of the interface powder on microwave-welded joints was observed. SS304 joints were developed by introducing interlayer powder having a particle size of 20, 30, 40, and 50 μm. The study concluded that the small size of the interlayer powder provided superior homogeneity to the joint region, thus, better interfacial properties were observed for a smaller size of the interface region [9]. Further study was extended to a nano-scale particle size of the interlayer powder and concluded that microwave-welded joint with nano-scale interlayer powder possesses better joint properties [73]. Exposure time for joining any joint under microwave radiation is one of the other influencing parameters that is important to define the joint strength. A study was carried out to determine the exposure time for joining several grades of SS (202, 304, and 316) using Ni powder as interface powder [67]. To reduce the overall cost of the microwave joining process, authors joined the SS304 and duplex stainless steel without filler material [10, 53]. SS202-SS202 lap joint were joined through SMHH for an exposure time of 15 to 25 min. The joint was characterized by mechanical and vibration texting. For the exposure time of 9 min sound joint was produced [68].

Microstructural analysis of the SS304 butt joint was examined through low-end optical microscopy and high-end orientation microscopy, i.e., electron back-scattered diffraction (EBSD). The post-EBSD data analysis was accomplished on TSL-Orientation imaging microscopy (OIM) software. Initially, slaying surfaces of the joints were cleaned and grinded, to avoid unwanted impurities in the joint interlayer region. The interlayer region was prepared by mixing Ni powder of 20 μm and epoxy. A mixture of Ni powder and epoxy formed a thick paste which was later molded to the thickness of approx. 1 mm. A Graphite separator sheet was used at the top of the joint interlayer region. Further, on the top of the graphite sheet, activated charcoal was used that acted as susceptor material. This whole arrangement was enclosed in refractory bricks, i.e., lower refractory brick and upper refractory brick. In addition, the upper brick provided masking of the specimen. In the lower brick, the specimen, along with the interlayer, was placed. The upper brick carried the susceptor powder in the drill hole, specifically targeting the interlayer region. This process is referred to as SMHH. The complete arrangement of the bricks, along with the specimen, was placed inside the cavity of the microwave applicator. The operating frequency of the microwave applicator was 2.45 GHz, with an operating power output of 900 W.

In lieu of the discussed state-of-the-art literature herein and by consideration of research and technological gaps; thus, the basic objective of the present attempt was to focus on the development of metallic pipe joint using microwave energy and subsequently its microstructural characterization through optical microscopy and EBSD analysis.

Microstructural examination plays a vital role in understanding the material's behaviour and response to the application of mechanical stresses, environmental conditions, etc. It becomes of utmost importance in terms of materials joining as the response of base material chemistry is sensitive to the degree of heating involved, and a reasonable estimate could be drawn on the corresponding mechanical behaviour of the obtained joint. Optical microscopy was used for the preliminary investigation of the joint region. The microstructure shown in Fig. 8.6 indicates complete melting of interface material and good metallurgical bonding with base metal for both the SS304 butt joint and as well as SS304 pipe joint.

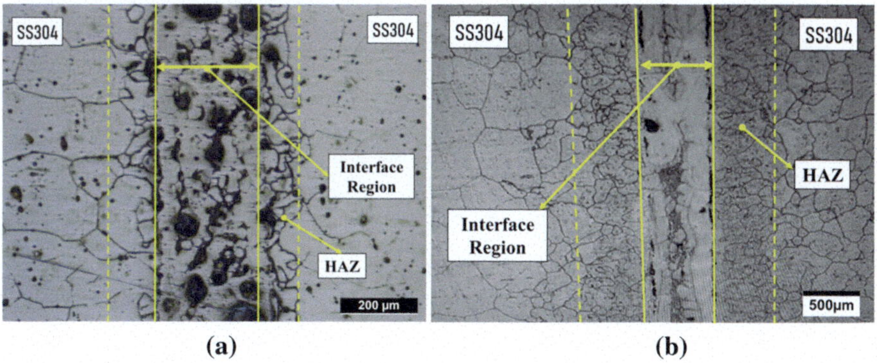

Fig. 8.6 Typical optical microstructure of joined specimen **a** SS304 Butt joint **b** SS304 pipe joint

3 EBSD of SS Pipe Joint

EBSD analysis for microwave-welded stainless-steel pipes is presented in this section. A comparative EBSD analysis is studied for two different regions: base metal and HAZ regions. Figure 8.7 presents the referencing area for the EBSD analysis.

3.1 Morphological Characteristics of Crystallographic Orientations

The post-processing of EBSD data yields concrete information in terms of the behavior of crystallographic orientations, local strain within the microstructure, susceptibility to deformation, the contributing fractions of grain boundaries and so on that could be analyzed to estimate mechanical attributes [45, 47]. Herein a preliminary post-EBSD data analysis of base metal and microwave joined specimen of SS304 has been presented to develop an initial understanding. Figure 8.8 represents the color-coded inverse pole figure (IPF) maps of base metal and joined specimen of SS304. IPF describes the crystallographic orientation of the grains as per the coded scale included in the bottom of Fig. 8.8. The IPF maps indicate <001>, <101>, and <111> planes in red, green, and blue color, respectively. The columnar grains of the joined specimen predominate in <101> plane as exhibited by green color (Fig. 8.8b), while fine grains were observed in base metal as indicated by Fig. 8.8a. Further, microstructural grain twinning appeared to exist in the joined specimen (Fig. 8.8b) which will increase the dislocation density of the grains resulting in improved ductility. As SS304 is austenitic steel by large weight fraction of nickel, twins are generally observed due to dislocation interactions that tend to thicken by

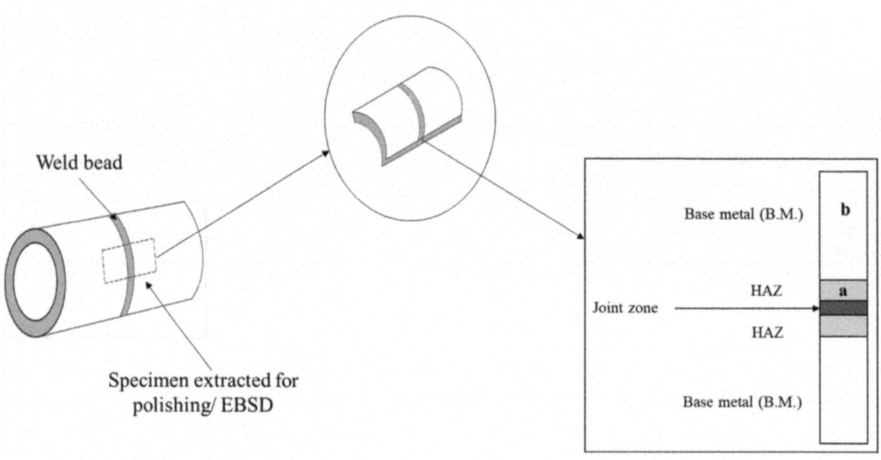

Fig. 8.7 Extraction of specimen for EBSD analysis

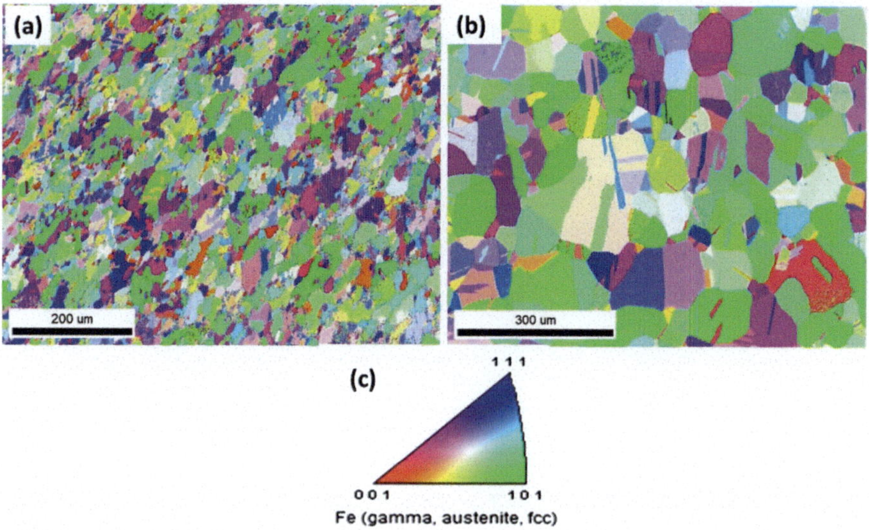

Fig. 8.8 IPF maps of base metal and joined specimen of SS304 **a** Base metal **b** Joined specimen **c** IPF scale

overlapping of other nuclei formed on the (1 1 1) planes parallel to its habit plane [71].

3.2 Interpretations from Pole Figure (PF) Maps

PF maps interpret the texture and contour intensity of the materials used to develop the joint. It is evident from PF maps (Fig. 8.9a,b) that grain boundaries tend to be random in base metal, whereas they have preferred orientation in a joined specimen. It signifies that obtained preferred orientation in the case of a microwave-joined specimen of SS304 should exhibit reliable mechanical properties.

3.3 Distribution of Grain Size and Grain Boundary Character

The average grain size of joined specimen has been observed to be (45–55 μm) compared to that of base metal (15–20 μm), as illustrated by statistical charts in Fig. 8.10a, b. In the case of a joined specimen, the area fraction occupied by large grains was relatively bigger than the base metal of SS304. It indicates that microwave heating led to the development of homogenous microstructure as desired to achieve better mechanical properties.

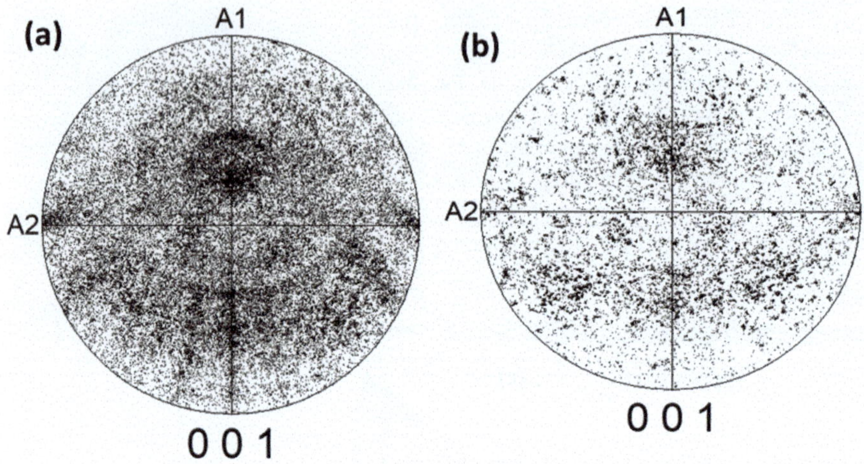

Fig. 8.9 PF maps of base metal and joined specimen of SS304 **a** Base metal **b** Joined specimen

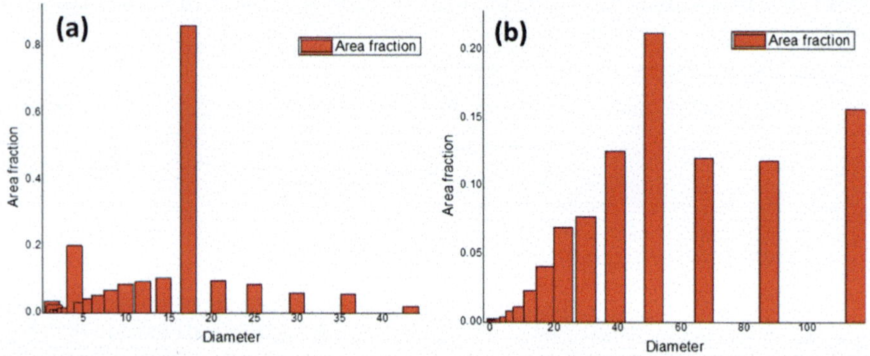

Fig. 8.10 Distribution of grain size in base metal and joined specimen of SS304 **a** Base metal **b** Joined specimen

Grain boundaries regarding misorientation angle between 0º to 15º are considered low angle grain boundaries (LAGB). In contrast, misorientation angles greater than 15º are assumed to be high-angle grain boundaries (HAGB) [46, 48]. Correspondingly in the grain boundary distribution map (Fig. 8.11a, b); dark blue lines represent HAGBs, whereas red and green lines depict LAGBs. The presence of relatively large HAGBs in the joined specimen will provide improved interference with the slip direction when compared to relatively large LAGBs in base metal due to increased misalignment in the grain boundary region, resulting in a greater change in slip direction. In other words, improved fractions of HAGBs in the microwave-joined specimen of SS304 will favor the microstructure to retard bulk deformation due to increased resistance in developing micro-cracks which otherwise will relatively quickly propagate with existing large fractions of LAGBs in base metal.

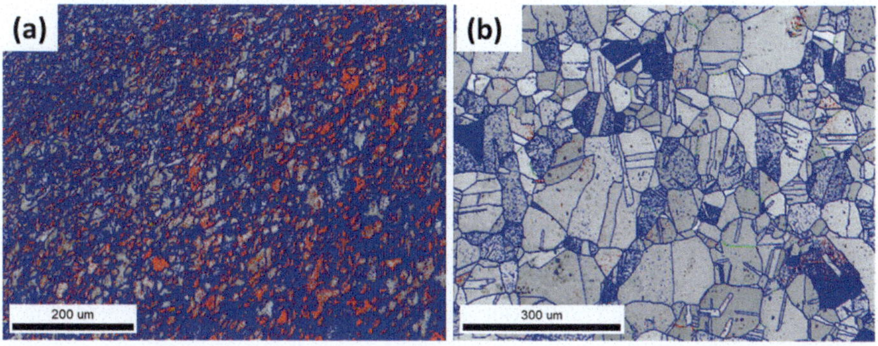

Fig. 8.11 Grain boundary maps of base metal and joined specimen of SS304 **a** Base metal **b** Joined specimen

3.4 Variation in Local Microstructural Strain

Kernel average misorientation (KAM) value characterizes the changes in the local misorientation and microstructure caused by geometrically necessary dislocations (GNDs), giving a variation and distribution of local strain within the microstructure. Also, KAM maps can reveal areas with an increased density of defects. KAM maps show the color-coded description of variation in microstructural strain present in base metal and microwave-joined specimens of SS304 (Fig. 8.12a, b). The increased KAM values were observed along grain boundaries, indicating the formation of slip lines at different locations. The higher KAM value sometimes indicated the dislocations inside the specimen. Further, grains in microwave-joined specimen exhibited small KAM values, i.e., less strain, indicating an improved ductility when compared to base metal, where fine grains revealed relatively large KAM values reflecting increased strain. Overall, the evolved microstructure in microwave joined specimen in terms of local strain distribution remarked less variation from KAM maps than base metal.

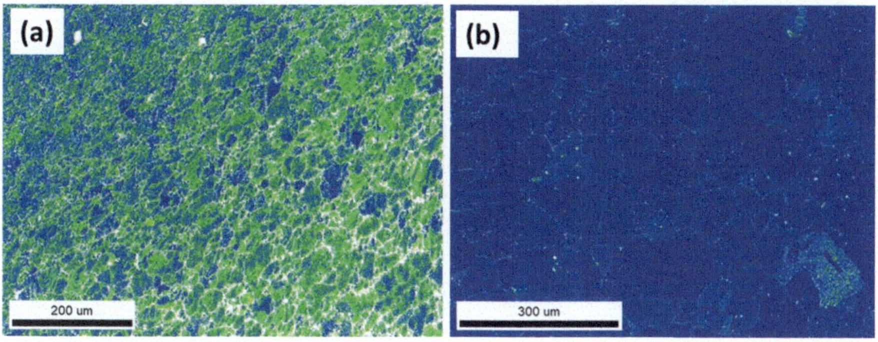

Fig. 8.12 KAM maps of base metal and joined specimen of SS304 **a** Base metal **b** Joined specimen

4 Summary

A new source of heat that provides various advantages over conventional heating sources like uniform heating, volumetric heating, eco-friendly, and energy efficient is interesting to explore. The MHH technique has been extensively used for joining different ceramics, composites, and metals. The type of interface (filler) powder, the particle size of the interface material, separator sheet, susceptor material, and exposure time are the involved parameters for microwave-processed joints. From the parametric review of the distinct reported studies, it was observed that particle size of the interface material and power out are the most influencing parameters to define the strength of the joints. Smaller particle size and low power output was recommended for better metallurgical properties and joint strength. The MHH technique was upgraded to SMHH to achieve better mechanical properties. Further, In SMHH, a conical vertical cavity shape with charcoal susceptor material was recommended for quick heating.

Microwave joining of mild steel material is lacking with good tensile strength properties. Mild steel is not suitable for heat treatments because microwave joining could not provide better tensile strength properties. Although, welding of Inconel is quite challenging with the conventional welding process due to its trend to initiate cracks. However, MHH was successfully used for joining different grades of Inconel effectively. The joining of different grades of Stainless steel revolves around the Ni or Ni-based filler material. Specific heat, atomic radius, density, and conductivity properties of Ni are almost similar to Fe or better than Fe. In the present study, SS 304 pipes have been successfully joined at an exposure time of 1080 s through a selective microwave hybrid heating process. Optical microscopy results revealed good metallurgical bonding between the filler material and base metal. EBSD results confirmed the formation of columnar grains near heat affected zone with an average grain size of 45–55 μm while base metal is 15–20 μm.

5 Future Scope

It is observed from the current study that, to establish the microwave joining process into industrial application, it is required to have a complete analysis of the joint: covering static, dynamic, and metallurgical analysis. Microwave joining of mild steel requires a more detailed study of the influencing parameters for tensile strength. Joining dissimilar materials like ceramics with metals could be possible with the MHH technique. This area of research still needs to be explored.

References

1. Ahmed A, Siores E (2001) Microwave joining of 48 % alumina-32 % zirconia-20 % silica ceramics. J Mater Process Technol 118:88–95
2. Aravindan S, Krishnamurthy R (1999) Joining of ceramic composites by microwave heating. Mater Lett 38:245–249
3. Azar AS, Ås SK, Akselsen OM (2012) Determination of welding heat source parameters from actual bead shape. Comput Mater Sci 54(1):176–182. https://doi.org/10.1016/j.commatsci.2011.10.025
4. Badiger RI, Narendranath S, Srinath MS (2015) Joining of Inconel-625 alloy through microwave hybrid heating and its characterization. J Manuf Process 18:117–123. https://doi.org/10.1016/j.jmapro.2015.02.002
5. Badiger RI, Narendranath S, Srinath MS (2018) Microstructure and mechanical properties of Inconel-625 welded joint developed through microwave hybrid heating. Proceedings of the Institution of Mechanical Engineers, Part B: Journal of Engineering Manufacture 232(14):2462–2477. https://doi.org/10.1177/0954405417697350
6. Badiger RI, Narendranath S, Srinath MS (2018) Optimization of parameters influencing tensile strength of inconel-625 welded joints developed through microwave hybrid heating. Mater Today: Proc 5(2):7659–7667. https://doi.org/10.1016/j.matpr.2017.11.441
7. Badiger RI, Srinath SNMS (2018) Optimization of process parameters by taguchi grey relational analysis in joining Inconel—625 through microwave hybrid heating. Metallogr Microstruct Anal 0123456789. https://doi.org/10.1007/s13632-018-0508-4
8. Badiger RI, Srinath SNMS, Hebbale AM (2019) Effect of power input on metallurgical and mechanical characteristics of Inconel-625 welded joints processed through microwave hybrid heating. Trans Indian Inst Met 72(3):811–824. https://doi.org/10.1007/s12666-018-1537-z
9. Bagha L, Sehgal S, Thakur A, Kumar H (2017) Effects of powder size of interface material on selective hybrid carbon microwave joining of SS304–SS304. J Manuf Process 25:290–295. https://doi.org/10.1016/j.jmapro.2016.12.013
10. Bagha L, Sehgal S, Thakur A, Kumar H, Goyal D (2019) Low cost joining of SS304-SS304 through microwave hybrid heating without filler-powder. Eng Res Express 1:025035
11. Bagha L, Shegal S, Thakur A (2016) Comparative analysis of microwave based joining/welding of SS304–SS304 using different interfacing materials. MATEC Web Conf 57:1–4. https://doi.org/10.1051/matecconf/20165703001
12. Bajpai PK, Singh I, Madaan J (2012) Joining of natural fiber reinforced composites using microwave energy: Experimental and finite element study. Mater Des 35:596–602. https://doi.org/10.1016/j.matdes.2011.10.007
13. Bansal A, Ansari S, Sharma S (2017) 3D electromagnetic field simulation of microwave joining of inconel-718. Int J Mech Eng Technol 8(7):1231–1237
14. Bansal A, Sharma AK, Das S (2013) Metallurgical and mechanical characterization of mild steel-mild steel joint formed by microwave hybrid heating process. Sadhana—Acad Proc Eng Sci 38(4):679–686. https://doi.org/10.1007/s12046-013-0142-4
15. Bansal A, Sharma AK, Das S, Kumar P (2016) On microstructure and strength properties of microwave welded Inconel 718/ stainless steel (SS-316L). Proceedings of the Institution of Mechanical Engineers, Part L: Journal of Materials: Design and Applications 230(5):939–948. https://doi.org/10.1177/1464420715589206
16. Bansal A, Sharma AK, Kumar P, Das S (2012a) Application of electromagnetic energy for joining of Inconel 718 plates. I-Manager's J Mech Eng 2(4):18–24. https://doi.org/10.26634/jme.2.4.1986
17. Bansal A, Sharma AK, Kumar P, Das S (2012) Joining of mild steel plates using microwave energy. Adv Mater Res 585:465–469. https://doi.org/10.4028/www.scientific.net/AMR.585.465
18. Bansal A, Sharma AK, Kumar P, Das S (2014) Characterization of bulk stainless steel joints developed through microwave hybrid heating. Mater Charact 91:34–41. https://doi.org/10.1016/j.matchar.2014.02.005

19. Bansal A, Vasudev H, Sharma AK (2019) Investigation on the effect of post weld heat treatment on microwave joining of the Alloy-718 weldment Investigation on the effect of post weld heat treatment on microwave joining of the Alloy-718 weldment. Mater Res Express 6
20. Bhattacharya M, Basak T (2016) A review on the susceptor assisted microwave processing of materials. Energy 97:306–338. https://doi.org/10.1016/j.energy.2015.11.034
21. Bhoi NK, Singh H, Pratap S (2019) Microwave material processing : a clean, green, and sustainable approach. In: Sustainable Engineering Products and Manufacturing Technologies (Issue 1). Elsevier Inc. https://doi.org/10.1016/B978-0-12-816564-5.00001-3
22. Chandrasekaran S, Ramanathan S, Basak T (2012) Microwave material processing—a review. Fluid Mechan Trans Phenom J 58(2):330–363. https://doi.org/10.1002/aic
23. Chandrasekaran S, Ramanathan S, Basak T (2013) Microwave food processing—a review. Food Res Int 52(1):243–261. https://doi.org/10.1016/j.foodres.2013.02.033
24. Chen J, Wu CS, Chen MA (2014) Improvement of welding heat source models for TIG-MIG hybrid welding process. J Manuf Process 16(4):485–493. https://doi.org/10.1016/j.jmapro.2014.06.002
25. Clark DE, Folz DC, West JK (2000) Processing materials with microwave energy. Mater Sci Eng, A 287:153–158
26. Das S, Mukhopadhyay AK, Datta S, Basu D (2009) Prospects of microwave processing: an overview. Bull Mater Sci 32(1):1–13
27. David SA, DebRoy T (1992) Current issues and problems in welding science. Science 257(5069):497–502. https://doi.org/10.1126/science.257.5069.497
28. Debroy T, David SA (1995) Physical processes in fusion welding. Rev Mod Phys 67(1):85–112. https://doi.org/10.1103/RevModPhys.67.85
29. Dehghan SF, Mehrifar Y (2019) Occupational exposure to fumes and gases during different arc welding processes. Int J Occup Hygiene 136–145
30. DuPont JN, Marder AR (1995) Thermal efficiency of arc welding processes. Welding J 74(12):406-s
31. Dwivedi SP, Sharma S (2014) Effect of process parameters on tensile strength of 1018 mild steel joints fabricated by microwave welding. Metall Microstruct Anal 3:58–69. https://doi.org/10.1007/s13632-013-0109-1
32. Farias RM, Teixeira PRF, Vilarinho LO (2022) Variable profile heat source models for numerical simulations of arc welding processes. Int J Therm Sci 179(April):107593. https://doi.org/10.1016/j.ijthermalsci.2022.107593
33. Gamit D, Mishra RR, Sharma AK (2017) Joining of mild steel pipes using microwave hybrid heating at 2.45 GHz and joint characterization. J Manuf Process 27:158–168. https://doi.org/10.1016/j.jmapro.2017.04.028
34. Gamit DN (2018) Modeling and simulation of microwave metallic pipe joining process. IJRAR-Int J Res Anal Rev 5(3):988–991
35. Gery D, Long H, Maropoulos P (2005) Effects of welding speed, energy input and heat source distribution on temperature variations in butt joint welding. J Mater Process Technol 167(2–3):393–401. https://doi.org/10.1016/j.jmatprotec.2005.06.018
36. Grohmann S, Langhans G, Reindl A, Sidarava V, Zaeh MF (2020) Investigation of reactive bimetallic Ni-Al particles as a heat source for microwave-assisted joining. J Mater Process Technol 282(December 2019):116637. https://doi.org/10.1016/j.jmatprotec.2020.116637
37. Gupta P, Kumar S (2014) Investigation of stainless steel joint fabricated through microwave energy. Mater Manuf Processes 29(8):910–915. https://doi.org/10.1080/10426914.2014.892975
38. Haque KE (1999) Microwave energy for mineral treatment processes — a brief review. Int J Mineral Process 57:1–24
39. Heinze C, Schwenk C, Rethmeier M (2012) Effect of heat source configuration on the result quality of numerical calculation of welding-induced distortion. Simul Model Pract Theory 20(1):112–123. https://doi.org/10.1016/j.simpat.2011.09.004
40. Horikoshi S (2018) Materials processing by microwave heating. In: Microwave Chemical and Materials Processing, pp 321–381

41. Jilla A (2019) Evaluation of total fume and heavy metal emission factors applicable to gas metal arc welding
42. Kah P, Latifi H, Suoranta R, Martikainen J, Pirinen M (2014) Usability of arc types in industrial welding. Int J Mech Mater Eng 9(1):1–12. https://doi.org/10.1186/s40712-014-0015-6
43. Kandagal ZB, Akkimaradi VG, Sonnad AN (2019) A review on green technology material processing through microwave energy. Int J Trend Sci Res Develop 3(2):402–405. https://doi.org/10.31142/ijtsrd21353
44. Khajuria A, Akhtar M, Bedi R, Kumar R, Ghosh M, Das CR, Albert SK (2020) Influence of boron on microstructure and mechanical properties of Gleeble simulated heat-affected zone in P91 steel. Int J Press Vessels Pip 188(November):104246. https://doi.org/10.1016/j.ijpvp.2020.104246
45. Khajuria A, Akhtar M, Bedi R, Kumar R, Ghosh M, Das CR, Albert SK (2020) Microstructural investigations on simulated intercritical heat-affected zone of boron modified P91-steel. Mater Sci Technol (United Kingdom) 36(13):1407–1418. https://doi.org/10.1080/02670836.2020.1784543
46. Khajuria A, Bedi R, Kumar R (2019) Investigation of impression creep deformation behavior of boron-modified P91 steel by high-end characterization techniques. Springer Singapore. https://doi.org/10.1007/978-981-13-6287-3_10
47. Khajuria A, Kumar R, Bedi R (2019) Effect of boron addition on creep strain during impression creep of P91 steel. J Mater Eng Perform 28(7):4128–4142. https://doi.org/10.1007/s11665-019-04167-z
48. Khaled D El, Novas N, Gazquez JA, Manzano-agugliaro F (2017) Microwave dielectric heating : applications on metals processing. Renew Sustain Energy Rev October, 0–1. https://doi.org/10.1016/j.rser.2017.10.043
49. Kik T (2020) Heat source models in numerical simulations of laser welding. Materials 13(11):1–24. https://doi.org/10.3390/ma13112653
50. Kiyoshima S, Deng D, Ogawa K, Yanagida N, Saito K (2009) Influences of heat source model on welding residual stress and distortion in a multi-pass J-groove joint. Comput Mater Sci 46(4):987–995. https://doi.org/10.1016/j.commatsci.2009.05.002
51. Krivtsun IV, Khaskin VY, Korzhik VN, Ziyi L (2015) Industrial application of hybrid laser-arc welding (Review). Paton Welding J 2015(7):41–46. https://doi.org/10.15407/tpwj2015.07.07
52. Kumar P, Ingle A, Jhavar S (2019) Parametric review of microwave-based materials. Integr Med Res x *x*:1–21. https://doi.org/10.1016/j.jmrt.2019.04.004https://doi.org/10.1016/j.jmrt.2019.04.004
53. Kumar V, Sehgal S (2020) Materials today: proceedings joining of duplex stainless steel through selective microwave hybrid heating technique without using filler material. Mater Today: Proc 28:1314–1318. https://doi.org/10.1016/j.matpr.2020.04.509
54. Lin B, Li H, Chen Z, Zheng C, Hong Y, Wang Z (2017) Sensitivity analysis on the microwave heating of coal: A coupled electromagnetic and heat transfer model. Appl Therm Eng 126:949–962. https://doi.org/10.1016/j.applthermaleng.2017.08.012
55. MacLeod JS, Harris MA, Tjepkema M, Peters PA, Demers PA (2017) Cancer risks among welders and occasional welders in a national population-based cohort study: canadian census health and environmental cohort. Saf Health Work 8(3):258–266. https://doi.org/10.1016/j.shaw.2016.12.001
56. Mandal AK, Sen R (2017) An overview on microwave processing of material: a special emphasis on glass melting. Mater Manuf Processes 32(1):1–20. https://doi.org/10.1080/10426914.2016.1151046
57. Mehrifar Y, Zamanian Z, Pirami H (2019) Respiratory exposure to toxic gases and metal fumes produced by welding processes and pulmonary function tests. Int J Occup Environ Med 10(1):40–49. https://doi.org/10.15171/ijoem.2019.1540
58. Mehrifar Y, Zeverdegani SK, Rismanchian M (2020) Chemical pollutants in the respiratory zone of welders: determination of concentrations and hazard analysis. Work 67(3):591–598. https://doi.org/10.3233/wor-203272

59. Mendez PF, Eagar TW (2001) Welding processes for aeronautics. Adv Mater Processes 159(5):39–43
60. Moslemi N, Gohari S, Abdi B, Sudin I, Ghandvar H, Redzuan N, Hassan S, Ayob A, Rhee S (2022) A novel systematic numerical approach on determination of heat source parameters in welding process. J Market Res 18:4427–4444. https://doi.org/10.1016/j.jmrt.2022.04.039
61. Pal M, Sehgal S, Kumar H, Goyal D (2020) Use of nickel fi ller powder in joining SS304-SS316 through microwave hybrid heating technique. Metal Powder Report 1–6. https://doi.org/10.1016/j.mprp.2020.10.001
62. Patel DK, Bhoi NK, Singh H (2022) Microwave heating capabilities of different susceptor material: experimental and simulation study. Silicon 14(12):6621–6635. https://doi.org/10.1007/s12633-021-01426-4
63. Pathania A, Gupta D (2014) A review: Microwave energy for materials processing. Int J Sci Res Develop 2(09):9–14
64. Podder D, Mandal NR, Das S (2014) Heat source modeling and analysis of submerged arc welding. Weld J 93(5):183–192
65. Popović O, Prokić-Cvetković R, Burzić M, Lukić U, Beljić B (2014) Fume and gas emission during arc welding: hazards and recommendation. Renew Sustain Energy Rev 37:509–516. https://doi.org/10.1016/j.rser.2014.05.076
66. Rawat S, Samyal R, Bedi R, Bagha AK (2022) Comparative performance of various susceptor materials and vertical cavity shapes for selective microwave hybrid heating (SMHH). Phys Scr 97(August):125704. https://doi.org/10.1088/1402-4896/ac9e7d
67. Samyal R, Bagha AK, Bedi R (2019) An experimental study to predict the exposure time for microwave based joining of different grades of stainless steel material. Mater Today Proc 27:2449–2454. https://doi.org/10.1016/j.matpr.2019.09.217
68. Samyal R, Bagha AK, Bedi R (2021) Evaluation of modal characteristics of SS202-SS202 lap joint produced using selective microwave hybrid heating. J Manuf Process 68(PB):1–13. https://doi.org/10.1016/j.jmapro.2021.07.018
69. Samyal R, Bedi R, Bagha AK (2023) Investigation of mechanical properties of microwave welded SS304-SS202 lap joints. Phys Scr. https://doi.org/10.1088/1402-4896/acb5d4
70. Sharma AK, Mishra RR (2018) Role of particle size in microwave processing of metallic material systems. Mater Sci Technol (United Kingdom) 34(2):123–137. https://doi.org/10.1080/02670836.2017.1412043
71. Shen YF, Li XX, Sun X, Wang YD, Zuo L (2012) Twinning and martensite in a 304 austenitic stainless steel. Mater Sci Eng A 552:514–522. https://doi.org/10.1016/j.msea.2012.05.080
72. Singh S, Gupta D, Jain V, Sharma AK (2015) Microwave processing of materials and applications in manufacturing industries: a review. Mater Manuf Processes 30(1):1–29. https://doi.org/10.1080/10426914.2014.952028
73. Soni P, Sehgal S, Kumar H, Singh AP (2018) Joining of SS316-SS316 through microwave hybrid heating by using Nickel nano-powder. Int J Appl Eng Res 13(8):6446–6449
74. Srinath MS, Sharma AK, Kumar P (2010) A novel route for joining of austenitic stainless steel (SS-316) using microwave energy. Proc IMechE B J Eng Manuf 225:1083–1091. https://doi.org/10.1177/2041297510393451
75. Srinath MS, Sharma AK, Kumar P (2011) A new approach to joining of bulk copper using microwave energy. Mater Des 32(5):2685–2694. https://doi.org/10.1016/j.matdes.2011.01.023
76. Srinath MS, Sharma AK, Kumar P (2011) Investigation on microstructural and mechanical properties of microwave processed dissimilar joints. J Manuf Process 13(2):141–146. https://doi.org/10.1016/j.jmapro.2011.03.001
77. Srinath MS, Sharma AK, Kumar P (2011) Microwave processing of metallic joints and their characterisation. I-Manager's J Mech Eng 1(1):21–25
78. Sun J, Wang W, Yue Q (2016) Review on microwave-matter interaction fundamentals and efficient microwave-associated heating strategies. Materials 9(1):1–25. https://doi.org/10.3390/ma9040231
79. Tamang S, Aravindan S (2019) 3D numerical modelling of microwave heating of SiC susceptor. Appl Therm Eng 162(August):114250. https://doi.org/10.1016/j.applthermaleng.2019.114250

80. Tamang S, Aravindan S (2021a) Effect of susceptors on joining of AA6061-T6 to AZ31B by microwave hybrid heating. J Mater Eng Performance 31(2):1130–1139. https://doi.org/10.1007/s11665-021-06220-2
81. Tamang S, Aravindan S (2021b) Effect of susceptors on joining of AA6061-T6 to AZ31B by microwave hybrid heating. J Mater Eng Perform Ref 31. https://doi.org/10.1007/s11665-021-06220-2
82. Tamang S, Aravindan S (2022) Joining of dissimilar metals by microwave hybrid heating: 3D numerical simulation and experiment. Int J Thermal Sci 172(PA), 107281. https://doi.org/10.1016/j.ijthermalsci.2021.107281
83. Tenkate TD (1998) Optical radiation hazards of welding arcs. Rev Environ Health 13(3):131–146. https://doi.org/10.1515/REVEH.1998.13.3.131
84. Thakur A, Bedi R, Bagha AK, Rao PS (2022) Microstructural and mechanical properties of mild steel pipes joined using selective microwave hybrid heating. J Manuf Process 82(September):848–859. https://doi.org/10.1016/j.jmapro.2022.08.024
85. Yeo SH, Neo KG (1998) Inclusion of environmental performance for decision making of welding processes. J Mater Process Technol 82(1–3):78–88. https://doi.org/10.1016/S0924-0136(98)00022-3

Development and Characterization of Microwave Sintered SiC Reinforced 3003 Aluminium Alloy

Vinay Kumar Soni, Raj Kumar Sahu, Agnivesh Kumar Sinha, and Royal Madan

Abstract Densification behaviour, hardness and microstructural evolution of microwave sintered 3003 aluminium alloy reinforced with varying SiC composition were investigated. The sintering was performed in a 2.45 GHz multimode microwave furnace at a temperature ranging from 524 °C to 557 °C. The study deals with the compacted 3003 aluminium alloy powder interacting with the microwaves and defines its density, hardness and grain size variation for different heating rates and SiC composition during the compaction and sintering process. The microhardness and grain size of a prepared sample were calculated by Vickers microhardness and optical microscopy and their grain size are measured using image analysis software. An increase and decrease in grain size were observed with heating rate and addition of SiC in 3003 Al alloy respectively, resulting in good mechanical properties. Optimum conditions for the heating rate and percentage of SiC were suggested which is beneficial for Aluminium based composite manufacturing industries. With the addition of SiC, the sintered density increases by 0.6%, 3% and 1.31% for heating rates 20 °C, 30 °C and 40 °C respectively.

Keywords 3003 aluminium alloy · Microwave Sintering · Heating rate · Density · Hardness · Grain size

---

V. K. Soni
Department of Mechanical Engineering, Columbia Institute Engineering and Technology, Raipur, Chhattisgarh 493111, India

R. K. Sahu
Department of Mechanical Engineering, National Institute of Technology, Raipur, Chhattisgarh 492010, India

A. K. Sinha
Department of Mechanical Engineering, Rungta College of Engineering and Technology, Bhilai, Chhattisgarh 490024, India

R. Madan (✉)
Department of Mechanical Engineering, Graphic Era (Deemed to be University), Dehradun, Uttarakhand 248002, India
e-mail: royalmadan6293@gmail.com

M. K. Singh et al. (eds.), *Composite Materials Processing Using Microwave Heating Technology*, Composites Science and Technology,
https://doi.org/10.1007/978-981-97-2772-8_9

1 Introduction

Powder metallurgy (P/M) is a solid phase manufacturing process in which the powder is compacted by applied external pressure with the help of presses to obtain the green compact, resulting in an 80 percent dense component. Further, the compact is sintered at the proper temperature i.e., 0.6–0.8T_m (melting temperature) to produce an approximately 100 percent dense component [4, 7]. A sintering is a process in which a green compact is heated from room temperature to recrystallization temperature. In the conventional sintering process, the compact is kept for a certain period of time at a slower heating rate for the formation of fine grains. However, this process is costly and time taking because the mode of heat transfer is due to conduction, convection and radiation. To overcome this issue a new technique, known as microwave sintering was developed in which, in contrast to conventional heating where layer-wise heating of material was carried out from outer surface to inner surface; here, the heating process was performed volumetrically as materials coupled with microwaves and absorbs heat. Due to this heating technique, it has several benefits such as improved diffusion processes, decreased energy consumption, very quick heating rates, and processing times [13]. In the powder metallurgy method, the compaction load, sintering temperature, and time play a vital role in the development of microstructure. The effects of these can be computed by performing microstructural characterization in the porosity, grain size, & other mechanical attributes such as density, modulus, hardness, etc. [8, 9].

The influence of microwave sintering on the attributes of (Ti, W) C particulate reinforced 6061 Al alloy composites were studied, they found that with the rise in the sintering temperature, the porosity first increased and then decreased. The mechanical attributes like as yield strength and hardness increased with reinforcement content and sintering temperature [19]. The densification of microwave-sintered W–Cu composites was significantly increased when sintered at higher temperatures or for soaking period [16]. Sintering at a lower temperature and for a shorter period, complete densification of the WC–Co alloy was achieved [1]. The particle size also has a significant effect on physical and mechanical attributes, a fine powder shows improved densification and so the improved properties [2]. The effect on attributes of Al–Mg-Si–Cu alloy for conventional & microwave heating was investigated. Higher microwave sintering temperatures and shorter sintering times led to an inhomogeneous microstructure with lower hardness and electrical conductivity [14]. To investigate the impact of the variations in the dielectric attributes during the heating process, a 2D simulation of the electromagnetic field distribution inside a microwave oven and the accompanying thermal distribution was conducted [15]. In the instance of microwave sintering, grain refinement led to a rise in hardness and fracture toughness [3].

With decrease in the particle size from micro to nano, the sintering temperature drops significantly thereby enhancing the specimen's bulk properties [5, 6]. Mechanical properties of optimized volume fraction of AA7075/SiC/graphite hybrid composite were studied for change in temperature from 400 to 550 °C by a step increase of 50 °C. The properties first increase with an increase in temperature till the temperature reaches a critical value of 500 °C, beyond this temperature the properties decrease [10, 11]. Literature review reveals that in comparison to traditional sintering: microwave sintering can improve densification and grain growth [18], and microwave sintering of W–Ni–Fe alloy shows a reduction in process time by approximately 75% thereby grain coarsening was achieved [17]. Higher density and finer average grain size was achieved by microwave sintering which manifested improved mechanical attributes [12].

The present research aims to investigate the following: the influence of heating rate and SiC on the grain size, microstructures, physical and mechanical attributes of sintered PM 3003 compacts. Microstructural characterization has been performed for 3003 aluminium alloy for 2 different temperatures (524 °C and 557 °C) and three different heating rates are 20 °C/min, 30 °C/min, and 40°C/min. The composite SiC composition is kept as 0 wt% and 5 wt% and their sintering temperature and heating rate on the densification of Aluminium with and without SiC reinforcement were seen. Vickers's hardness test was performed for the above-fabricated samples. The average grain size was calculated using image analysis software. The mechanical properties first improve with temperature rise up to a critical temperature beyond this the properties degrades because of the development of larger grains.

2 Experimental Procedure

Elemental composition for the preparation of alloy is illustrated in Table 9.1. Appropriate proportions of elements in powder form were milled in a Centrifugal ball mixer (model: T2C, supplier: Bachoffen, Switzerland) for 5 h. Milled elemental powders were pressed at 370 MPa and 410 MPa to make cylindrical pellets (diameter 25 mm, height 10 mm) utilizing a Universal testing machine (model: UTN-100, supplier: Fuel Instruments & Engineers Pvt. Ltd.) with different wt% of SiC particulate, i.e., 0 wt% and 5 wt% by cold compression forming. Compaction was done at two different magnitudes of pressure (370 MPa and 410 MPa) for 2 min using lithium stearate as a die wall lubricant.

Table 9.1 Material composition

Material	% Weight
Al	97
Cu	0.2
Fe	0.6
Mn	1.5
Zn	0.1
Si	0.6

Sintering of green compacts was carried out in a microwave furnace. Microwave sintering was done in a multimode cavity 2·45 GHz, 10 kW commercial microwave furnace (VB Ceramic Consultants Chennai). Argon atmosphere (99.99%) was maintained within the microwave furnace for heating the green compacts. Green compacts were heated till 524 °C and 557 °C at three different rates i.e. at 20 °C/min, 30 °C/min, 40 °C/min. Figure 9.1 shows the schematic of a typical microwave furnace. Microwave power controls the heating temperature. An infrared thermometer is utilized for recording and monitoring the temperature of specimens.

The density of sintered compacts was evaluated using the Archimedes principle. Three experiments were performed for each set of conditions and their average was recorded. And to determine the efficiency of sintering, densification is evaluated by Eq. (1).

$$\rho = \frac{\rho_{s}\_\rho_{g}}{\rho_{t}\_\rho_{g}} \tag{1}$$

where, ρ, ρ_g, ρ_s *and* ρ_t are the densification, green density, sinter density and theoretical density of specimen, respectively [10].

A metallographic study of sintered samples was prepared wherein etching was performed using Keller's reagent for the 30 s. Subsequently, with the help of an

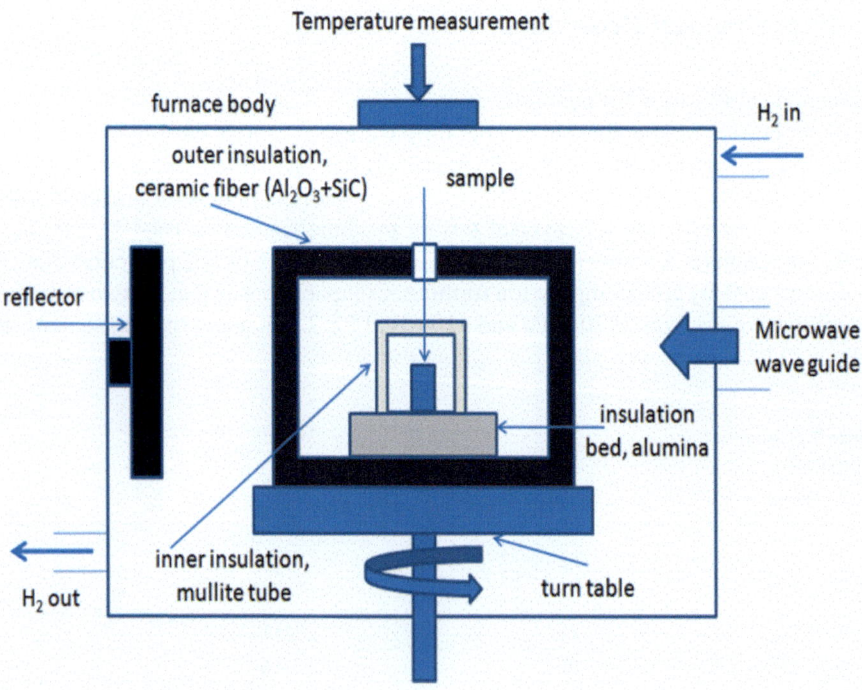

Fig. 9.1 Microwave furnace for sintering

optical microscope (supplier: Zeiss, Germany) sintered samples were investigated for their microstructures. An image analyser was used for determining the grain size. Three micrographs were obtained at 500X magnification for each specimen. And an average of three-grain sizes was recorded for enhanced accuracy. The hardness of sintered specimens was determined by Vicker's micro-hardness apparatus (Model: Hmv-2 T Adw, Supplier: Shimadzu, Germany) wherein load of 10 g was applied in each case for 5 s.

3 Results and Discussion

3.1 Macro-Scale Morphology of the Sintered Samples

3003 Al alloy specimens with and without SiC particulates were successfully prepared by using a microwave furnace. Figure 9.2 illustrates the heating behavior of 3003 Al alloy with sintering time. It manifests that 3003 Al alloy interacted well with microwave resulting in achieving samples without macro scale cracks or porosity. Samples of 3003 Al alloy and 3003 Al/SiC were prepared for heating temperatures of 524 °C and 557 °C. Sintering temperature was selected such that deformation or bubbling could be prevented due to melting.

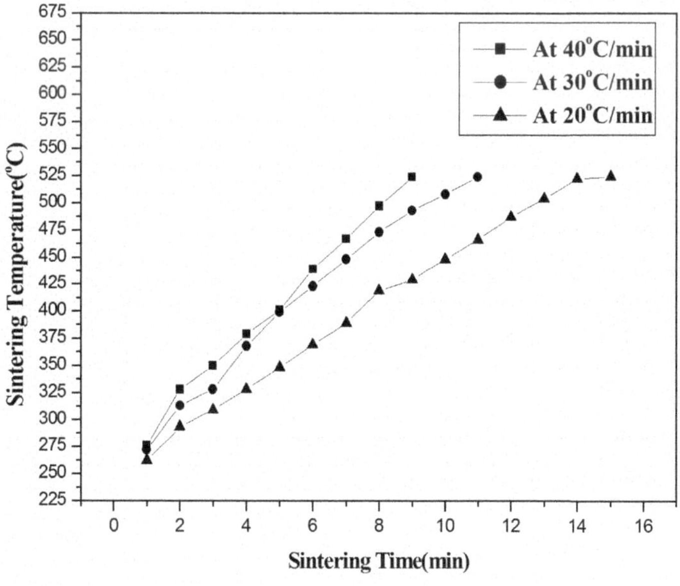

Fig. 9.2 Effect of heating rate on the thermal profiles of 3003 Al Alloy compacts

3.2 Heating Profiles and Densification Response

The heating rate in microwave sintering plays key role in governing the mechanical attributes of 3003 al alloy. Thermal heating profiles of 3003 Al alloy compacts for 3 different heating rates that are 20 °C/min, 30 °C/min and 40°C/min are shown in Fig. 9.2. 3003 Al alloy compacts couple well at 20 °C/min with a microwave. Therefore, 3003 Al alloy compacts after sintering show much finer mechanical properties than at other heating rates.

Figure 9.3 illustrates the influence of heating rate and SiC on density of sintered compacts of 3003 Al alloy. It is observed that, on increasing the rate of heating from 20 to 30 °C /min, sintered density of 3003 Al alloy with SiC increases due to the consolidation effect on the grains, which decrease the porosity. But, when rate of heating is further raised by 10 °C /min i.e., from 30 to 40 °C /min, porosity increases relatively due to the oxide formation, which in turn decreases the sintered density of 3003 Al alloy (with SiC). The results are in accord with the work done by [4].

It is also evident from Fig. 9.3 that the sintered density for 3003 Al alloy with SiC is greater for all the values of heating rates from the 3003 Al alloy without SiC. From this, it is inferred that SiC enhances the density by filling the voids of the alloy. However, on increasing the compaction load and sintering temperature, there is a significant amount of increase in the sintered density of 3003 Al alloy without SiC. This enhancement in density is due to the consolidation of 3003 Al alloy particles due to high compaction load and sintering temperature which reduces the voids between the particles and leads to high values of density. The values of

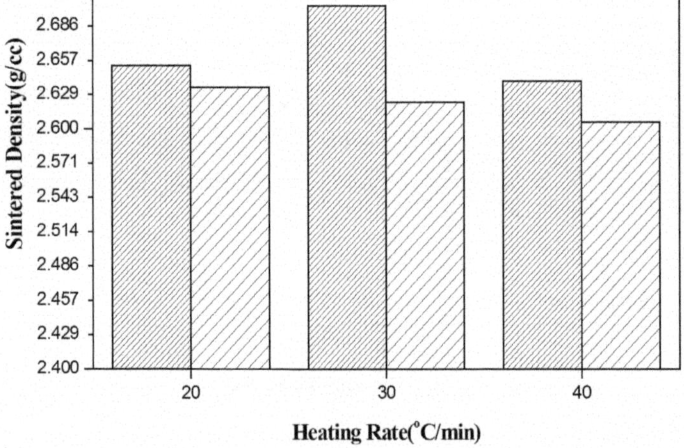

Fig. 9.3 Sintered density at 524 °C and 180 kN load for 3003Al alloy with and without SiC

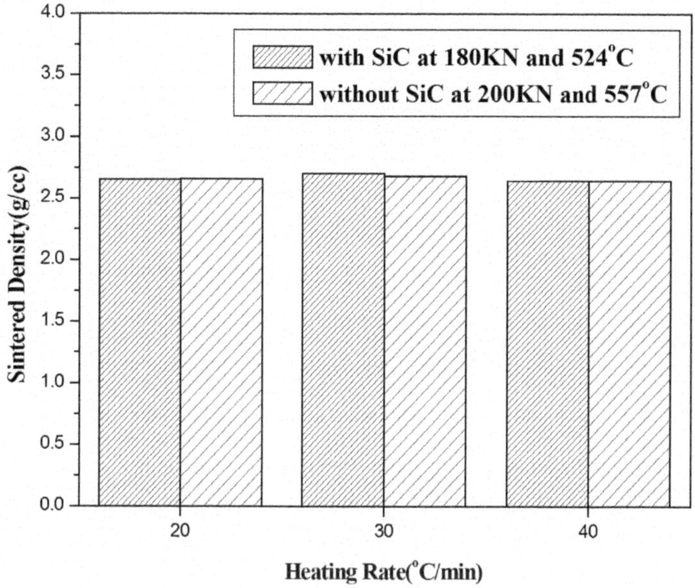

Fig. 9.4 Sintered density for 3003 Al alloy with SiC (at 524 °C and 180 kN load) and without SiC (at 557 °C and 200 kN load)

sintered density obtained here are at par with the values obtained for the density of 3003 Al alloy with SiC at 180 kN load and 524 °C sintering temperature as illustrated in Fig. 9.4.

From above, it can be concluded that on increasing the compaction load and sintering temperature, the cost of adding SiC can be skipped or saved. This will bring down the cost of development of alloy.

3.3 Microstructure Feature and Hardness of the Sintered Samples

Grain size measurement is carried out in image analysis software. The mean grain size number is obtained from the image analysis software. This was done at 500X magnification. For each sample, three images were taken and their average value was considered. This was done to get accuracy in the value of grain size number (Table 9.2).

The influence of sintering temperature and heating rate on mean grain size number is illustrated in Fig. 9.5. It shows that an increase in heating rate increases the mean grain size of 3003 Al/SiC composites. However, the mean grain size number decreases with the heating rate in the case of 3003 Al alloy and it may be a good reason for a decrease in hardness of 3003 Al alloy. And, as the grain size decreases

Table 9.2 Mean grain size number

Sintering parameters	Heating rate (°C/min)	Image -1	Image -2	Image -3	Mean grain size number
524 °C, 180 kN without SiC	20	5.30	3.63	4.45	4.46
	30	3.33	3.52	3.33	3.39
	40	3.63	3.38	3.24	3.41
524 °C, 180 kN with SiC	20	3.73	4.33	3.12	3.72
	30	4.99	3.98	2.89	3.95
	40	10.45	10.40	10.33	10.39
524 °C, 200 kN without SiC	20	5.73	5.36	5.50	5.53
	30	5.50	4.29	2.78	4.19
	40	3.24	5.21	4.33	4.26

dramatically with the addition of SiC in 3003 Al alloy, it results in achieving good mechanical properties.

The grain size calculation was carried out with the help of microstructures taken for 524°C sintering temperature, with SiC (Fig. 9.6) and without SiC (Fig. 9.7) for 20 °C/min, 30 °C/min and 40 °C/min respectively. In Fig. 9.8, the mean grain size number decreases with an increase in heating rate which indicates the enlargement of grains of 3003 Al alloy. The conclusion drawn from these results shows that the

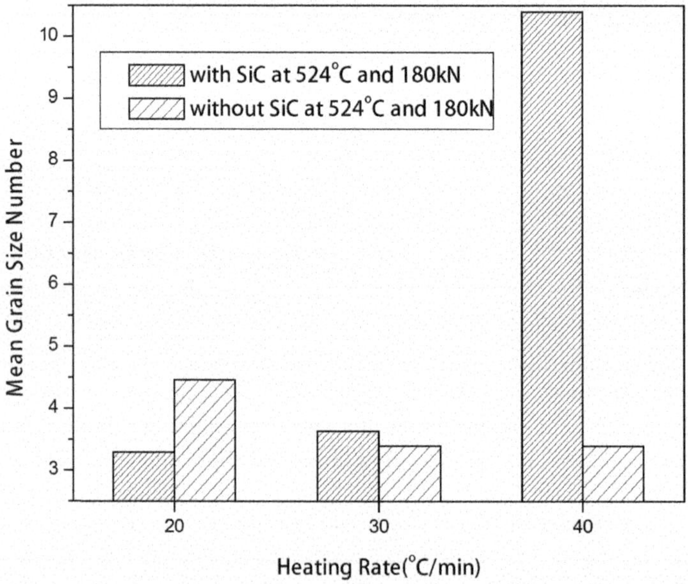

Fig. 9.5 Mean grain size number of 3003 Al-alloy with and without SiC at 180 kN load and 524 °C sintering temperature

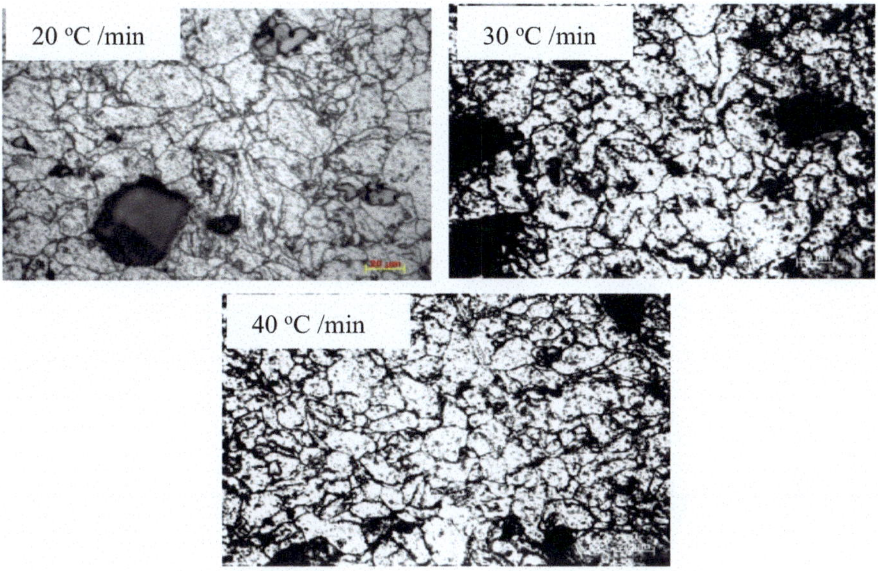

Fig. 9.6 Optical micrographs of 3003 alloy (with SiC) sintered at 524°C (pressed at 370 MPa)

maximum value of the heating rate exhibits the largest grain size. On comparing the grain size for 3003 Al alloy with SiC at 180 kN and 524 °C with 3003 Al alloy without SiC at 200 kN and 557 °C, the finest grain size at 40 °C/min heating rate was obtained in both cases. But, as far as hardness is concerned, the value still is higher at 200 kN and 557 °C without SiC. Therefore, it is not necessary to add SiC to get a good hardness value, just going up for high sintering temperature and compaction load will be enough. Hence, it saves the cost of materials (SiC). Figure 9.9 shows the optical micrographs of 3003 alloy (with SiC) sintered at 557 °C (pressed at 410 MPa).

From Fig. 9.10, it is inferred that the hardness decreases on increasing the heating rate because of the increase in grain size of 3003 Al alloy. And it can be ascribed that grain size here plays a vital role in improving the hardness. In Fig. 9.10, it is observed that the hardness of the 3003 Al alloy with SiC first increases, on raising the rate of heating from 20 to 30 °C/min, and then decreases at 40 °C/min. This increase in hardness value at 30 °C/min is due to the precipitation hardening of SiC particles in 3003 Al alloy which causes localized hardening.

On comparing the hardness values, due to the increase in sintering temperature and compaction load, the hardness of 3003 Al alloy is significantly increased as compared to the hardness values obtained at lower compaction load and sintering temperature. This means the effect of high sintering temperature and compaction load. However, the hardness values decrease with an increase in heating rate as shown in Fig. 9.11. Table 9.3 tabulates the consolidated physical and mechanical properties of 3003 Al alloy with and without SiC.

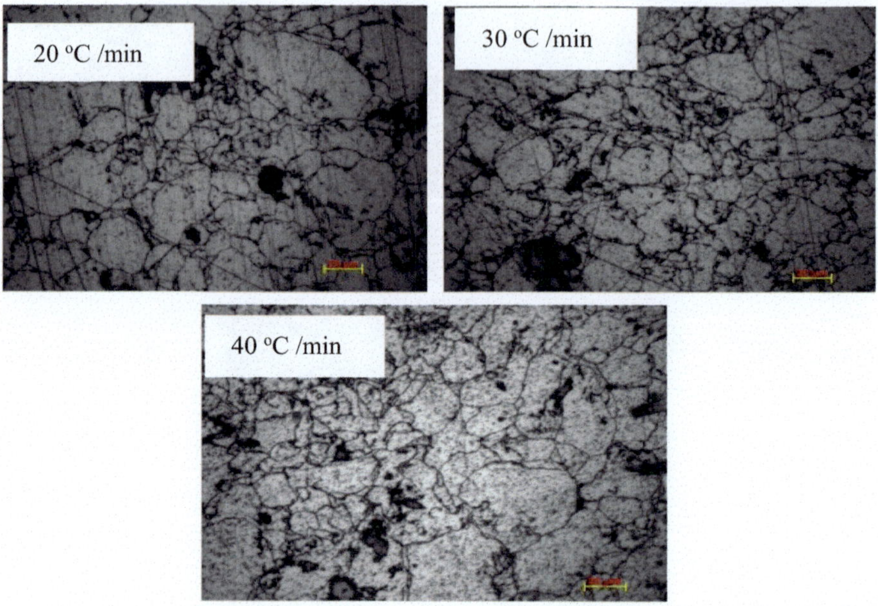

Fig. 9.7 Optical micrographs of 3003 alloy (without SiC) sintered at 524 °C (pressed at 370 MPa)

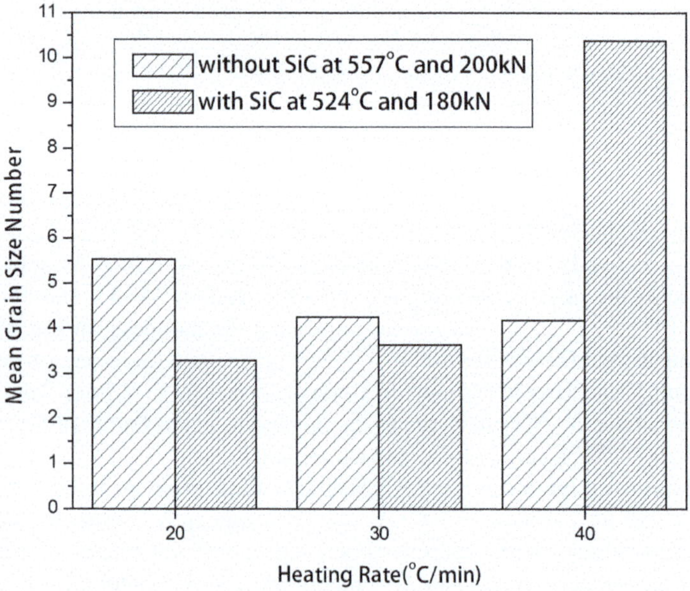

Fig. 9.8 Mean grain size number of 3003 Al-alloy with SiC (at 180 kN and 524 °C) and without SiC (at 200 kN and 524 °C 557 °C)

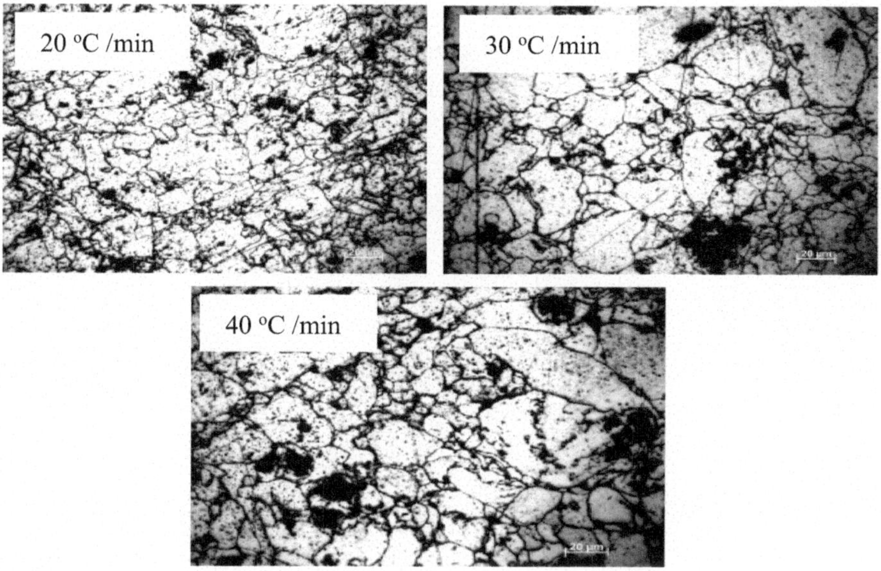

Fig. 9.9 Optical micrographs of 3003 alloy (with SiC) sintered at 557 °C (pressed at 410 MPa)

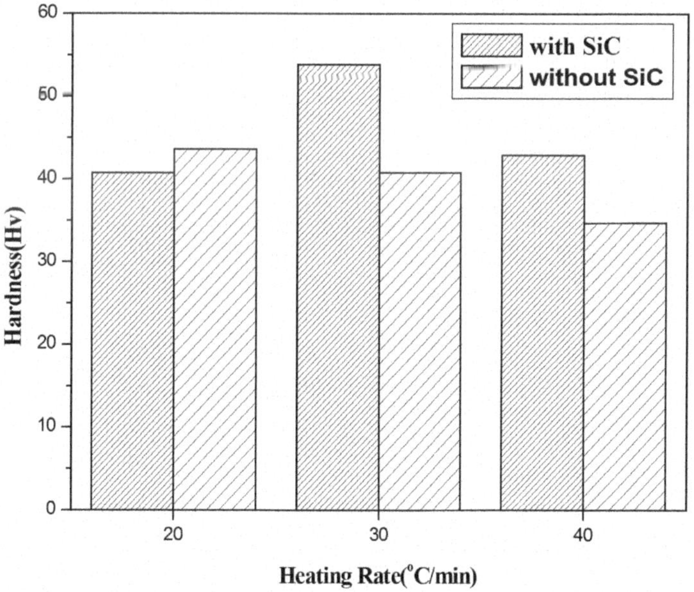

Fig. 9.10 Hardness at 524 °C and 180 kN load for 3003Al alloy with and without SiC

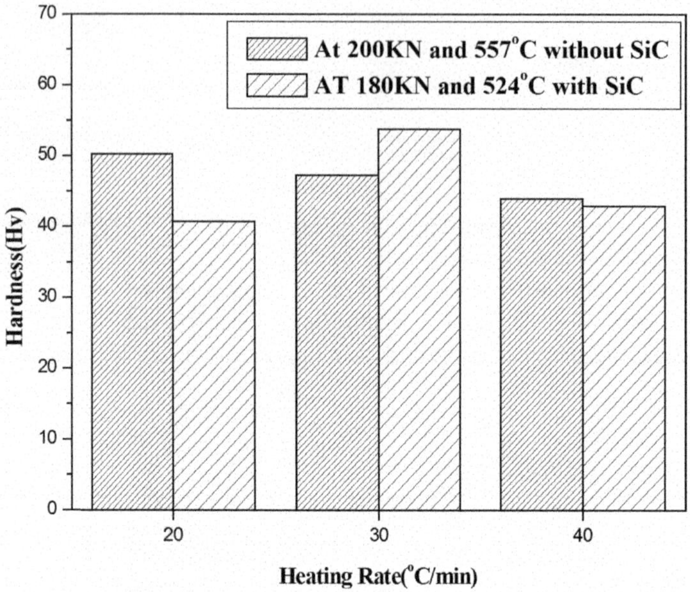

Fig. 9.11 Hardness for 3003 Al alloy with SiC (at 524 °C and 180KN load) and without SiC (at 557 °C and 200KN load)

Table 9.3 Physical and mechanical properties of 3003 Al alloy

Specimen	Heating rate (°C/min)	Sintering temp (°C/min)	Green density (g/cc)	Sintered density (g/cc)	Densification parameter	Porosity (in %)	Hardness (Hv)	Mean grain size
3003 Al (180 kN)	20	524	2.60	2.62	0.15	4.02	43.6	4.46
	30		2.58	2.63	0.34	3.66	40.8	3.39
	40		2.58	2.60	0.13	4.76	34.7	3.41
3003 Al + SiC (180 kN)	20	524	2.58	2.65	0.46	2.90	40.7	3.27
	30		2.61	2.70	0.75	1.09	53.2	3.63
	40		2.59	2.64	0.35	3.29	42.9	10.39
3003 Al (200 kN)	20		2.60	2.66	0.46	2.56	50.3	5.53
	30	527	2.58	2.68	0.67	1.83	47.3	4.26
	40		2.58	2.64	0.4	3.29	44	4.19

3.4 Validation

The values for 6061 Al alloy are taken from literature based on microstructure and mechanical properties of Al/(Ti, W)C composites prepared by microwave sintering [1] as shown in Table 9.4. From the data obtained for 3003 and 6061 Al alloy, it can

Table 9.4 Hardness and density for 3003 and 6061 Al alloy for validation

Properties	3003 Al alloy			6061 Al alloy		
	524 °C without SiC	524 °C with 5% SiC	557 °C Without SiC	520 °C without (Ti, W) C	520 °C with 10% (Ti, W) C	560 °C with (Ti, W) C
Hardness (Hv)	40.8	53.8	47.32	50	53	45
Density (g/cc)	2.63	2.7	2.68	2.63	2.75	2.66

be concluded that the trend followed by 3003 al alloy is almost the same as that in the case of 6061 Al alloy. Also, the hardness obtained by the addition of SiC in 3003 Al alloy is greater than the hardness value obtained in the case of 6061 Al alloy by addition of (Ti, W)C. In the case of density, the values for 3003 Al alloy were found relatively greater than the 6061Al alloy.

4 Conclusions

The present study manifested that 3003 Al and 3003 (Al/SiC) alloys can be successfully consolidated through a microwave sintering route. It is evident from the results that sinter density for 3003 Al alloy reinforced with SiC is greater for all the values of heating rates compared to 3003 Al alloy. From this, it is inferred that SiC enhances the density by filling the voids of the alloy. On raising the rate of heating from 20 to 30 °C/min, the sinter density of 3003 Al alloy increases due to the consolidation effect on the grains which decreases the porosity. On the other hand, when the rate of heating is further raised from 30 to 40 °C/min, porosity percentage increases relatively due to the oxide formation which in turn decreases the sinter density.

The hardness of the 3003 Al alloy increases with the SiC composition. The hardness increases with the rise in the rate of heating (from 20 to 30 °C/min) and then decreases at 40 °C/min. This increase in hardness value at 30 °C/min is due to the precipitation hardening of SiC particles in 3003 Al alloy which causes localized hardening. The hardness values decrease with further increase in the heating rate (at 40 °C/min) due to increases in grain size. When the heating rate is 40 °C/min the size of grain is the largest when compared to the size at the lower temperatures. This is because of the fast heating and insufficient sintering time for alloying elements during microwave sintering which results manifestation of inhomogeneous microstructure. Thus, the results reveal that the grain size plays a very important role in enhancing the hardness of composites.

References

1. Bao R, Yi J (2014) Densification and alloying of microwave sintering WC–8wt%Co composites. Int J Refract Metal Hard Mater 43:269–275. https://doi.org/10.1016/j.ijrmhm.2013.12.010
2. Bhushan RK (2021) Effect of SiC particle size and weight % on mechanical properties of AA7075 SiC composite. Adv Compos Hybrid Mater 4:74–85. https://doi.org/10.1007/s42114-020-00175-z
3. Demirskyi D, Agrawal D, Ragulya A (2013) Comparisons of grain size-density trajectory during microwave and conventional sintering of titanium nitride. J Alloy Compd 581:498–501. https://doi.org/10.1016/j.jallcom.2013.07.159
4. Ertugrul O, Park H-S, Onel K, Willert-Porada M (2014) Effect of particle size and heating rate in microwave sintering of 316L stainless steel. Powder Technol 253:703–709. https://doi.org/10.1016/j.powtec.2013.12.043
5. Gamit DN, Chudasama MK (2020) Size-effect in microwave processing of engineering materials—A review. JMES 14:6770–6788. https://doi.org/10.15282/jmes.14.2.2020.18.0530
6. Khadem SA, Nategh S, Yoozbashizadeh H (2011) Structural and morphological evaluation of Al–5vol.%SiC nanocomposite powder produced by mechanical milling. J Alloy Compd 509:2221–2226. https://doi.org/10.1016/j.jallcom.2010.10.188
7. Madan R, Bhowmick S (2020) A review on application of FGM fabricated using solid-state processes. Adv Mater Process Technol 6:608–619. https://doi.org/10.1080/2374068X.2020.1731153
8. Madan R, Bhowmick S (2022) Fabrication, microstructural characterization and finite element analysis of functionally graded Al-Al_2O_3 disk using powder metallurgy technique. Mater Today Commun 32:103878. https://doi.org/10.1016/j.mtcomm.2022.103878
9. Madan R, Bhowmick S (2022) Fabrication and microstructural characterization of Al-SiC based functionally graded disk. AEAT. https://doi.org/10.1108/AEAT-03-2022-0096
10. Manohar G, Maity SR, Pandey KM (2022) Microstructural and mechanical properties of microwave sintered AA7075/Graphite/SiC hybrid composite fabricated by powder metallurgy techniques. Silicon 14:5179–5189. https://doi.org/10.1007/s12633-021-01299-7
11. Manohar G, Pandey KM, Maity SR (2021) Effect of sintering mechanisms on mechanical properties of AA7075/B4C composite fabricated by powder metallurgy techniques. Ceram Int 47:15147–15154. https://doi.org/10.1016/j.ceramint.2021.02.073
12. Menezes RR, Kiminami RHGA (2008) Microwave sintering of alumina–zirconia nanocomposites. J Mater Process Technol 203:513–517. https://doi.org/10.1016/j.jmatprotec.2007.10.057
13. Oghbaei M, Mirzaee O (2010) Microwave versus conventional sintering: a review of fundamentals, advantages and applications. J Alloy Compd 494:175–189. https://doi.org/10.1016/j.jallcom.2010.01.068
14. Padmavathi C, Upadhyaya A, Agrawal D (2011) Effect of microwave and conventional heating on sintering behavior and properties of Al–Mg–Si–Cu alloy. Mater Chem Phys 130:449–457. https://doi.org/10.1016/j.matchemphys.2011.07.008
15. Santos T, Valente MA, Monteiro J, Sousa J, Costa LC (2011) Electromagnetic and thermal history during microwave heating. Appl Therm Eng 31:3255–3261. https://doi.org/10.1016/j.applthermaleng.2011.06.006
16. Shu-dong L, Jian-hong Y, Ying-Li G, Yuan-dong P, Li-ya L, Jun-ming R (2009) Microwave sintering W-Cu composites: analyses of densification and microstructural homogenization. J Alloy Compd 473:L5–L9. https://doi.org/10.1016/j.jallcom.2008.05.038
17. Upadhyaya A, Tiwari SK, Mishra P (2007) Microwave sintering of W–Ni–Fe alloy. Scripta Mater 56:5–8. https://doi.org/10.1016/j.scriptamat.2006.09.010
18. Yang JH, Song KW, Lee YW, Kim JH, Kang KW, Kim KS, Jung YH (2004) Microwave process for sintering of uranium dioxide. J Nucl Mater 325:210–216. https://doi.org/10.1016/j.jnucmat.2003.12.003

19. Zheng RR, Wu Y, Liao SL, Wang WY, Wang WB, Wang AH (2014) Microstructure and mechanical properties of Al/(Ti, W)C composites prepared by microwave sintering. J Alloy Compd 590:168–175. https://doi.org/10.1016/j.jallcom.2013.12.016

Application of Microwaves in Development of Metal Matrix Composites

M. Shashank Lingappa

Abstract Need for tailor made properties of materials desired for corresponding applications has increased demand in development of composite materials. Metal Matrix Composites (MMC) is renowned among all the composites and finds wide range of applications in everyday applications, owing to higher functional temperatures, strength, stiffness, radiation resistance, etc. Hence, developing metal matrix composites in an energy-efficient and environmentally greener route becomes prime responsibility. In this regard, the present work is a successful attempt made to develop an Al–5%SiC metal matrix composite by microwave stir casting technique. The process involves melting of the commercially available bulk pure aluminum by microwave hybrid heating, further addition of reinforcement particles, and further stirring the mixture of aluminum and silicon carbide to obtain a homogeneous distribution of the reinforcements. The developed metal matrix composite is characterized for the phases, microstructure, tensile properties and hardness.

1 Introduction

Microwave processing is one of the most popular and efficient methods for treating materials. In the frequency range of 300 MHz to 300 GHz, microwaves are electromagnetic radiations with wavelengths between 1 mm and 1 m. The frequency of microwave processing varies depending on the application. The Federal Communications Commission (FCC) has designated a number of frequencies for industrial processing, including 915 MHz, 18 GHz, and 2.45 GHz [1]. Most currently employed conventional processing techniques have a variety of drawbacks, such as a lengthy processing time, an uneven microstructure, and material and energy loss [2]. As technology advanced, non-traditional energy sources usage has increased exponentially. Nonetheless, there are similar issues with processing time and energy requirements

M. S. Lingappa (✉)
Department of Mechanical Engineering, Malnad College of Engineering, Hassan, Karnataka, India
e-mail: shashanklingappam@gmail.com

M. K. Singh et al. (eds.), *Composite Materials Processing Using Microwave Heating Technology*, Composites Science and Technology,
https://doi.org/10.1007/978-981-97-2772-8_10

that can be addressed. Microwave processing is used to mitigate the drawbacks of conventional processing and improve the materials' mechanical and physical qualities. The method of heat transfer varies between conventional processing and microwave processing. In contrast to microwaves, which interact with materials at the molecular level and cause volumetric heating, conventional heating involves the transfer of heat from a material's external surface to its internal core [3]. Due to the numerous advantages, it offers in a variety of industries, microwave processing has a wide range of applications. Microwaves have been utilized for a number of decades in a variety of applications, including radar, telecommunication, food processing, and others [4, 5]. Metals, rubber, and ceramics have all recently been processed using microwave technology. The advantages of using microwaves for material processing include;

i. reduction in the processing time
ii. ease of heat controlling
iii. operator safety
iv. environmental cleanliness
v. rapid heating
vi. precise heating control
vii. reduced thermal gradients

2 Theory of Microwave–Material Interaction

Microwaves are one type of electromagnetic radiation that has magnetic and electrical components that are orthogonal to one another. During their interaction, these elements are crucial in producing heat at the molecular level inside the material. Yet, the parameters of energy absorption are greatly influenced by the material properties. The specific method through which microwaves interact with materials is unknown. Some of the theories, nevertheless, might be true.

Coupling of microwaves with materials is defined by certain relations, called Maxwell's equations as given by the relations (1) to (4), which shows the time-dependent electric and magnetic field changes with the updated material properties [6].

$$\nabla \times E = j\mu\omega H \tag{1}$$

$$\nabla \times E = -j\omega\varepsilon_0\varepsilon^* E \tag{2}$$

$$\nabla(\varepsilon E) = 0 \tag{3}$$

$$\nabla H = 0 \tag{4}$$

where, H—magnetic field (Am^{-1}), E—electric field (Vm^{-1}), ε_0—permittivity in air (Fm^{-1}), ε^*—permittivity of material (Fm^{-1}), μ—magnetic permeability (Hm^{-1}), ε—permittivity (Fm^{-1}), ω—angular frequency (s^{-1}), j—imaginary unit

Complex permittivity of non-magnetic material which is given by $\varepsilon^* = \varepsilon' - j\varepsilon''$ is the key parameter that illustrates the intensity of generation of heat during the interaction of materials with microwaves. The Dielectric Constant, which is the real component of the complex permittivity equation, represents the amount of electric energy that may be stored in the heated material. The capacity of the heated material to disperse microwaves is indicated by the imaginary component of the equation known as the Dielectric Loss Factor. In order to determine the distribution of the electric and magnetic fields within the subject material and to ensure a faster rate of heating inside the microwave applicator, these equations that control the generation of heat in the material during interaction with microwaves can be solved with the proper boundary conditions.

3 Fundamentals of Microwave Processing

Characteristic nature of microwave processing that makes it one of the most attractive processing techniques is the inside-out heating due to interaction with materials at the molecular level. Microwaves penetrate into materials and generate heat at the molecular level, while conventional heating takes place from the outside to the inside nature of the material. This nature of conventional heating results in temperature differences between layers of the material, causing energy and material losses. Heat transfer in microwave and conventional processing methods is schematically shown in Fig. 10.1 [7].

Processing of materials by microwave energy is a rapidly emerging area owing the various advantages. However, the mechanism of microwave and material interaction is still an unknown concept and has been presented by various researches in their own perspective. Hence, the present chapter is an attempt made on describing the mechanism of melting metallic materials using microwave energy. Discussion on material interaction with microwaves is also presented.

The properties of a material have a significant role in heating up by interaction with microwaves. The interaction of microwaves differs from one material to the

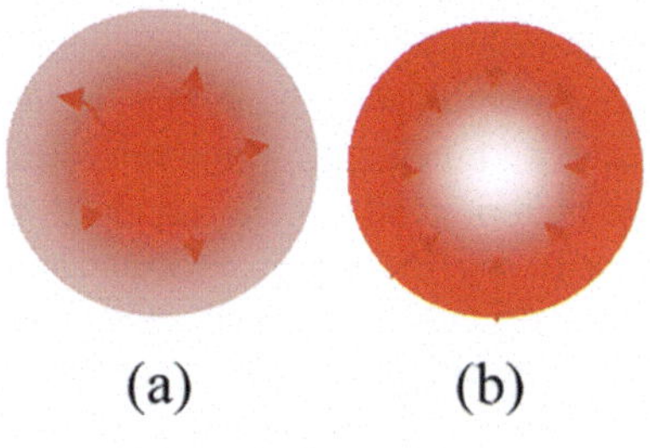

Fig. 10.1 Heat transfer in **a** microwave processing **b** conventional processing

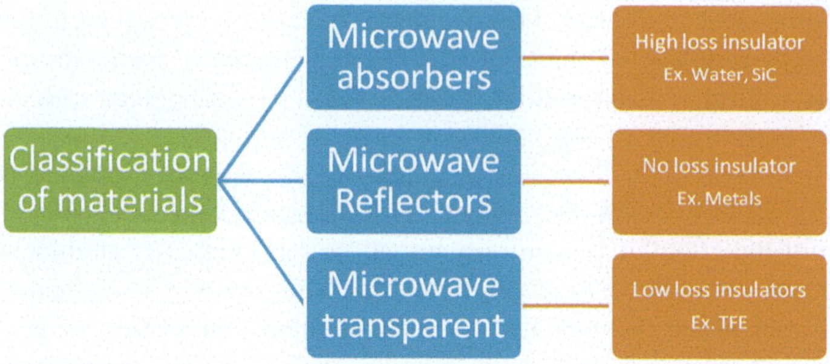

Fig. 10.2 Classification materials depending upon interaction with microwaves

other. Based on this interaction of microwaves, materials are broadly categorized into three main groups as shown in Fig. 10.2. Materials that allow microwaves to pass through them without any interaction are called transparent materials. Materials that absorb microwaves and get heated up are called absorbers and those materials which are conductive and reflect microwaves from their surface owing to very limited depth of penetration are called microwave reflectors.

The microwave absorbing properties of the materials vary from one material to another when in contact with microwaves due to variations in the strength of the electric (E) and magnetic (M) fields along the thickness of the material [6]. Figure 10.3 provides an example of how the strength of the electric and magnetic fields varies with the thickness of the material. Low-loss insulators and microwave transparent materials fit into the first group since the reduction in electric and magnetic field strength is minimal. In materials of the second type, referred to as microwave absorbers, the strength of the electric and magnetic fields diminishes as the dielectric loss factor rises. Some materials in this category have exceptionally high energy absorption rates per unit volume. The third category of materials includes those with extremely high conductivities, with very little change in electric and magnetic field intensity up to a few micrometers in thickness. This category includes materials made of metal. The amount of energy absorbed by a material from microwaves per unit volume is determined by its dielectric loss factor, which is also affected by the electric field component of the microwaves. A higher loss factor results in greater energy absorption per volume of material and less penetration depth.

Materials can be heated in different ways by microwave interaction. Some of the techniques used most often are briefed in this section. Three types of material heating by microwaves are (a) direct heating (b) selective heating and (c) hybrid heating [1]. Direct heating method is used in such places where the candidate material is compatible with absorbing microwaves readily. Heat is generated from the centre of the material and transferred to the surface which causes thermal differences, giving way to overheating and formation of hotspots in the material and further causing thermal runaway. Further, these thermal instabilities in the material may give rise to non-uniform properties and cracking. Another special type of masking arrangement for heating the material with direct interaction of microwaves is called as selective

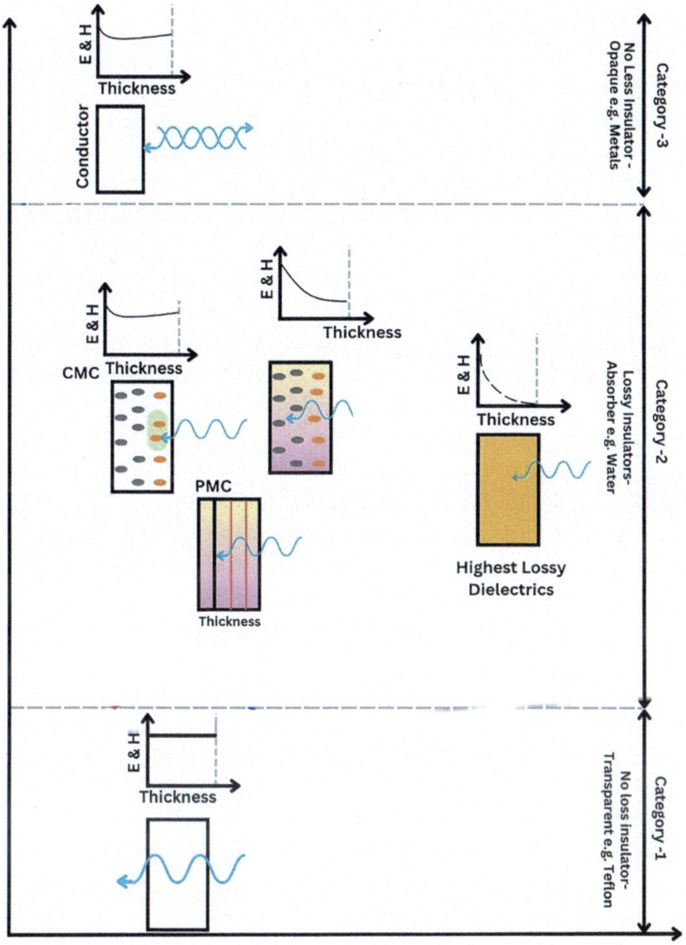

Fig. 10.3 Variation in electric and magnetic field of materials during microwave interaction

heating. Here, whole region of the material is covered except where microwave exposure is required. This technique is widely used in microwave joining, where only the joint area is exposed to microwaves, while leaving the rest of the material unaffected. Final type of heating technique used for processing of material through microwave irradiation is the hybrid heating. The drawbacks of direct microwave heating and conventional heating can be overcome by this method of heating. Masking systems and a microwave-absorbing substance (susceptor) are used to produce hybrid heating.

4 Microwave Processing of Metallic Materials

The mechanism involved in processing of materials in microwave environment is still not completely known. Various materials behave in different manner with microwaves, leading to various types of possibilities. Type of microwave interaction required is prioritized based on the final applications [8–12]. Hence, materials are classified into various groups depending on their interaction with microwaves.

Anyhow, microwaves pose a major problem during the processing of metallic materials. Microwaves are reflected from the surface of the metallic materials due to the shallow skin depth at room temperature as shown in the Fig. 10.4a. This also causes arcing/sparking at the surface or corners of the metallic materials. Skin depth is the penetration depth of the microwaves into the material from its surface where the intensity falls to 1/μ of its value at the surface. However, microwaves are known to be absorbed even by metallic materials after certain elevated temperatures, known as the 'Critical Temperature (T_C) as shown in Fig. 10.4b [13]. Hence, the difficulty lies in bringing the metallic materials to critical temperature. This can be achieved by using a susceptor (microwave absorbers) material along with the candidate material. Such a technique of using an external source for initial heating of the candidate metallic materials during microwave processing is called as Microwave Hybrid Heating (MHH). Various phases involved in MHH are presented in the subsequent section.

Some of the metal processing areas where microwaves are being used are sintering, cladding, joining and melting. These areas can be further sub-categorized into Powder Metal systems, Powder Metal–Bulk Metal systems and Bulk Metal–Bulk Metal systems. Sintering, cladding and joining applications will have at least one powdered metal system and are already established. However, effective processing of bulk metals in microwave environment still needs to be developed. In this regard, much of the concentration has been made on the processing of bulk metallic materials (melting).

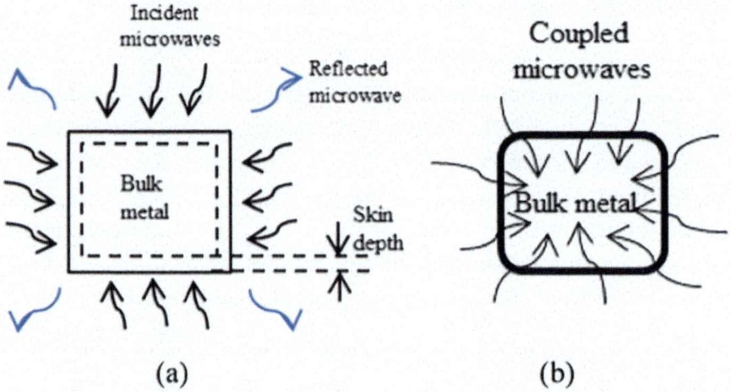

Fig. 10.4 Behavior of bulk metals and microwaves at **a** ambient temperature **b** higher temperature

5 Microwave Hybrid Heating

Microwave material processing is gaining importance because of various advantages that are offered. The nature of heat transfers in conventional and even direct microwave heating lead to thermal non-uniformities within the material, which leads to formation of temperature gradients resulting in substandard microstructure in the processed material. In order to eliminate the said drawback, a novel processing technique called hybrid heating provides a solution by heating the material evenly all over by interacting with them at the molecular level, thereby resulting in rapid uniform heat generation, as shown in Fig. 10.5. Hybrid heating consists of a unique setup that includes microwave absorbing material (susceptor) and masking objects. The heating process is usually achieved in three steps starting with susceptor heating, conventional heating of candidate material by heated susceptor and direct microwave heating of material (beyond critical temperature). The initial heating of the candidate material by susceptor takes place through the mode of conduction, where the heat is transferred from the surface of the material to its core (as in conventional heating). As the temperature of the candidate material passes beyond its critical temperature, microwaves directly couple and heat the material from inside to outside. The process is also called Microwave Hybrid Heating (MHH) and found its application mainly in processing of metals. Hence, the candidate material undergoes two-way heating, which is theoretically uniform heating of the material, as shown in Fig. 10.5.

However, the same interaction nature does not hold good with bulk metallic materials. This is because of the lower depth of penetration in metallic materials. Microwaves are capable of penetrating only to a certain depth inside the material where the value of current density falls to 1/e of its value at the surface. Hence, the interaction of microwaves and the generation of heat depend on the depth up to which microwaves can penetrate into the material, which is called as the depth

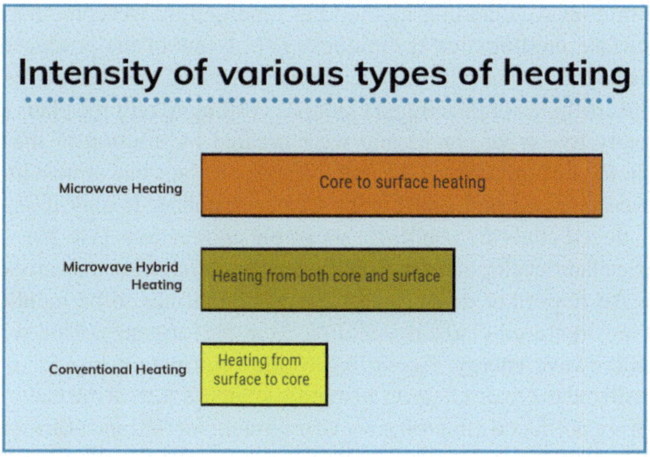

Fig. 10.5 Conventional, microwave and hybrid heating

of penetration and it differs based on the size of the material. Effective area of microwave heating is the portion of material with which microwave couples, resulting in uniform heating. Most of the powdered metal particles in the submicron and nano segment have revealed complete and efficient absorption of microwaves resulting in volumetric heating and thereby improving the microstructure and enhancing the mechanical properties of the processed material. Available literature indicates that the effective skin depth for processing of metallic materials is only 100 μm at the microwave frequency of 2.45 GHz. Materials having thickness of more than 100 μm are considered to be coarse particles for microwave processing, which gets heated in a non-uniform manner. In coarser particles/materials, uniform heating by microwave penetration is possible only up to skin depth, while the rest of the region is heated by conduction of heat from the skin depth to its core. The same heat is transferred to the neighboring particles. Hence, uniform heating can be achieved effectively with decreasing particle size [14].

6 Application of Microwaves in Processing of Metallic Materials

6.1 Microwave Cladding

Surface modification is essential when a material needs to be resistant to external harm [15, 16]. Surface modification includes enhancing a material's hardness, resistance to corrosion or erosion, or combining the two. Surface modification techniques include thermal spraying, laser cladding, TIG surface-based welding, and others. However, there are a number of drawbacks to these conventional techniques, including significant deformation and energy consumption. As a new technology with significant benefits, the microwave cladding method has emerged to overcome the limitations of existing surface modification techniques [17]. A few of the produced claddings have been discussed, along with their outcomes. Several methodologies were used to describe the attributes of developed clads. According to reports, clads that formed were completely free from cracks and were created by dilution of the underlying material. It should be emphasized that the microwave clads had almost little porosity when they were created and had higher wear resistance. It may be said that the microwave-induced clads are useful for wearable applications [18–20].

Nanometric clad development and characterization have been discussed by Zafar and Sharma. An important discovery of the work that has to be mentioned is the WC-12Co nanostructured clad's dry sliding wear performance. This was achieved employing microwave energy. According to the information provided, WC-12Co powder was utilized to create clads on austenitic stainless steel at the nano- and micro-scale. Clads were produced after microwave treatment for 540 and 600 s, respectively. In nanometric clads, nanocarbides were found to be dispersed as clusters, but in micrometric clads, carbides were found to be distributed as skeleton-like structures in the metal matrix. The nanometric clad performed 54% better in terms of wear resistance than the micrometric clads due to the higher volume fraction and uniform distribution of nanocarbides in the clad [21–23].

Hebbale and Srinath presented a microwave irradiation technique for creating nickel-based coatings on austenitic stainless steel (SS-304). Because nickel has a great resistance to erosion, it was employed as a cladding material. Clads were created utilizing a home microwave applicator with a 900 W power output and a 2.45 GHz operating frequency. Mechanical and metallographic characteristics of the generated clads were determined. A microstructural analysis indicated a crack-free clad surface and a homogeneous distribution of metallic carbides. Also, it is claimed that the distribution of intermetallic and metal carbides has boosted the created clad's hardness [24, 25]. Similarly, martensitic stainless steel was used to develop cobalt cladding (AISI-420). A study of the microstructure and erosion of developed clad was conducted. The factors influencing the erosion rate were investigated using the Taguchi Orthogonal Array. Ultimately, it was discovered that the mass loss was considerably impacted by the slurry speed [26–29].

6.2 Microwave Joining

Most manufacturing and assembly sectors depend heavily on permanently joining two or more materials. The limits of traditional joining methods like welding, brazing, and soldering are the materials required and processing times. An alternate method that uses microwaves as a source of energy for connecting materials can overcome these restrictions. The ability to join materials that are different from one another is a major benefit of microwave joining [30].

By overcoming the challenge of microwave reflection, Sharma et al. showed that metallic materials may also be joined using microwaves. It has been established that microwave metal joining is an environmentally friendly method that can be done even in a home microwave oven. Joints with a substantial deal of strength were acquired.

Srinath et al. discovered that joints between bulk copper plates and coins developed as a result of microwave irradiation. Joints were made by exposing the substrate and interface material to microwaves for exposure durations ranging from 300 to 900 s in a 900 W microwave oven. It is important to note that the particles were better integrated. Jointly formed products are only 1.92% porous and have a hardness of roughly 84% of the basic material. Similar work was done by the same authors employing microwave irradiation to bind bulk stainless steel (SS-316). The interface layer's metallurgical bonding as a result of improved particle fusion made it possible to join. A higher level of hardness in the interface layer was discovered during the joint's characterization, which may be the result of metallic carbide precipitation. Just 0.78% of the joint's surface was minimally porous, according to porosity measurements. The consequence of extensive field research is the blending of various metals. A multimode microwave applicator operating at 2.45 GHz was used to connect stainless steel (SS-316) and mild steel. Microwave hybrid heating

(MHH) techniques were used to generate the junction. At the joint region, Cementite and metal carbide development were observed, and the porosity was exceptionally low (0.58%) [3, 31, 32].

Moreover, Badiger et al. claimed that couplings between Inconel-625 could be produced. Microwave applicator of 900 W was used to apply microwave radiation to link commercially available Inconel-625 plates. The interface material was a powder made of nickel. A microstructural analysis of the joint showed that nickel, chromium, and molybdenum carbides had formed. At the joint, a hardness reading of 350 ± 10 Hv was averaged out. Joint fractographs demonstrated that a mixed mode of deformation was to blame for the failure. [33–37].

7 Microwave Melting and Development of MMC

It was believed that microwave melting of metals was impossible due to the lower skin depth of metallic materials at room temperature, which prevents microwaves from permeating the metal. Yet, after undertaking considerable research in this area, it was conceivable to use microwave energy to melt metallic materials.

According to Clark et al. metallic materials can absorb microwaves at greater temperatures. Later, by exposing them to microwaves, a variety of metals and alloys were melted. The Oak Ridge Y-12 National Security Complex (NSC) invented microwave melting [38]. However, there were no reports on characterization. Theoretical and practical investigations on the melting of copper, lead, aluminum, and tin at various microwave power levels were later published by Chandrasekaran et al. A multimode microwave applicator was employed to achieve the purpose. Metals in the form of granules were used for melting purposes. A lumped parameter model was also developed to support the experimental results [39].

Gowtham et al. were the first to report on the bulk melting of non-ferrous metals like tin (60Sn40Pb). To melt tin alloy, a 900 W household microwave oven was employed. The melting of tin alloy in a conventional muffle furnace was also compared by the authors [40]. According to reports, bulk tin may be melted using a microwave in less time than it would have taken using a traditional procedure. Mishra and Sharma also reported on in-situ casting of bulk aluminum using the MHH method. A 1400 W microwave applicator operating at 2.45 GHz was used to melt the AA 7039 alloy. The theory behind how microwave irradiation heats and melts the candidate material is also reported. Aluminum that has been cast exhibits dense casting and minimal porosity of less than 2% [41].

The nature of microwave interaction with the materials and the results that followed serve as evidence that microwave processing of metallic materials is superior to conventional processing techniques.

7.1 Microwave Stir Casting Process to Develop MMC

Metal casting is one of the major economical manufacturing processes used in industries from centuries. Metal casting forms the basis for development of metal matrix composites. Conventional route of metal casting involves furnace charging, high temperature maintenance for bringing the raw metals into molten state and casting into specific desired shapes. However, there exist some drawbacks in conventional casting techniques like formation of thermal differences, higher consumption of energy and longer processing time [42]. Development of new techniques for overcoming the drawbacks of conventional processing is essential. Further, call for environmentally cleaner production processes for green manufacturing and assisting in meeting the global energy crisis makes it feasible for developing metal matrix composites by microwave stir casting technique. This process uses microwave energy as a source of heat for melting of the base metal and further stirring the molten metal added with reinforcement particles for homogenous dispersion. This technique helps in achieving melting process in an efficient and environmentally cleaner route. However, processing metallic materials in a microwave field is a difficult task, owing to lower skin depth. Hence, the microwave hybrid heating technique is employed in order to process metallic materials in the microwave field. The basis for the higher efficiency of microwave processing is the nature by which heat is transferred. Heat generated by microwaves is transferred from inside of the material to its surface. In contrary to the conventional heating processes, microwaves interact with the materials at the molecular level, generating heat energy at much faster rates leading to shorter processing time. Further, some of the researchers have reported the time, energy, and cost savings involved in the microwave processing of materials. Also, the microstructure development and the mechanical properties exhibited by microwave-processed materials are far much superior to those processed by conventional counterparts.

Aluminium–Silicon Carbide (Al–SiC), one of the most widely renowned and widely used MMC is opted for the feasibility study of MMC development through microwave stir casting route.

In connection with the same, base material aluminum in bulk form is bought to the molten state by Microwave Hybrid Heating technique, using an industrial furnace irradiating microwaves from a 3.3 kW magnetron at 2.45 GHz frequency. Further, silicon carbide particles are introduced into to the molten metal matrix, followed by uniform stirring. The developed metal matrix composite is characterized for metallographic and mechanical properties.

8 Mechanism of Metal Melting by Microwave Irradiation and Development of MMC

Our focus will be on the melting of bulk metallic materials using microwave energy because the majority of the aforementioned applications have already been the subject of specialized research. Metals are known to have thinner skin because they are efficient heat- and electricity-conductors. As illustrated in Fig. 10.6a, when microwaves are bombarded on metallic objects at room temperature, they are reflected, decreasing the depth of the skin. Freely moving electrons that create a magnetic field that is pointed in the opposite direction from the electromagnetic waves being transmitted are what cause reflection. When the temperature rises and the strength of the produced magnetic field weakens, the penetration depth of electromagnetic radiation increases. The term "Critical Temperature (T_C)" refers to the temperature at which the depth of penetration of microwaves into metal reaches a specific raised level and is sufficient for absorption. To reach the critical temperature, the candidate material must receive heat energy from a separate source. A substance that produces heat energy is referred to as a "susceptor" and it is capable of absorbing microwaves and transforming them into heat energy. According to Fig. 10.7b, heat energy from the susceptor is transmitted to the candidate material. As the temperature reaches the critical point, the candidate material starts directly absorbing microwaves, which causes molecule-level heating (Fig. 10.6c). From this point on, the candidate material is heated in two directions until total melting is accomplished; the first direction is conduction by the susceptor, and the second direction is direct heating by microwaves. Because of the fluctuating electromagnetic (EM) irradiation, friction between moving charges results in direct heating (Fig. 10.6d). Microwave Hybrid Heating is the common name for this technique of heating metals (MHH). MHH is maintained until total melting is attained (Fig. 10.6e and f).

Commercially available pure aluminum in bulk form is taken as a base material. Composition of aluminum used in the present work is given in the Table 10.1. Silicon Carbide powder with particle size varying in the range 200–1200 μm is used as a reinforcement material, a typical X-Ray Diffraction spectrum is presented in the Fig. 10.7. Distinctive peaks of the XRD spectrum corresponds to SiC phase.

The microwave stir casting method is used to create metal matrix composites. Bulk metal is heated using a microwave hybrid system to melt it, and additional reinforcement is manually added and stirred into the molten metal to ensure even distribution. A schematic representation of the experimental setup is as shown in Fig. 10.8.

Commercially available aluminum in bulk form is taken in a clay-graphite crucible, and placed on SiC (susceptor) block inside a microwave transparent refractory casket as shown. This whole setup acts as a heating chamber and is placed inside the microwave furnace chamber. Temperature of the material before the experiment is recorded. Maximum temperature is restricted to 700 °C using a Eurotherm temperature controller. The candidate material is subjected to microwave exposure. At room temperature, microwaves are reflected by aluminum (skin depth ~ 1.67 μm) and hence

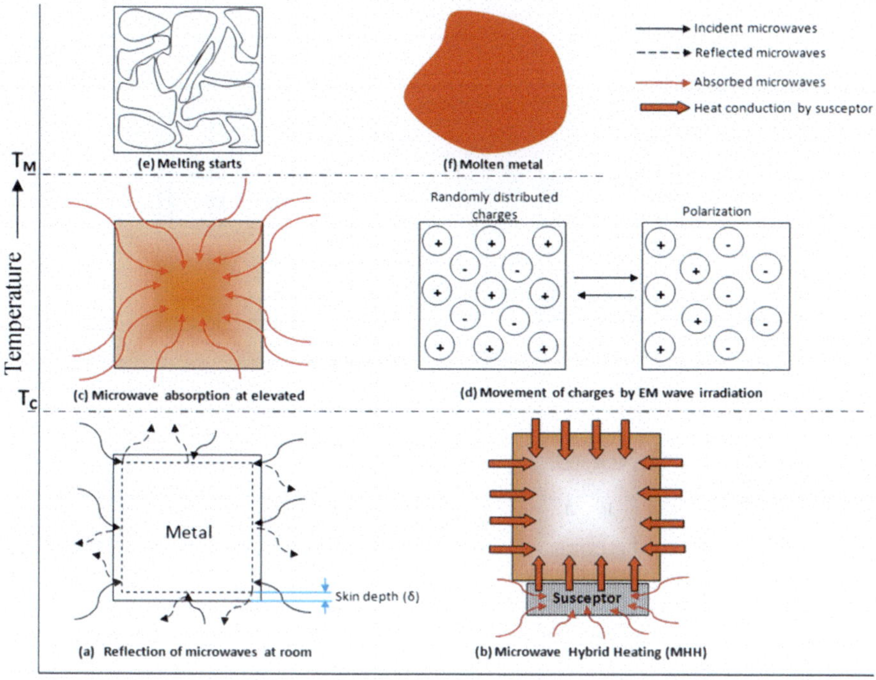

Fig. 10.6 Mechanism of melting of bulk metal by MHH

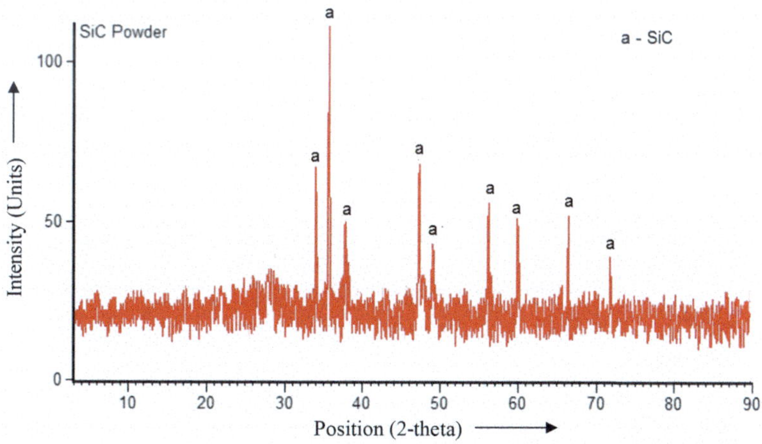

Fig. 10.7 XRD spectrum of SiC powder

Table 10.1 Chemical composition of the commercially available pure aluminum

Elements	Cu	Mg	Si	Fe	Mn	Zn	Ti	Al
Wt (%)	0.01	0.015	0.35	0.01	0.008	0.05	0.015	Balance

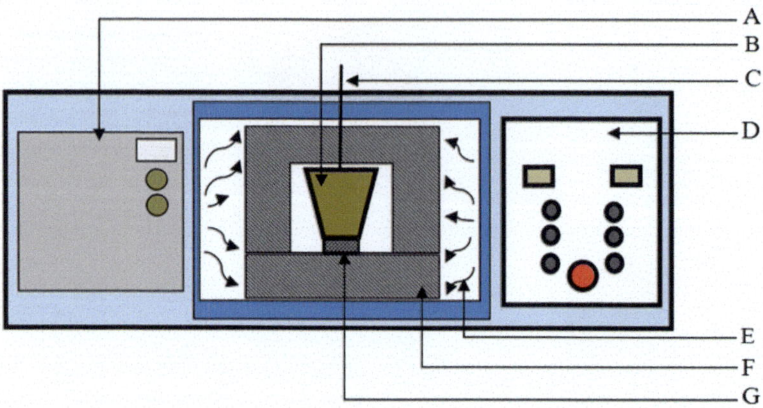

Fig. 10.8 Schematic representation of the experimental setup **A** Chiller **B** Crucible **C** Stirrer **D** Control panel **E** Microwaves **F** Refractory insulation **G** Susceptor

don't get affected. On the contrary, susceptor material gets heated up initially and transfers the generated heat energy to the candidate material, which gets heated up. When the temperature rises over the critical point, microwaves pair with the candidate material and cause uniform molecular heating, which lasts until complete melting is achieved. SiC powder (5% wt.) is manually incorporated into the metal matrix once the candidate material has reached the liquid condition. Heating is further continued and the mixture is stirred at a speed of 50 rpm for duration of 02 min. Aluminum-SiC in molten state is cast by pouring into a sand mold previously prepared. The developed Al–5%SiC MMC is shown in the Fig. 10.9.

Al–5% SiC MMC developed by microwave stir casting technique is sectioned using a low speed saw (Model: Baincut LSS) as per requirement and subjected to various characterization techniques like X-Ray Diffraction, microstructural study,

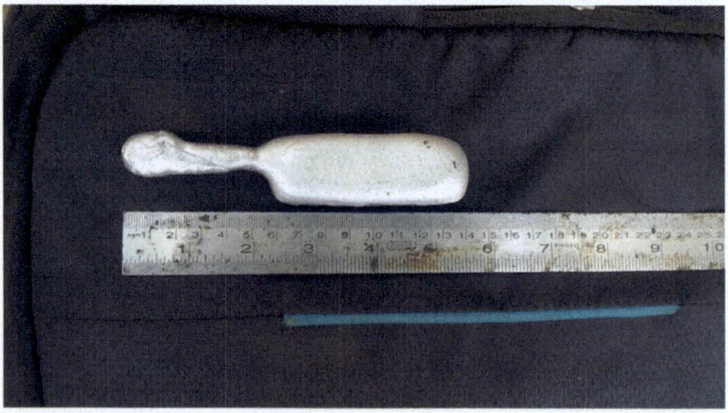

Fig. 10.9 Al–5% SiC metal matrix composite developed by microwave stir casting technique

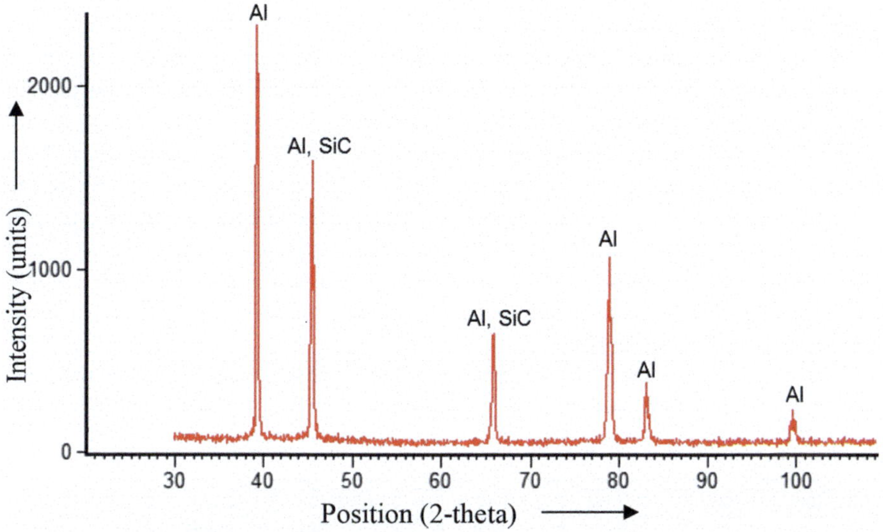

Fig. 10.10 XRD spectrum of the developed Al–5% SiC composite

Energy Dispersive X-Ray Spectroscopy (EDS), tensile test and microhardness study. Detailed discussion on the characterization carried out is presented in this section.

8.1 X-Ray Diffraction (XRD) Analysis

Microwave stir cast Al–5% SiC is subjected to XRD at a rate of scanning equal to 1°/ minute between the range 30°–110° using XPert[3] Powder Diffractometer. A typical XRD pattern obtained is shown in Fig. 10.10. Six high intensity distinctive peaks are obtained at various 2-theta positions. Maximum intensity peak is observed at 39.5° corresponding to aluminum, with remaining peaks at 47.47°, 66.13°, 78.9°, 83.26° and 99.74°. SiC also shows its presence at 47.47° and 66.13° along with aluminum. Formation of oxide during processing of aluminum in atmospheric conditions is inevitable and is evident by EDS study. However, the peak corresponding to oxides doesn't exist as the percentage is comparatively very less.

8.2 Microstructural Analysis and Elemental Study

Assessing the microstructure developed and distribution of reinforcements during the microwave stir casting of Al–5% SiC is of vital importance. Uniform microstructure lead to invariable propertied throughout the material, while homogeneous distribution of reinforcement result in better resistance to wear and erosion all over. In this regard, the microstructure of the as-cast MMC is analyzed by using TESCAN Vega 3 scanning electron microscope integrated with EDS attachment. Microstructure of Al–5% SiC MMC is shown in Fig. 10.11. Grains and grain boundaries are clearly revealed, with nearly uniform structure. SiC particles are observed to be segregated at the grain boundaries, giving definite hardness and strength to the developed MMC. Obtaining proper distribution of the reinforcement particles plays a chief role in the development of the metal matrix composites. Hence, area mapping of the developed composite is carried out and the obtained spectrum is shown in the Fig. 10.12, which clearly shows the presence of aluminum and silicon carbide. Furthermore, oxide also attests to its existence, which may be related to conducting the experiment under atmospheric circumstances.

Hardness of the microwave stir cast Al–5%SiC is assessed by using indentations made by the diamond indenter of Highwood Vicker's Microhardness Tester

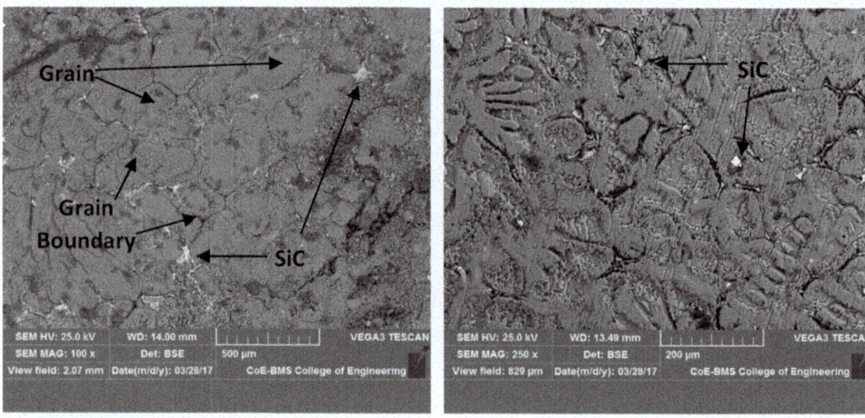

Fig. 10.11 Microstructure of the developed Al–5% SiC metal matrix composite

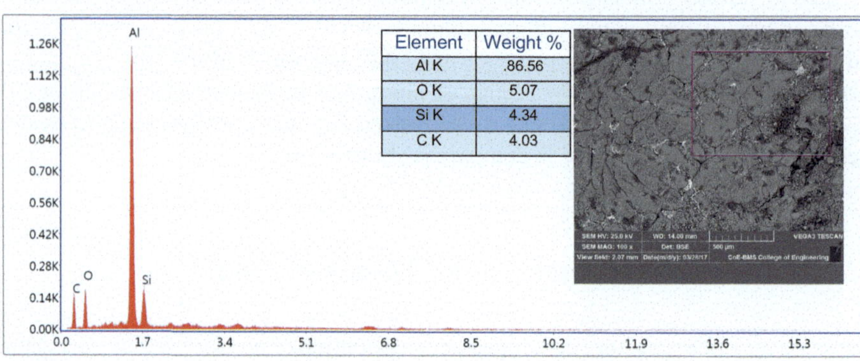

Fig. 10.12 Area selected for EDS analysis of Al–5% SiC and its corresponding spectrum

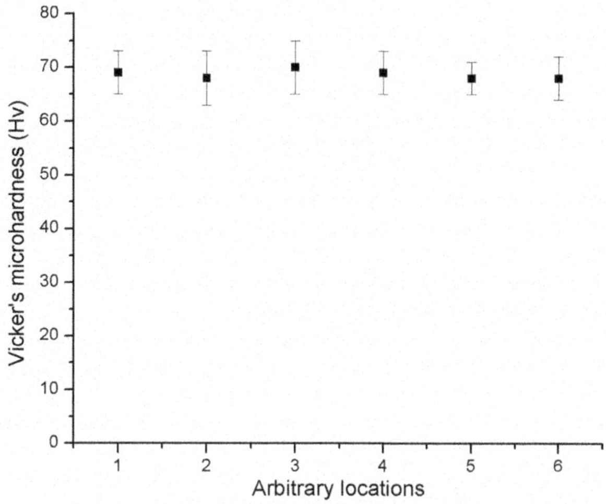

Fig. 10.13 Distribution of microhardness of the developed Al–5%SiC MMC

HWMMT-X7 for the load and dwell time of 500 g and 10 s, respectively. Indentations were made at six different locations on the vertically sectioned surface of the stir cast MMC. Distribution of microhardness values of the developed MMC is represented in the Fig. 10.13. The average hardness exhibited by the developed MMC is measured to be 68.23 ± $3H_V$ Better hardness of the developed MMC could be due to the addition of reinforcements.

8.3 Tensile Properties

Tensile strength plays a very important role in determining whether the material is suitable for a particular application or not. Taking that to consideration, tensile properties of the Al–5% SiC MMC is acquired by carrying out tensile test using Mecmesin Multitest 10-i micro universal testing machine. Tensile load is applied at a uniform strain rate of 1 mm per minute on the specimens prepared according to ASTM E8/M-09, as shown in the Fig. 10.14a. Graph of stress against strain is plotted and is presented in the Fig. 10.14b, which shows the curve related to maximum tensile strength and its corresponding ultimate tensile strength (UTS) and fracture point. Microwave stir cast Al–5% SiC MMC exhibits an average tensile strength of 72.38 MPa with an elongation of ~9%.

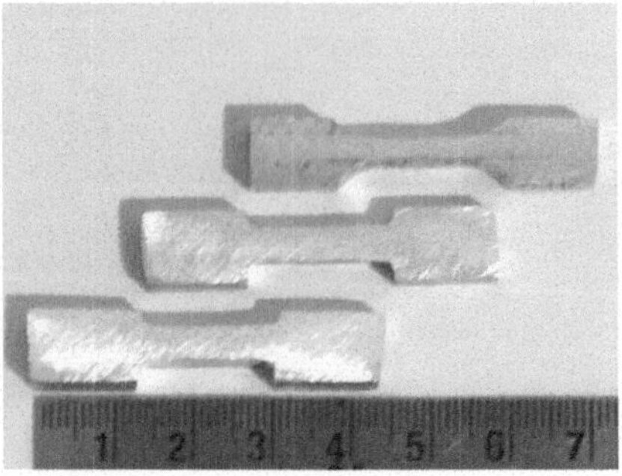

(a)

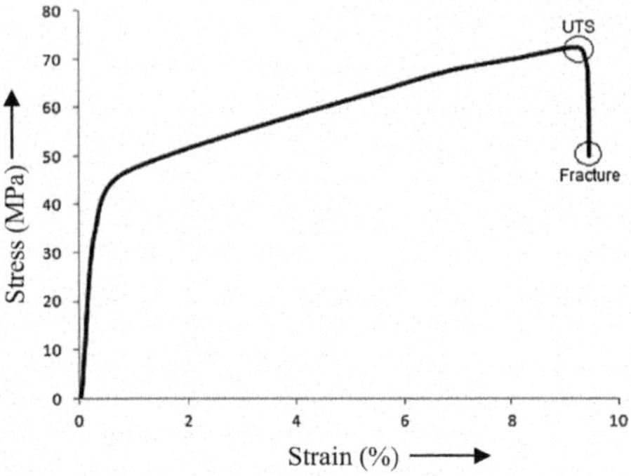

(b)

Fig. 10.14 **a** Tensile specimens prepared as per ASTM E8/M-09 **b** Strain–stress curve obtained during tensile test of the developed Al–5%SiC MMC

9 Future Aspects in Microwave Processing of Metallic Materials

We have come across a number of methods for treating metallic materials that use microwaves as a heat source. However, there still exists a gap where microwaves can be effectively used. Some of such areas are listed below;

i. A conceptual description of microwave materials processing has not been reported yet in the form of mathematical concepts and equations, which is a great opportunity to carry out the future work.
ii. Even though a lot of progress has taken place in the area of microwave materials processing, exact temperature monitoring at various regions of the material being processed and the ways of regulating the same hasn't reported till date.
iii. Insulation around the workpiece becomes very crucial while processing materials as the heat around would damage the furnace walls and even might spoil the magnetron. Hence, optimization of insulation and susceptor layer thickness for processing of metals has to be studied.
iv. Depending upon the positioning of materials for processing inside the microwave cavity, temperature gradient formation could be an issue, which might lead to improper heating and results in variation in processing time. Hence, optimization of materials processing time happens to be a core research area in future.
v. Microwaves have found applications in metal processing from many decades. Metallic materials can be processing in microwave environment only through hybrid heating. On the contrary, metallic materials can absorb microwaves at higher temperatures. However, the Physics involved during the interaction of microwaves with metals has been proposed and not practically proved. Hence, microwave-metal interaction phenomenon needs to be addressed.

References

1. Singh S, Gupta D, Jain V, Sharma AK (2015) Microwave processing of materials and applications in manufacturing industries: a review. 30(1):1–29. https://doi.org/10.1080/10426914.2014.952028
2. Moore AF, Schechter DE, Morrow MS (2003) Method and apparatus for melting metals. U.S Patent 20030089481
3. Srinath MS, Sharma AK, Kumar P (2011) A new approach to joining of bulk copper using microwave energy. 32(5):2685–2694. https://doi.org/10.1016/j.matdes.2011.01.023
4. Gupta M, Leong WW, Wong WL (2007) Microwaves and metals. John Wiley and Sons, Singapore
5. Metaxas AC, Meredith RJ (1983) Industrial microwave heating. Peter Peregrinus Ltd., London
6. Mishra RR, Sharma AK (2016) Microwave–material interaction phenomena: heating mechanisms, challenges and opportunities in material processing. 81:78–97. https://doi.org/10.1016/j.compositesa.2015.10.035
7. Shashank LM, Srinath MS, Amarendra HJ (2017) A review on microwave processing techniques and its potentials 145 – 151
8. Agrawal D (2010) Latest global developments in microwave materials processing. 14:3–8. https://doi.org/10.1179/143307510X12599329342926
9. Chandrasekaran S, Basak T, Ramanathan S (2011) Experimental and theoretical investigation on microwave melting of metals. 211:482–487. https://doi.org/10.1016/j.jmatprotec.2010.11.001
10. Chandrasekaran S, Ramanathan S, Basak T (2012) Microwave material processing—a review. 58:330–63. https://doi.org/10.1002/aic.12766
11. Clark DE, Sutton WH (1996) Microwave processing of materials 26:299–331. https://doi.org/10.17226/2266
12. Das S, Sharma AK (2012) Microwave drilling of materials. BARC. Newsletter 329:15–21
13. Shashank LM, Srinath MS, Amarendra H J (2017) Microstructural and mechanical investigation of aluminium alloy (Al 1050) melted by microwave hybrid heating. 4:076504. https://doi.org/10.1088/2053-1591/aa7aaf
14. Shashank LM, Srinath MS, Amarendra HJ (2017) An experimental investigation to find the critical (coupling) temperature in microwave hybrid heating of bulk metallic materials. 4:106521. https://doi.org/10.1088/2053-1591/aa931e
15. Singh B, Zafar S (2020) Understanding time-temperature characteristics in microwave cladding. 25:75–80. https://doi.org/10.1016/j.mfglet.2020.08.002
16. Kaushal S, Singh D, Gupta D, Jain V (2019) Processing of Ni–WC–Cr3C2-based metal matrix composite cladding on SS-316L substrate through microwave irradiation. 53(8):1023–1032. https://doi.org/10.1177/0021998318794846
17. Sarbjeet K, Singh D, Gupta D, Jain V (2019) Wear resistance improvement of austenitic 316 L steel by microwave-processed composite clads 141(4):041605. https://doi.org/10.1088/2053-1591/ab3674
18. Bansal S, Gupta D, Jain V (2021) Surface Modification of SS-316 To Enhance Cavitation Resistance Through Composite Microwave Clads 2141004. https://doi.org/10.1142/S0218625X21410043
19. Ajit HM, Anikethan B, Srinath MS (2019) Wear studies of composite microwave clad on martensitic stainless steel. 1:1–7. https://doi.org/10.1007/s42452-019-0214-0
20. Ajit HM, Badiger RI, Srinath MS, Naik GM (2020) An experimental investigation of microwave developed nickel-based clads for slurry erosion wear performance using Taguchi approach 9:293–304. https://doi.org/10.1007/s13632-020-00644-3
21. Zafar S, Sharma AK (2016) Structure-property correlations in nanostructured WC–12Co microwave clad, 370:92–101. https://doi.org/10.1016/j.apsusc.2016.02.114
22. Zafar S, Sharma AK (2016) Prediction of tribological behaviour of WC-12Co nanostructured microwave clad through ANN 11:333–340. https://doi.org/10.2474/trol.11.333

23. Zafar S, Sharma AK (2016) Abrasive and erosive wear behaviour of nanometric WC–12Co microwave clads. 346:29–45. https://doi.org/10.1016/j.wear.2015.11.003
24. Hebbale A, Srinath MS (2016) Microstructural investigation of Ni based cladding developed on austenitic SS-304 through microwave irradiation. 5(4):293–301. https://doi.org/10.1016/j.jmrt.2016.01.002
25. Hebbale A, Srinath MS (2016) Microstructure and experimental design analysis of nickel based clad developed through microwave energy. 8:257–259. https://doi.org/10.1016/j.pisc.2016.04.044
26. Hebbale A, Srinath MS (2017) Fuzzy prediction of slurry erosive behavior of cobalt based clad developed through microwave energy. 4(2):1804–1811. https://doi.org/10.1016/j.matpr.2017.02.023
27. Hebbale A, Srinath MS (2017) Taguchi analysis on erosive wear behavior of cobalt based microwave cladding on stainless steel AISI-420 99:98–107. https://doi.org/10.1016/j.measurement.2016.12.024
28. Hebbale A, Srinath MS (2018) Microstructural studies of cobalt based microwave clad developed on martensitic stainless steel (AISI-420) 71:737–743. https://doi.org/10.1007/s12666-017-1206-7
29. Hebbale A, Srinath MS (2019) Characterization of cobalt based microwave clad developed on SS-355 895:259–264. https://doi.org/10.4028/www.scientific.net/AMM.895.259
30. Srinath MS, Sharma AK (2011) Investigation on microstructural and mechanical properties of microwave processed dissimilar joints 13(2):141–146. https://doi.org/10.1016/j.jmapro.2011.03.001
31. Srinath MS, Sharma AK, Kumar P (2011) A novel route for joining of austenitic stainless steel (SS-316) using microwave energy 225(7):1083–1091. https://doi.org/10.1177/2041297510393451
32. Srinath MS, Murthy PS, Sharma AK, Kumar P, Kartikeyan MV (2012) Simulation and analysis of microwave heating while joining bulk 4(2):152–158. https://doi.org/10.4314/ijest.v4i2.11
33. Badiger RI, Narendranath S, Srinath MS (2015) Joining of Inconel-625 alloy through microwave hybrid heating and its characterization 18:117–123. https://doi.org/10.1016/j.jmapro.2015.02.002
34. Badiger RI, Narendranath S, Srinath MS (2018) Optimization of parameters influencing tensile strength of Inconel-625 welded joints developed through microwave hybrid heating 5 (2):7659–7667. https://doi.org/10.1016/j.matpr.2017.11.441
35. Badiger RI, Narendranath S, Srinath MS (2018) Microstructure and mechanical properties of Inconel-625 welded joint developed through microwave hybrid heating 232(14):2642–2477. https://doi.org/10.1177/0954405417697350
36. Badiger RI, Narendranath S, Srinath MS (2019) Optimization of process parameters by Taguchi grey relational analysis in joining Inconel-625 through microwave hybrid heating 8:92–108. https://doi.org/10.1007/s13632-018-0508-4
37. Badiger RI, Narendranath S, Srinath MS, Hebbale AM (2019) Effect of power input on metallurgical and mechanical characteristics of Inconel-625 welded joints processed through microwave hybrid heating 72:811–824. https://doi.org/10.1007/s12666-018-1537-z
38. Clark DE, Sutton WH (1996) Microwave processing of materials 26(1):299–331. https://doi.org/10.1146/annurev.ms.26.080196.001503
39. Chandrasekaran S, Basak T, Ramanathan S (2011) Experimental and theoretical investigation on microwave melting of metals 211(3):482–487. https://doi.org/10.1016/j.jmatprotec.2010.11.001
40. Gouthama TR, Harisha G, Manjunatha YR, Mohana Kumara SM, Srinath MS, Shashank Lingappa M (2016) Melting of tin using muffle furnace and microwave energy and its characterization 149(1):012100. https://doi.org/10.1088/1757-899X/149/1/012100
41. Mishra RR, Sharma AK (2016) On mechanism of in-situ microwave casting of aluminium alloy 7039 and cast microstructure 112:97–106. https://doi.org/10.1016/j.matdes.2016.09.041
42. Shashank LM, Srinath MS, Amarendra HJ (2018) Melting of bulk non-ferrous metallic materials by microwave hybrid heating (MHH) and conventional heating: a comparative study on energy consumption 40:1–11. https://doi.org/10.1007/s40430-017-0921-7

Focus on Carbon Fiber Reinforced Composites Manufacturing and Properties

Mohit Kumar and Jarnail Singh

Abstract From last few decades, carbon fiber reinforced polymer (CFRP) composites are emerging as the most promising composite for many industrial applications comprising of aerospace industries, automotive industries, marine industries and renewable energy. A huge consideration is given to these composite materials over old conventional materials because of its exceptional properties such as high strength/stiffness to their weight ratio and excellent corrosion resistance properties. Different polymers *i.e.,* thermoplastics, thermosetting and elastomers can be used as matrix for fabrication of composites. Out of these, the main emphasis is given to the thermoplastic polymers due to its advantage of having good flexibility, recyclability, easy processing and less manufacturing time. In this chapter, the different manufacturing techniques such as filament winding, vacuum assisted resin transfer molding (VARTM) compression molding, injection molding and pultrusion process and the curing processes of carbon fiber reinforced composites using thermal curing mechanism and radiation curing is discussed. The properties of CFRP composites *i.e.,* mechanical strength and modulus properties, electromagnetic shielding effectiveness, thermal stability, thermal conductivity and the surface modification effect on CFRP composites are thoroughly discussed. Overall, this chapter focuses and summaries the ongoing research and development and future prospective of these composite materials.

Keywords Carbon fiber · Polymers · FRP composites · Mechanical properties · Surface modification · Electrical properties

M. Kumar (✉) · J. Singh
Department of Mechanical Engineering, Chandigarh University, Mohali, Punjab, India
e-mail: mohit.dhiman349@gmail.com

J. Singh
University Centre for Research and Development, Chandigarh University, Mohali, Punjab, India

M. K. Singh et al. (eds.), *Composite Materials Processing Using Microwave Heating Technology*, Composites Science and Technology,
https://doi.org/10.1007/978-981-97-2772-8_11

1 Introduction

In this dynamic world, emerging new sustainable materials with exceptional properties and economically stable is necessary to meet the industrial expectations and prevention from environmental effects. The challenges such as low efficiency, heavy weights, high cost, recyclability and reliability should be countered by use of new materials. In contrast of these new lightweight materials have been developed that replace the old conventional materials to increase the strength and efficiency of the product [1]. Consequently, a huge consideration is given to the lightweight fibrous composite materials by the researchers and industries due to its high strength and stiffness to their weight ratios [2]. These fibrous composite materials are extensively used in applications of aerospace, civil structures, automobiles and marine sectors [3, 4]. Usually, the manufacturing of FRP composites, consists of a polymer (matrix) and fiber (reinforcement). Fiber reinforcements are preferred, as the strength of any material is much higher in fiber form than in bulk form [5]. FRP composite materials are orthotropic in nature, i.e., having different properties in different directions [5]. In these composites, matrix material surrounds and keep the fibers in the desired orientation and location whereas the fiber reinforcements contribute to improve the mechanical and physical properties of the matrix material [6]. Matrix part also acts as a shield for fibers against environmental surroundings. FRP composite materials have shown huge benefits over conventional materials with reference to low thermal expansion, good dimensional stability, erosion and wear resistance, enhanced fatigue life, lightweight, high strength properties, corrosion resistance, etc. [7, 8].

The fibers in the composite materials are the main strength builders on which the whole structural component is supported. The fibers reinforced in the matrix can be classified depending upon their fiber length [6]. Fibers that are continuously oriented in one direction are continuous fibers. These provide excellent directional tailoring properties and imparts high strength and modulus to the composite materials in the direction of fibers. Depending upon their fiber orientations these can be unidirectional, bidirectional and multidirectional. Fibers that impart isotropic properties i.e., no change in mechanical properties along the directions are the discontinuous fibers. Although the strength of these fibers is less than continuous fibers in a particular direction but can take load along any direction of the composite material. The combination of continuous fibers and discontinuous fibers can be termed as hybrid fibers.

The function of the matrix is to protect and support the fibers and to deliver the distributing load and transferring the load between the fibers. This load transfer mechanism imparts shear stress in the matrix that resists the pulling out of the broken fibers [6]. The matrix in the composite materials can be metallic, organic, ceramic, or carbon, but the main emphasis of this section is on the polymer matrix. Since polymer matrix is light in weight, have low processing temperatures and have excellent corrosive resistant properties, these are widely used with glass and carbon fiber reinforcements [9].

Polymers are the long molecular chain or network of repeating monomer units that are bonded together using chemical reactions [10]. Polymer matrix act as a continuous phase that holds the fiber in its place, with its properties influencing most of the degradative processes like chemical resistance, delamination, water absorption, etc., [11].

Generally, the polymer matrix is categorized in thermoset and thermoplastic polymers [12]. A three-dimensional cross-linked solid structure is obtained after the curing of thermoset polymers that claims solvent resistant properties and high thermal stability. Epoxies, polyesters, phenolic, vinyl esters and polyamides are some examples of thermoset polymers. Thermosets imparts good mechanical strength and low manufacturing cost. Thermoplastic polymers can be melted and reformed without altering its physical properties due to weak intermolecular bonds. The composites manufactured from thermoplastic polymers claim good impact resistance and damage tolerance but are less brittle than thermosets and required a large amount of processing energy and cost. Few examples of thermoplastic polymers are polyphenylene sulfide, polyether ether ketone, polycarbonates and polyetherimides.

Apart from all the fibrous materials, carbon fiber is one of the strongest and lightweight material that possesses excellent mechanical strength, good corrosion resistance and wear resistance properties. Moreover, carbon fibers have exceptional mechanical, electrical and thermal properties. In recent years, the products/components made from carbon fiber reinforced composites have potentially been used in different sectors such as automobile, aircraft, marine, civil, space and sports. Its applications also range in industrial products (i.e., tanks, power transmission shafts, step ladders), in medical industries (i.e., bone plates for physically challenged, implants) and in furniture manufacturing [13].

Figure 3.1 shows the demand of CFRP composites is setting huge standards globally from past few decades and is expected to reach upto 200 kiloton by 2050 [14, 15]. The global response at present confirms predominant use of CFRP in different application areas such as aerospace, pressure vessels, wind turbine blades, automotive components, construction, marine and others. The contribution of different application areas in shown in Fig. 3.2a and their cost analysis is presented in Fig. 2b. Japan, Europe, North America and Asia Pacific regions are the major contributors of CFRP composites (Fig. 3.2c). North America and Europe consumes around 40% of the CFRP composites. Companies like Boeing, Airbus, defense leads the sales in these regions. The global market can also be classified in terms of manufacturing techniques of CFRP composites as shown in Fig. 3.2d. The techniques like hand layup and prepregs are widely used in fabrication of sports and aerospace applications.

2 Manufacturing Techniques

Manufacturing of FRP composites consists of the preparation of fiber preforms and reinforcing these fibers with polymer matrix using different techniques. Here, fiber preforms signify weaving, braiding, stitching and knitting of fibers in a long sheet [17,

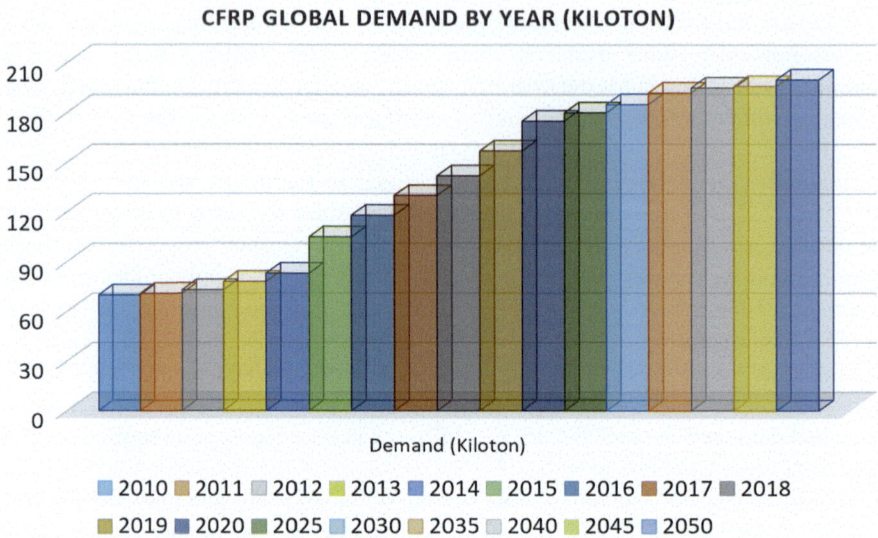

Fig. 3.1 CFRP composites global demand in terms of years [14, 15]

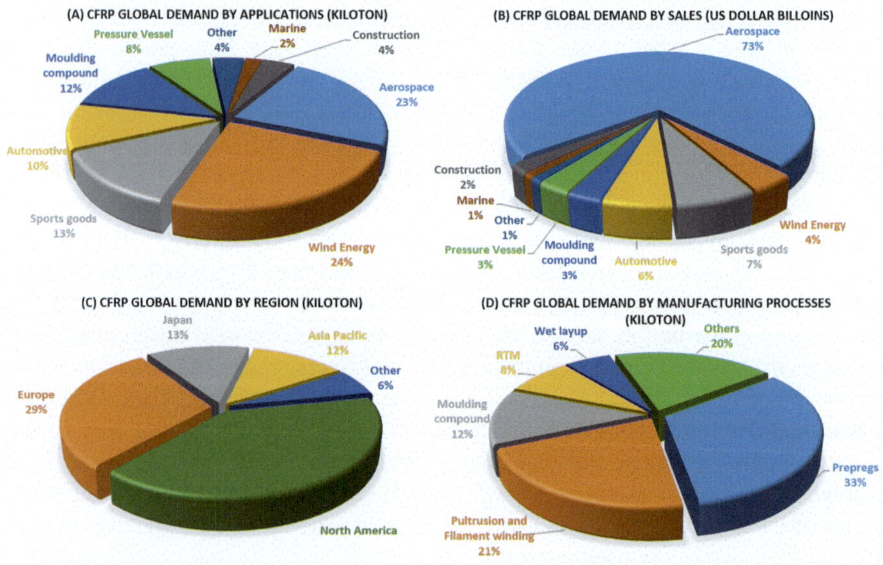

Fig. 3.2 CFRP composites global demand in terms of **a** applications, **b** sales, **c** region and **d** manufacturing processes [14, 16]

18]. The hand layup technique is the special technique and the primary phase for the manufacturing of FRPs [19]. Under this process, the desired fiber preforms are placed in a mold and the matrix material is applied onto the fibers through brush or roller. Rollers ensure the interaction between the successive layers of fibers and the matrix by forcing the matrix part into the fiber reinforcement. The combination of uncured matrix material and fibers is known as prepregs. These prepared composite prepregs are finally cured using different curing processes [19]. Manufacturing techniques are the primary step in the fabrication of composite materials. Some of the composite manufacturing techniques are discussed in the following sections [6].

2.1 Microwave Processing

The electromagnetic spectrum includes microwaves, which have frequencies between 300 MHz and 300 GHz. Generally speaking, there are three main components to a microwave furnace: the applicator, transmission lines, and microwave source, which is usually a magnetron. To put it briefly, the source generates the microwaves, which are then transmitted to the applicator through transmission lines, and the applicator then delivers the microwave energy to the item being heated. Applicators are essential to producing adequately cured composites because they control the distribution of microwaves within a material [20, 21]. Waveguides, single-mode cavities, and multi-mode cavities are examples of applicator types. One hot area will result from the predictable non-uniform fields produced by single-mode applicators. Multiple hot spots are produced by multi-mode applicators, which are utilized in household microwave ovens.

Thermosetting polymers and other dielectric materials are heated by microwaves through molecular-level electromagnetic field interaction. Across the mechanism of polarization, microwaves are able to pass through dense materials and transmit energy. Different types of polarization exist. Electrons are displaced from their state of equilibrium with respect to the nucleus during the process of electronic polarization. When distinct types of atoms in a molecule move in relation to one another, it results in atomic polarization. Molecular dipoles are rotated to coincide with the one being used electromagnetic field in order to achieve dipole polarization. Friction caused by the material's reluctance to this reorientation produces thermal heat. The accumulation of charges at a material contact is the basis for interfacial (Maxwell–Wagner) polarization. In composite materials, the two most important pathways for energy transfer are interfacial polarization and dipole polarization [22]. The degree of polarization is ultimately determined by a material's dielectric characteristics. The conductive portion of the material response is quantified by the dielectric loss factor, whereas the capacitive portion of the dielectric response is quantified by the dielectric constant. Microwave penetration depth is minimal in materials like carbon fibers because of their high dielectric loss coefficients. Consequently, it is claimed that these substances serve as microwave reflectors. Conversely, materials possessing a low dielectric loss factor have a high penetration depth, hence minimizing their absorption of microwave radiation. Thus, the most efficient way to heat materials in the lower half of the conduction range, such thermosetting polymers, is by using microwaves to transfer energy. This is unlike the case with traditional heating, where energy is transferred most efficiently into materials with high conductivity (high dielectric loss). During the curing process, thermosetting resins crosslink, which modifies the resin's molecular dielectric characteristics. The dielectric characteristics exhibit a direct correlation with the variations in resin viscosity. The viscosity of the resin is quite low at the start of the curing process, and because of its dielectric characteristics, which allow its dipoles to be somewhat liberated from motion and coordinate with the electromagnetic field, the resin couples well with microwaves. The resin gets increasingly viscous during crosslinking, which lowers the dielectric loss factor. At this point, the material will be more susceptible to microwave heating because the thermosets will turn more visible to microwaves since the molecular dipoles are less free to orient oneself to the electromagnetic field. Using in situ cure monitoring to maximize the usage of microwaves throughout processing allows for the exploitation of variations in dielectric characteristics during the curing process [23]. In order to achieve this, Zhang et al. [24] created an off-line dielectric measurement technique that may be used in the microwave curing procedure. Microwaves are the most suitable electromagnetic radiation band for heating and curing thermosetting polymers, while other bands may also be employed. This is so that, unlike additional electromagnetic waves like gamma radiation, which pose a serious health risk, microwaves may successfully penetrate the polymer.

2.2 *Filament Winding*

Filament winding is a technique to wind the fiber on the mandrel after passing it through liquid resin and then cured. Fibers can be wrapped in different orientations to have strength and stiffness properties in different directions. The filament winding technique can be used to produce pressure vessels, power and transmission poles, aircraft fuselages, pipes, etc.

2.3 *Compression Molding*

Compression molding is the primary process that is used to manufacture high volume complicated FRP composite products. This molding operation begins with the hand layup technique to form prepregs inside the open molds followed by compression of these prepregs under high temperature and pressure. A component of low void content and high fiber volume fraction is produced using this technique. Application of compression molding can be found in automotive industries which include, leaf springs, rod wheels and bumpers. A schematic diagram of the manufacturing of the composite part using compression molding is shown in Fig. 3.3 [6].

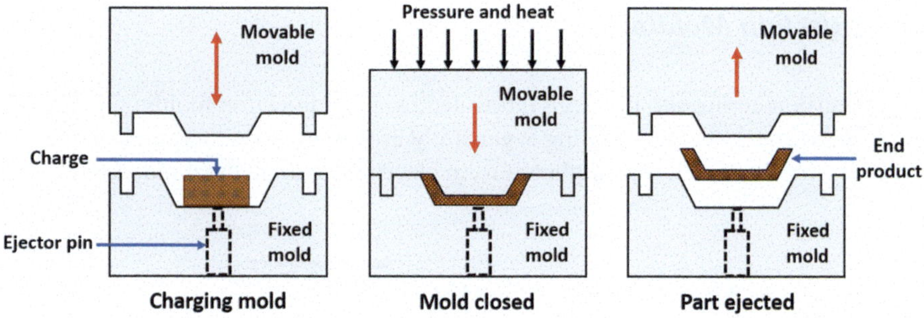

Fig. 3.3 Compression molding [6]

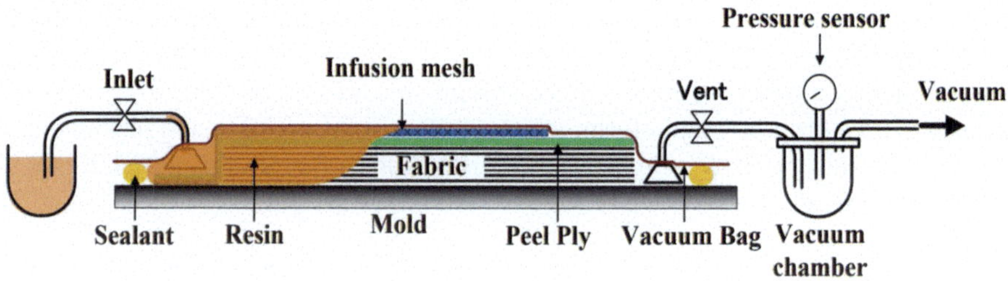

Fig. 3.4 Vacuum-assisted resin transfer molding (VARTM) [6]

2.4 Vacuum-Assisted Resin Transfer Molding (VARTM)

VARTM is the resin transfer molding technique in which fiber preforms are placed on the molds and resin is transferred through perforated tubes over the fibers under vacuum. This technique leaves no traces of excess air in the composite that makes it popular in many applications like wind turbine blades and boat hulls. Figure 3.4 shows the VARTM process for the preparation of a composite laminate [6].

2.5 Pultrusion Process

In pultrusion process, the stand of continuous fibers is pulled through the resin bath and then cured in a heating die. A uniform cross-section and with relatively long lengths allow this technique to produce beams, flat sheets, solid and hollow tubes, etc.

2.6 Injection Molding

High precision composite parts are fabricated using an injection molding technique in very less time. Injection molding is generally used to produce thermoplastic based composite products that find applications in electronic and medical industries.

2.7 Additive Manufacturing (FDM)

Additive manufacturing, commonly known as "3D printing" or "rapid prototyping," is a modern manufacturing technology in which layers of material are placed at a time (successively) until a final object is achieved. FDM (Fused Deposition Modelling) technology has recently become one of the most extensively used fast prototyping technologies for a variety of applications. This process is based on fused fibre material deposition on a drop-down platform, which allows for the creation and introduction of novel materials, such as composites. ABS is the most often used material in FDM, followed by PC, PLA, PPSF, ULTEM9085, and mixes thereof.

3 Properties of CFRP Composites

Due to wide range of applications and excellent properties of CFRP composites, the developers and researchers have shown tremendous amount of interest in reviewing these composites. Additionally, by proper combination of materials and methods, the properties of the CFRP composites can be enhanced. E.g., proper selection of fiber length, proper orientation of fibers in the matrix, use of surface modification techniques etc. Each selection parameter has important role in fabricating the composites properties and further set the prospect of using the material in specific industries. For many applications the mechanical strength of CFRP composites is the major requirement. Meanwhile, for sectors like electronics, automotive, or energy storage electromagnetic shielding effectiveness and electrical conductivity are the major requirements. Specially, when it is used in aeroplane wings to shield them from lightning strikes. From previous literature, some of the properties of CFRP composite properties was explored and shown in Fig. 3.5 also discussed in the following sections.

3.1 Mechanical Strength and Modulus of CFRP

Due to several inherent limitations in terms of mechanical characteristics, the polymer matrix impairs its usage in structural applications. To deal with this problem, carbon

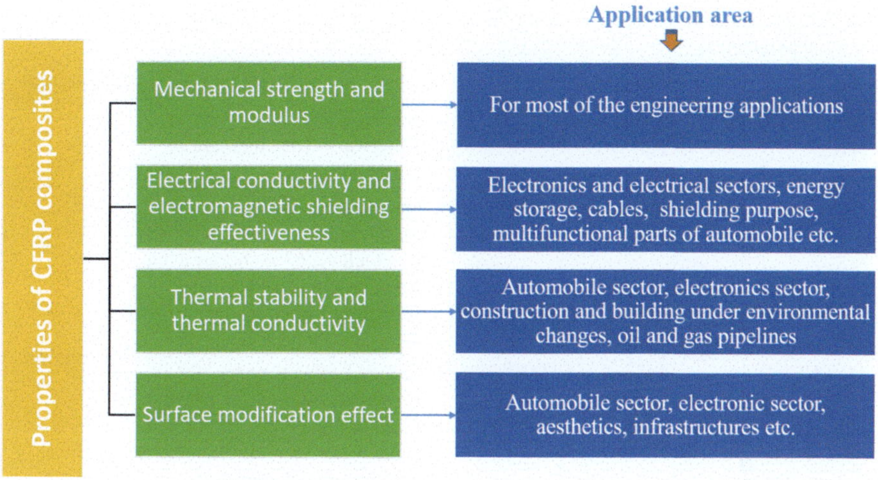

Fig. 3.5 Properties of CFRP and their application areas

fibers (CF) can be fused with matrix. These fibers are available in short or continuous form and commonly known as short CFRPs and continuous CFRPs. Currently, the CFs are used extensively because of its excellent mechanical properties. These composites remarkably enhanced the mechanical properties in terms of strength, modulus, and resistance properties along with electrical and thermal properties. Apart from structural applications, these composites have great usage in aerospace, automotive as well as manufacturing industries for decades due to mentioned remarkable properties. However, their properties will differ depending upon the following influential parameters:—fiber breakage and loading of fibers [25], the length to diameter ratio (aspect ratio) of the reinforcing materials [26], Interfacial stress shifting and dispersion homogeneity, Distinguishing structure and chirality of nanofiller [27], void content of the composite structure [28], waviness of the fibers, interaction between the matrix/reinforcement which is accelerated due to physical and chemical properties [29], fiber length of composites [30], processing parameters such as temperature, pressure, environmental conditions, orientation and stacking sequence of the fibers [31], fiber packing geometry [32], type of polymers and manufacturing process used [33]. The mechanical properties of CFRP composite material depending upon the various influential parameters are shown in Table 3.1.

3.2 *Electromagnetic Shielding Effectiveness of CFRP*

Effective radiation shielding systems are required for preventing EMI and electromagnetic pollution. EMI shielding is the practice of limiting the effects of electromagnetic interference through the use of conductive or magnetic barriers [49]. Through

Table 3.1 Mechanical properties of CFRP composite material depending upon the various influential parameters

Material	Methodology	Mechanical properties	References
WE91-2/CFRP prepreg	Microwave (MW) curing	• Industrial MW designed for composites. 10% increase in tensile strength	[34]
Bismaleimide (BM)/CF	Microwave (MW) curing	• Comparable void and interlaminar shear stress (ILSS) to autoclave, 63% shorter cycle times, modest reductions in flexural and fiber volume portion	[35]
Epoxy/CF	Microwave (MW) curing	• Reduced cycle time by 39%, increased compression strength by 22%, and enhanced porosity. Evidence of increased fiber to resin interfacial adhesion	[36]
CF/polyetherimide	Hand layup technique	• Reduced wear at the fiber's 0° orientation • However, when the orientation angle increased, the mechanical properties deteriorated	[37]
CFRP/epoxy steel	Pultrusion technique	• Adhesive J133 tested best in terms of toughness and tensile strength	[38]
CNTs/continuous CFR polyethylene terephthalate	Single screw extruder	• At 1.0 wt.% of CNT and 56 wt.% of CF-content, the tensile strength and the tensile modulus was 1728.7 MPa and 25.1 GPa, respectively	[39]
Thin CF/epoxy	Autoclave	• Unidirectional hybrid laminates have far greater ultimate tensile and compression strengths • There is an almost 50% gain in tensile strength	[40]
CF/PLA	Fused Deposition Molding (FDM)	• As the printing temperature rises, the material's tensile properties improve	[41]
CF/epoxy hybrid	Hand lay-up method followed by compression molding	• Composites with a filler (Al_2O_3) loading of 1.75 wt.% showed the highest increase in tensile strength (29.54%) and modulus (2.42%)	[42]

(continued)

Table 3.1 (continued)

Material	Methodology	Mechanical properties	References
Modified CF/epoxy and carbon micro and nanofiller	Hand lay-up	• 0.3 wt.% single-wall CNTs maximized static tensile strength by 8.6% and flexural strength by 14% • Fatigue durability also enhanced greatly. Milled carbon fibers improved static loading by 3%–9% but degraded fatigue properties	[43]
CF/epoxy	Vacuum resin transfer molding technique	• The incorporation of carbon nanofiller increases the material's tensile strength by over 60% and its interlinear shear strength by around 16%	[44]
CFRPs with nano-modified matrix	Dissolver dispermat	• In comparison to composites containing either GNPs or MWCNTs in equal proportions, the second hybrid showed up to a 45% improvement in mode I fracture toughness (G_{IC}) • The results of Mode II indicated that the second hybrid had a G_{IIC} rise of up to 25%, while the identical contents used separately had increases of 18% and 13%, respectively	[45]
Graphene oxide-incorporated woven CF/epoxy	VARTM process	• Tensile strength, Young's modulus, toughness, flexural strength, and flexural modulus of GO (0.2 wt%)/CF/epoxy composites were increased by 34, 20, 83, 55, and 31%, respectively, when compared to the CF/epoxy composite	[46]
Pineapple leaf fiber (PALF)/Carbon hybrid laminate	Vacuum infusion process	• Interior carbon plies and ply orientation [0°, 90°] produced the maximum tensile strength (187.67 MPa) and modulus (5.23 GPa) • The laminate with exterior carbon plies and similar ply orientation had the maximum flexural strength (289.46 MPa) and modulus (4.82 GPa)	[47]
CF/epoxy	Compression molding method	• Maximum flexural strength is demonstrated by the composite with the unidirectional fiber layers oriented at 0°	[48]

either internal consumption and dissipation of radiation power or outward reflection of field waves, shielding materials can block and dampen the radiated signal [50]. The shield needs to have electrical conductivity because electromagnetic radiation is reflected off of moving charge carriers, however great conductivity is not necessary therefore a volume resistivity of 1 Ω-cm is permissible [51]. Thus, metal sheets, metal screens, networks, metal foam, and volumetric conductive paints containing metal particles are the conventional shielding materials [51]. Shielding Effectiveness (SE) is a popular metric used to describe the degree to which a material may act as a shield. Schelkunoff's equations can be used to determine the SE for metals given their permeability to magnetic fields and electric conductivity [52]. A SE value above 100 dB indicates that the material is practically impenetrable, while a value of 30 dB is regarded as adequate for the majority of commercial and consumer applications [53], while a value of 40 dB is required for the production of electronic housing and for electronic appliances [54].

Recent years have seen research into potential replacements to standard metals for EMI shielding applications. Carbon fiber-reinforced plastic (CFRP) is widely utilized in the aerospace sector due to its excellent mechanical strength (such as a Young's modulus and relatively low density) and outstanding chemical inertness. Since the polymeric matrices is non-conductive, the shielding efficiency of these composites relies on the metallic filler utilized; the composites in this class have a mean conductivity of 10^4 S/m, which is 3 times lower than metals [55]. Filler with a high aspect ratio, high conductivity, and tiny unit size all contribute to better electrical connectivity and hence greater shielding efficiency. Carbon materials, such as carbon fibers, colloidal graphite, and flexible graphite, have been the subject of extensive research on their shielding efficiency by Chung [51]. By comparing the SE of polyether sulfone (PES)-matrix composites that were filled with isotropic pitch-based carbon fibers at different percentages of quantity filler material and different filament lengths, Li and Chung [44] discovered that the SE of composites improved with longer filaments and that a filler volume fraction of at least 40% was required to achieve a SE of >50 dB [56]. Fouladgar et al. provided an explanation for the nonlinear behavior of the SE as a function of frequencies of CFRP sheet containing 10% of fiber concentration as a result of the permissible intricate structure of the material [57]. Various research looked into the correlation between SE and layer count, fiber density, and aspect ratio. The SE has been demonstrated to be irrespective of the fiber volume proportion for with higher frequencies exceeding 8 dB [58], and it was shown that increasing the sum of several layers of the composite material was associated with an increase in SE. Similarly, decreasing the filler concentration was shown to be associated with a progression regarding greater frequency ranges of the maximum SE. Absorption loss increases with fiber concentration, which is correlated with improved EMI shielding; this trend holds true for increasing fiber proportions and sample thickness as well [59].

3.3 Thermal Stability and Thermal Conductivity of CFRP

Due to their relatively low thermal conductivity (TC), CFRPs are not ideal for use in thermal management [60]. Therefore, elevating the thermal stability of CFRP laminates relies heavily on the development of through-thickness TC. Direct heat conduction pathways through the thickness of CFRP can be improved by using 3-D woven fabrics or z-pin fillers [61]. However, making 3D fabric that's woven is a difficult and expensive procedure. However, z-pin fillers weaken composite laminates' structure and diminish their mechanical characteristics. CFRP laminates' TC is severely constrained by the polymer matrix's low TC. Matrix TC is typically improved by including high TC fillers. Although the out-of-plane TC was raised by 40% when Kandare et al. [62] incorporated silver nano-inclusions and graphene into CFRP, the maximum conductivity wasn't increased above 0.6 W/m K. Blending fillers and polymer is feasible, however significant improvements in TC often call for a high filler content, which might raise blend viscosity, cause nanoparticle aggregation, and result in insufficient impregnation of fibers. Altering filler inclination in polymers has already been extensively explored for its potential to reduce filler amounts and enhance TC in a desired direction. The thermally conductive fillers were formed into a foam structure with 3-D linked networks using expendable template approaches, such as ice template, metal template, or polymer template [63]. However, there are still difficulties in producing CFRPs with fillers that have an oriented structure. Still, several fillers (thermally conductive) were affixed to the surfaces of fiber in order to provide a heat conduction channel bridging adjacent fibers [64, 65]. Fibers' structure and strength were compromised due to the chemical treatment of their surfaces. The difficulty and expense of producing such composites stems from the complexity of their production. It was suggested that the interlayers of CFRP laminates be modified due to the presence of resin rich zones there. However, there is still potential for development because the sandwich layer remodeling technique unable to manage the polymeric matrix studded with carbon fibers inside lamina. To obtain metal-level through-plane TC, the layered architecture of graphene paper was mechanically manipulated to create a wrinkled structure consisting predominantly of graphene (vertical) and an extremely thin capping of horizontally graphene layers (both sides) [65]. By intertwining highly conductive films (HCFs) like Copper (Cu) film and exceedingly orientated graphite film (HOGF), Zeyu Ouyang et al. dramatically increased the lateral thermal conduction (TC) of unidirectional CFRP laminates. In comparison to CFRP, the maximum thermal conductivity of HCF-CFRPs are as high as 49.19 W/m K (at 1.97 vol% HOGF for HOGF-CFRP), and 21.45 W/m K (at 1.74 vol% Cu film for Cu-CFRP).

In contrast, SiC nanowires (SiCNWs) are commonly used as reinforcements for epoxy-based composites because of their exceptional performance in terms of resilience, thermal stability, and thermal conductivity (390 W/mK) [66]. Here, a hierarchical reinforcing structure was formed by using electrophoretic deposition (EPD) of SiCNWs onto the CF fiber to improve thermal characteristics. SiCNWs and CF were then grafted with graphene using in-situ chemical vapor deposition (CVD). The

TC of SiCNWs/Graphene augmented epoxy composites improved with increasing CVD duration, owing in part to the combinatorial effect of the two nanomaterials [67].

Glass fibre-modified epoxy composites (GFEC) were reinforced with two types of Si-based nanoparticles (SiC and SiO_2) by Kwon et al. Compared to SiO_2 GFEC, SiC GFEC was reported to have greater heat resistance [68]. With the concurrent incorporation of nanoplatelets of graphene and silver nanoparticles/nanowires at a combined loading of 1 vol%, the through-thickness thermal conductivity was found to be improved by around 40%, as discovered by Kandare et al. [62], who investigated the combined mechanical interaction of conductive filler materials (graphene nano-platelets and silver nanowires). Han et al. [69] found that adding 0.80% carbon black to epoxy composites increased their thermal conductivity by up to 210% compared to the unmodified epoxy composites. Incorporating several types of nano-filler (including clay, silica, graphene, and CNT) into the thermoplastics matrices has improved the resilience and thermal properties of CF composites. It has been shown that adding inorganic particle and fiber fillers can significantly enhance thermal properties. Due to its high value of aspect ratio, diameter in the nanoscale range, and thermal resistivity [70], the inexpensive natural material nano-clay has been described by several research as one of the prospective choices for nanocomposite and CFRP preparations. Longer carbon fibers (CFs) endowed the composites with superior thermomechanical capabilities and thermal stability, as determined by an investigation into the effect of the fiber's length on the thermal characteristics of minimal carbon fiber/polypropylene composites conducted by Rezaei et al. [71].

The thermal properties of improved polymer composites is thus concluded to rely mostly on the microstructure and intrinsic features of the polymeric matrix and fillers [72]. Increasing the filler's percolation in the polymer matrix increases the thermal conductivity channel, which is why it's important to construct a three-dimensional thermal conductivity network. Despite the many attempts to boost polymer composites' performance, the problems of interfacial transition and size mismatch across nanofiller and matrix are frequently disregarded. This means that modified polymer composites do not have adequate thermal characteristics. Realizing a controlled interface that also meets mechanical and functional requirements is therefore still a significant issue.

3.4 Surface Modification Effect on CFRP

Carbon fibers (CFs) are frequently employed in polymer composites as reinforcing material. Ninety percent of all commercial CFs are made from polyacrylonitrile using a tried-and-true method that yields a highly durable product with an exceptional strength-to-weight ratio [73]. Due to the adaptability of both materials and the exceptional combination to create high-performance structural materials [74], CFs as reinforcement are a topic of active study. Physically blending two components at high shear to achieve adequate dispersion of CFs in the polymer matrix, then curing under precise circumstances, is a common method for fabricating CF-based composites [75]. However, their inert surface prevents bonding with matrix materials. Several investigators have hypothesized that this is due to insufficient adhesion and bonding between the fiber and matrix. When CFs are employed without any sort of surface treatment, the resulting composites have poor ILSS (interlaminar shear strength). Therefore, carbon fibers have not reached their full potential as a material for composite's reinforcement. The interfacial adhesion between the polymeric matrix and the inorganic filler is crucial to the effectiveness of polymer composites. The efficiency with which the polymer matrix transfers loads to the CF can be greatly affected by the interfacial adhesion between them [76]. High-performance characteristics and resilience to different chemical and physical influences necessitate modulation of the CF surface, which is both vital and challenging.

Oxidation, coating with sizing agents, nanomaterial deposition, and chemical grafting are just some of the methods used to alter CFs in order to improve interfacial adhesion through chemically bonded electro-grafting, surface wetting, van der Waals interactions, and interlocking mechanically [73]. By increasing the surface energy for better wetting, increasing the roughness of the surface to produce more effective mechanical interlocking, and incorporating chemically active groups on the fiber surfaces, all surface treatments increase the interfacial shear strength (IFSS). Chemical treatment has been the focus of many reported methods for enhancing interfacial adhesion. Improvements in fiber-resin matrix adhesion have been attributed to treatments which improve the area of the surface and surface functional groups that are acidic [77]. This generally improves the CFs' wettability and ILSS. Oxidative etching of CFs using micro plasma generated by microwave irradiation has been reported by Yuan et al. to increase the interfaces shear resistance of the CF/epoxy composite [78]. However, considerable oxidation of the CF surface generated defects and irregularities, which reduced the overall tensile characteristics of the resultant composites. Amino-functionalizing of CFs is an efficient and non-destructive way to produce strong interfacial adhesion, which improves stress transfer in the CF/epoxy combination and results in enhanced mechanical characteristics. According to the research, modulus of elasticity of epoxy composites can be increased by 50% when amino-functionalized carbon-based fillers are used [79]. Zabihi et al. recently reported that the fracture strength and modulus of epoxy composites were found to be increased by 19 and 10%, respectively after successful amino-functionalization of CFs [80]. Another study claimed an advanced strategy that utilized polyamide grafting and coating of carbon nanotubes/graphene oxide that combines covalent adhesion with mechanical interlocking [81]. Covalent bonding between carbon fibers (CFs) and the epoxy matrix improves the composites' interfacial adhesion, as observed by Zoriana Demchuk et al. [73]. This results in better resistance to thermal ageing and improved final mechanical characteristics.

Most of the research has centered on the impact of treatment with chemicals on interfacial adhesion. The effect of the interfacial contact that occurs between treated carbon fillers and the polymer matrix is studied using both experimental research and molecular dynamics simulations. While polymer composites have been extensively studied, very few work has been done for better mechanical characteristics and long-term attributes.

4 Future Prospect

In upcoming years, CFRP composite teams and researchers will have more focus on fabrication procedures, recycling methods, cost reduction, and property improvement. While numerous types of thermoplastic polymers may be used as matrix for CFRP composites, these materials have appealing properties for a variety of industrial applications. Continuous advancements in process ability and engaging CFRP from a cost-effective standpoint, as well as synergies within industrial sectors, will pave the road for high-volume production that businesses require to fulfil progressive expectations. Advancements are also needed to create efficient, cost-effective, and simple CFRP materials. Innovative concepts are also essential for the high manufacturing rate of such materials. In terms of CF, the global CF manufacturing business is expected to expand dramatically over the next 10 years. This huge development will be driven by a variety of causes, including a strong demand for lightweight materials and imaginative design that requires composite technology solutions [33]. CF's prospects for the future are unquestionably bright. Nevertheless, the incorporation of CF into vast markets and a variety of typical applications is mostly dependent on the skills of the producers. If both polymer and CF producers continue to study new uses and create innovative and low-cost technologies, CFRP will have a broad potential. The aviation, defense, automobile, sustainable sources, sports goods, and infrastructure sectors are among the key CFRP consumption sectors [82]. For fact, aeroplane manufacturers have been concentrating on decreasing total aircraft weight in order to boost product efficiency. The adoption of CFRP aids in significantly reducing total weight. Increasing demand for CFRP as a metal substitute for steel and aluminium is predicted to drive growth in these kind of applications [83]. Furthermore, the usage of CFRP-based components in aeroplanes and vehicles is rising in comparison to metallic-based parts, which might reduce weight, greenhouse gas emissions, and energy consumption. Perhaps, CFRP composites have been renowned for automotive body panel applications together with structural and non-structural components such as fuel tanks, bumpers, hoods and seat structures [14, 84, 85].

CFRP also have a tremendous future in pipelines in the oil and gas sectors. Morozov et al. [86] reviewed the progress in pipelines for oil and gas applications based on thermoplastic polymers alternative to metallic pipes due to its inherent advantages of corrosion and fatigue resistivity.

Additionally, CFRP may be used as an alternative to conventional composite materials in the construction of wind turbine blades. Mishnaevsky et al. [87] and Prabhakaran et al. [88] discover the suitability of CFRP/thermoplastic resin for future turbine blades and related challenges from these materials.

Despite these potential benefits, CFRP has certain downsides and limitations. One of the hurdles is the high cost of virgin CF, which has reduced significantly in recent years. This further increase the fabrication cost of CFRP material. However, it costs less than CF/thermosetting polymers as it consumes a smaller amount energy. These are some of the obstacles to the use of composites as a high-demand material in a variety of sectors. Researchers have concentrated their efforts on developing recycling technologies for CF and CFRP composites, which are predicted to reduce costs by 50% [89].

As a whole, CFRP composite materials are becoming a more popular lightweight material class. Various researcher and R & D departments are actively working on the better association between manufacturing technique, structural properties of interfacial connectivity of CFRP that will be led to an improvement of material properties, which further provide more flexibility in the design for several applications. Yet, with the enormous demand of expanding sectors, the prospects for development, and the support of setting standards for testing and utilizing CFRP composite, more high-efficiency CFRP products will be produced.

5 Conclusions

Developments in CFRP composite material research have been fully considered in this broad review, with an emphasis on composite abilities such as mechanical properties, electrical properties, thermal properties, and surface modifications. The remarkable features of CFRPs are the key reason for additional research and development, e.g., as compared to the neat polymer qualities, these properties improved dramatically with the inclusion of CF as a reinforcement, paving the way for CFRP products in many manufacturing sectors. Moreover, surface modification of the CF is required to enhance the interfacial connection between the CF and the matrix. The oxygen content on the CF's surface can be increased using either a chemical or physical modification approach. In terms of polarity, increasing oxygen makes the CF surface more similar to the matrix. As compared with untreated polymer/CF composites, the modifications enhanced the filler/matrix bond and had a significant favorable influence on the mechanical capabilities of the composite. A range of surface modification strategies for CF were considered before to the production process.

In general, the use of CF as a reinforcement increased the characteristics of diverse composites greatly when compared to the qualities of the neat polymer. Yet, there is a wide range of such enhancements. This might be related to a variety of variables, such as process parameters, type of polymer, CF type, manufacturing technique, fiber orientation and surface modification processes, which result in interfacial adhesion and dispersion statistics. All of these factors are required to achieve the desired qualities, notably mechanical properties, and to comprehend the links between the modification techniques and the mechanical properties of the final CFRP composites. As a result, this evaluation serves as a necessary stepping stone for fully realising the potential of CFRPs in the industrial business.

References

1. Sonnenschein R, Gajdosova K, Holly I (2016) FRP composites and their using in the construction of bridges. Proc Eng 161:477–482
2. Pathak H, et al. (2015) Thermo-elastic failure simulation of 3-D orthotropic composites by XFEM. In: AIP Conference Proceedings 2015. AIP Publishing LLC
3. Soutis C (2015) Introduction: Engineering requirements for aerospace composite materials. Polymer composites in the aerospace industry. Elsevier, pp 1–18
4. Selvam JDR et al (2018) Microstructure and mechanical characterization of in situ synthesized AA6061/(TiB2+ Al2O3) hybrid aluminum matrix composites. J Alloy Compd 740:529–535
5. Mallick PK (2007) Fiber-reinforced composites: materials, manufacturing, and design. CRC press
6. Kumar M et al (2020) Performance of mechanical joints prepared from carbon-fiber-reinforced polymer nanocomposites under accelerated environmental aging. J Mater Eng Perform 29:7511–7525
7. Sen R, Mullins G (2007) Application of FRP composites for underwater piles repair. Compos B Eng 38(5–6):751–758
8. Chohan JS et al. (2020) Manufacturing techniques and applications of polymer matrix composites: a brief review. Adv Mater Process Technol 1–11
9. Tang LG, Kardos JL (1997) A review of methods for improving the interfacial adhesion between carbon fiber and polymer matrix. Polym Compos 18(1):100–113
10. Ebewele RO (2000) Polymer science and technology. CRC press
11. Dang Z-M et al (2012) Fundamentals, processes and applications of high-permittivity polymer–matrix composites. Prog Mater Sci 57(4):660–723
12. Brandrup J et al (1999) Polymer handbook, vol 89. Wiley, New York
13. Su DH (2014) Application of fiber reinforced composites for sports instruments. In: Applied Mechanics and Materials. Trans Tech Publ
14. Zhang J et al (2020) Current status of carbon fibre and carbon fibre composites recycling. Compos B Eng 193:108053
15. Lin G (2016) Global carbon fibre composites market report; ATA Carbon Fiber Tech. Guangzhou Co., Ltd.: Germany
16. Wang Y et al (2023) A review on new methods of recycling waste carbon fiber and its application in construction and industry. Constr Build Mater 367:130301
17. Chung DD (2016) Carbon composites: composites with carbon fibers, nanofibers, and nanotubes. Butterworth-Heinemann
18. Balasubramanian K, Sultan MT, Rajeswari N (2018) Manufacturing techniques of composites for aerospace applications. Sustainable Composites for Aerospace Applications. Elsevier, pp 55–67
19. Elkington M et al (2015) Hand layup: understanding the manual process. Adv Manuf: Polym Compos Sci 1(3):138–151
20. Singh MK et al. (2021) Microwave processing of polymer composites. Advanced Welding and Deforming 351
21. Arora G (2022) Microwave irradiation manufacturing of polymer composites: a sustainable manufacturing technique. Sustainable advanced manufacturing and materials processing. CRC Press, pp 9–32
22. Thostenson E, Chou T-W (1999) Microwave processing: fundamentals and applications. Compos A Appl Sci Manuf 30(9):1055–1071
23. Zhou J et al (2018) A multi-pattern compensation method to ensure even temperature in composite materials during microwave curing process. Compos A Appl Sci Manuf 107:10–20
24. Zhang L, Li Y, Zhou J (2018) Anisotropic dielectric properties of carbon fiber reinforced polymer composites during microwave curing. Appl Compos Mater 25:1339–1356
25. Özturk S (2010) Effect of fiber loading on the mechanical properties of kenaf and fiberfrax fiber-reinforced phenol-formaldehyde composites. J Compos Mater 44(19):2265–2288

26. Vallittu PK (2015) High-aspect ratio fillers: fiber-reinforced composites and their anisotropic properties. Dent Mater 31(1):1–7
27. Sayam A et al (2022) A review on carbon fiber-reinforced hierarchical composites: Mechanical performance, manufacturing process, structural applications and allied challenges. Carbon Lett 32(5):1173–1205
28. de Almeida SFM, ZdSN Neto (1994) Effect of void content on the strength of composite laminates. Compos Struct 28(2):139–148
29. Sethi S, Ray BC (2015) Environmental effects on fibre reinforced polymeric composites: evolving reasons and remarks on interfacial strength and stability. Adv Coll Interface Sci 217:43–67
30. Chollakup R et al (2011) Pineapple leaf fiber reinforced thermoplastic composites: Effects of fiber length and fiber content on their characteristics. J Appl Polym Sci 119(4):1952–1960
31. Sakib S et al (2023) Effect of fibre orientation and stacking sequence on properties of hybrid composites. Mater Sci Technol 1–13
32. Ahmadian H et al (2019) Effects of shape and misalignment of fibers on the failure response of carbon fiber reinforced polymers. Comput Mech 63:999–1017
33. Alshammari BA et al (2021) Comprehensive review of the properties and modifications of carbon fiber-reinforced thermoplastic composites. Polymers 13(15):2474
34. Kwak et al. (2011) Curing of composite materials using the recently developed hephaistos microwave. In: 18th international conference on composite materials
35. Xu X et al (2016) Effect of microwave curing process on the flexural strength and interlaminar shear strength of carbon fiber/bismaleimide composites. Compos Sci Technol 123:10–16
36. Xu X et al (2016) Improvement of the compressive strength of carbon fiber/epoxy composites via microwave curing. J Mater Sci Technol 32(3):226–232
37. Sharma M, Rao I, Bijwe J (2009) Influence of orientation of long fibers in carbon fiber–polyetherimide composites on mechanical and tribological properties. Wear 267(5–8):839–845
38. Ke L et al (2020) Effects of elevated temperatures on mechanical behavior of epoxy adhesives and CFRP-steel hybrid joints. Compos Struct 235:111789
39. Qiao L et al (2022) Mechanical properties, melting and crystallization behaviors, and morphology of carbon nanotubes/continuous carbon fiber reinforced polyethylene terephthalate composites. Polymers 14(14):2892
40. Yan B et al (2020) Experimental study on the mechanical properties of laminates made of thin carbon fiber plies. Compos Struct 245:112336
41. Dou H et al (2020) Effect of process parameters on tensile mechanical properties of 3D printing continuous carbon fiber-reinforced PLA composites. Materials 13(17):3850
42. Shahabaz S et al (2023) Effect of Al_2O_3 and SiC nano-fillers on the mechanical properties of carbon fiber-reinforced epoxy hybrid composites. J Compos Sci 7(4):133
43. Burkov MV, Eremin AV (2021) Mechanical properties of carbon-fiber-reinforced epoxy composites modified by carbon micro-and nanofillers. Polym Compos 42(9):4265–4276
44. Bisht A, Dasgupta K, Lahiri D (2019) Investigating the role of 3D network of carbon nanofillers in improving the mechanical properties of carbon fiber epoxy laminated composite. Compos A Appl Sci Manuf 126:105601
45. Kostagiannakopoulou C et al (2017) Synergy effect of carbon nano-fillers on the fracture toughness of structural composites. Compos B Eng 129:18–25
46. Adak NC et al (2018) Static and dynamic mechanical properties of graphene oxide-incorporated woven carbon fiber/epoxy composite. J Mater Eng Perform 27:1138–1147
47. Hashim MKR et al (2021) The effect of stacking sequence and ply orientation on the mechanical properties of pineapple leaf fibre (Palf)/carbon hybrid laminate composites. Polymers 13(3):455
48. Tarfaoui M, Choukri S, Nême A (2008) Effect of fibre orientation on mechanical properties of the laminated polymer composites subjected to out-of-plane high strain rate compressive loadings. Compos Sci Technol 68(2):477–485
49. Håkansson E, Amiet A, Kaynak A (2006) Electromagnetic shielding properties of polypyrrole/polyester composites in the 1–18 GHz frequency range. Synth Met 156(14–15):917–925

50. Thomassin J-M et al (2013) Polymer/carbon based composites as electromagnetic interference (EMI) shielding materials. Mater Sci Eng R Rep 74(7):211–232
51. Chung, D., *Electromagnetic interference shielding effectiveness of carbon materials.* carbon, 2001. **39**(2): p. 279–285.
52. Schulz RB, Plantz V, Brush D (1988) Shielding theory and practice. IEEE Trans Electromagn Compat 30(3):187–201
53. Bigg D (1987) The effect of chemical exposure on the EMI shielding of conductive plastics. Polym Compos 8(1):1–7
54. Jou W et al (2001) Electromagnetic shielding of nylon-66 composites applied to laser modules. J Electron Mater 30:1287–1293
55. Balageas D, Fritzen C-P, Güemes A, *Structural health monitoring*. Vol. 90. 2010: John Wiley & Sons.
56. Li L, Chung D (1994) Electrical and mechanical properties of electrically conductive polyethersulfone composites. Composites 25(3):215–224
57. Fouladgar J, Wasselynck G, Trichet D (2013) Shielding and reflecting effectiveness of carbon fiber reinforced polymer (CFRP) composites. In: 2013 international symposium on electromagnetic theory. 2013. IEEE
58. Jana P, Mallick A, De S (1992) Effects of sample thickness and fiber aspect ratio on EMI shielding effectiveness of carbon fiber filled polychloroprene composites in the X-band frequency range. IEEE Trans Electromagn Compat 34(4):478–481
59. Jana P, Mallick A, De S (1991) Electromagnetic interference shielding by carbon fibre-filled polychloroprene rubber composites. Composites 22(6):451–455
60. Cho J et al (2019) Enhanced mechanical and thermal properties of carbon fiber-reinforced thermoplastic polyketone composites. Compos A Appl Sci Manuf 126:105599
61. Yu G-C et al (2016) Thermal and mechanical properties of carbon fiber polymer-matrix composites with a 3D thermal conductive pathway. Compos Struct 149:213–219
62. Kandare E et al (2015) Improving the through-thickness thermal and electrical conductivity of carbon fibre/epoxy laminates by exploiting synergy between graphene and silver nano-inclusions. Compos A Appl Sci Manuf 69:72–82
63. Ouyang Z, Rao Q, Peng X (2022) Significantly improving thermal conductivity of carbon fiber polymer composite by weaving highly conductive films. Compos A Appl Sci Manuf 163:107183
64. Zhang J et al (2017) MgO nanoparticles-decorated carbon fibers hybrid for improving thermal conductive and electrical insulating properties of Nylon 6 composite. Compos Sci Technol 148:1–8
65. Dai W et al (2019) Metal-level thermally conductive yet soft graphene thermal interface materials. ACS Nano 13(10):11561–11571
66. Papanikolaou N (2008) Lattice thermal conductivity of SiC nanowires. J Phys: Condens Matter 20(13):135201
67. Wang T et al (2021) Simultaneous enhancement of mechanical and electrical/thermal properties of carbon fiber/polymer composites via SiC nanowires/graphene hybrid nanofillers. Compos A Appl Sci Manuf 145:106404
68. Kwon D-J et al (2017) Interfacial properties and thermal aging of glass fiber/epoxy composites reinforced with SiC and SiO_2 nanoparticles. Compos B Eng 130:46–53
69. Han S et al (2008) Enhancing the thermal conductivity and compressive modulus of carbon fiber polymer–matrix composites in the through-thickness direction by nanostructuring the interlaminar interface with carbon black. Carbon 46(7):1060–1071
70. Phonthammachai N et al (2011) Fabrication of CFRP from high performance clay/epoxy nanocomposite: Preparation conditions, thermal–mechanical properties and interlaminar fracture characteristics. Compos A Appl Sci Manuf 42(8):881–887
71. Rezaei F, Yunus R, Ibrahim N (2009) Effect of fiber length on thermomechanical properties of short carbon fiber reinforced polypropylene composites. Mater Des 30(2):260–263
72. Shi S et al (2022) Carbon fiber/phenolic composites with high thermal conductivity reinforced by a three-dimensional carbon fiber felt network structure. ACS Omega 7(33):29433–29442

73. Demchuk Z et al (2022) Unravelling the influence of surface modification on the ultimate performance of carbon fiber/epoxy composites. ACS Appl Mater Interfaces 14(40):45775–45787
74. Love LJ et al (2014) The importance of carbon fiber to polymer additive manufacturing. J Mater Res 29(17):1893–1898
75. Siddiqui NA et al (2011) Manufacturing and characterization of carbon fibre/epoxy composite prepregs containing carbon nanotubes. Compos A Appl Sci Manuf 42(10):1412–1420
76. Chi X et al (2021) Enhanced interfacial interactions of carbon fiber/epoxy resin composites by regulating PEG-E51 and graphene oxide complex sizing at the interface. Polym Adv Technol 32(9):3458–3473
77. Weitzsacker CL, Xie M, Drzal LT (1997) Using XPS to Investigate Fiber/Matrix Chemical Interactions in Carbon-fiber-reinforced Composites. Surface and Interface Analysis: An International Journal devoted to the development and application of techniques for the analysis of surfaces, interfaces and thin films 25(2):53–63
78. Yuan H et al (2012) Effect of surface modification on carbon fiber and its reinforced phenolic matrix composite. Appl Surf Sci 259:288–293
79. Zhang W et al (2019) Improving interfacial and mechanical properties of carbon nanotube-sized carbon fiber/epoxy composites. Carbon 145:629–639
80. Zabihi O et al (2016) One-step amino-functionalization of milled carbon fibre for enhancement of thermo-physical properties of epoxy composites. Compos A Appl Sci Manuf 88:243–252
81. Feng P et al (2021) Enhanced interfacial adhesion of carbon fiber/epoxy composites by synergistic reinforcement with multiscale "rigid-flexible" structure at interphase. Compos B Eng 225:109315
82. Khurshid MF et al (2020) Recent developments in the processing of waste carbon fibre for thermoplastic composites–A review. J Compos Mater 54(14):1925–1944
83. Kim Y et al (2020) Origami-inspired reforming method for carbon fiber-reinforced thermoplastics via simple thermal stitching. Compos B Eng 193:108043
84. Ishikawa T et al (2018) Overview of automotive structural composites technology developments in Japan. Compos Sci Technol 155:221–246
85. Ahmed A, Wei L (2016) Introducing CFRP as an alternative material for engine hood to achieve better pedestrian safety using finite element modeling. Thin-Walled Structures 99:97–108
86. Yu K et al (2017) A review of the design and analysis of reinforced thermoplastic pipes for offshore applications. J Reinf Plast Compos 36(20):1514–1530
87. Mishnaevsky L Jr et al (2017) Materials for wind turbine blades: An overview. Materials 10(11):1285
88. Durai Prabhakaran R (2014) Are reactive thermoplastic polymers suitable for future wind turbine composite materials blades? Mech Adv Mater Struct 21(3):213–221
89. Giorgini L et al (2020) Recycling of carbon fiber reinforced composite waste to close their life cycle in a cradle-to-cradle approach. Current opinion in green and sustainable chemistry 26:100368

Application of Microwave Processing in Orthopedic Implants

Amrinder Mehta and Hitesh Vasudev

Abstract In the not-too-distant future, microwave energy will be one of the energy sources that is utilized most frequently in processing various materials. This possibility exists because microwaves can produce high levels of heat. During the processing that is done with microwaves, electromagnetic energy is transferred directly to the material at the molecules' level through various interactions. During the processing of materials using a microwave, electromagnetic energy is transferred to the materials themselves directly and involves the exchange of their molecules. Because of the inherent benefits that it provides in terms of time and energy efficiency, significantly shorter process cycle times, improved considerably mechanical properties, and reduced environmental risks, the use of microwaves to assist in the processing of materials has recently become increasingly popular and is expected to continue to grow in popularity soon. Modern medical practices extensively use cutting-edge equipment and techniques made possible by technological progress. In several areas of medicine, biomaterials are utilized to replace or repair damaged biological tissues. Most biomaterials are processed by heating them using a variety of conventional and cutting-edge methods. However, these methods still require development so that necessary product attributes may be attained, and processing times can be decreased. It has been widely reported that treating materials with microwaves saves both time and energy without lowering product quality. Hence technology has considerable promise. However, there hasn't been a lot of research on using this method with biomaterials. New developments in the utilization of a wide variety of biomaterials in arthroplasty and their processing using microwave energy are discussed here.

Keywords Microwaves · Energy sources · Electromagnetic energy · Processing · Materials · Biomaterials

A. Mehta
Division of Research and Development, Lovely Professional University, Phagwara 144411, India
e-mail: amrindermehta896@gmail.com

H. Vasudev (✉)
School of Mechanical Engineering, Lovely Professional University, Phagwara 144411, India
e-mail: hiteshvasudev@yahoo.in

M. K. Singh et al. (eds.), *Composite Materials Processing Using Microwave Heating Technology*, Composites Science and Technology,
https://doi.org/10.1007/978-981-97-2772-8_12

1 Introduction

The growth of science and technology has made it possible to quicken the pace of multidisciplinary research in a range of fields, such as health, materials science, and manufacturing technology, which has ultimately led to an increase in the amount of goods that are manufactured. The field of medical science is one of the primary areas that has been one of the primary contributors to making significant strides toward improving the overall quality of life [32]. These advancements have been made possible because of advancements in pharmaceuticals, biomaterials, and other novel biomedical instruments (BMs). The key goals of research into advancements in BMs are to increase the amount of time that body transplants can be used and to make it easier to replace them. However, throughout the course of the previous 500 years, there has been a gradual but consistent improvement not only in the BMs themselves but also in the procedures that are used to process them. In recorded documents, dental implants made of stone, gold, and ivory are described for the first time probably in the sixteenth and seventeenth centuries. During that time, natural biomaterials like as animal and human bones were utilized in the process of tissue transplantation and repair in order to treat illness and injury [30]. This was done so that patients might be healed from their wounds and illnesses. In the second half of the eighteenth century, this field of study, which included the investigation of a variety of devices and implants designed for use within the human body, came to be known as implantology. Implantology embraced all these topics and more. It was primarily for the purpose of gaining experience in the replacement and integration of missing body components as well as the support of a damaged structure within the human body that these implants were put to use in a variety of medical procedures and traumas [43]. This was accomplished through the use of a variety of medical procedures and traumas.

The process of modifying prosthetic surfaces has a long and illustrious history. As seen in, four distinct generations have descended from it (Fig. 12.1). Scientists and medical professionals alike are intrigued by "New generation coatings" that either encourage more bone formation or have greater biological functionality than thermally sprayed HAp. These coatings are of particular interest [38].

As a direct result of this, composites and ceramics make superior candidates for implants that are designed for hard tissue, whilst polymers and new metals are superior candidates for implants that are intended for soft tissue. In the field of medicine, the process of fabricating implants for hard tissue is used by several different BMs. On the other hand, natural materials do not meet the requirements for biocompatibility, and the development of artificial materials that followed was impacted by interactions with the host tissues [41]. As shown in (Fig. 12.2) to maximize an implant's resistance to corrosion, the discharge of metal ions from a metallic component located in the harshest conditions of the body should be minimized. After being implanted, the stent in the patient's heart began to corrode.

The complexity of the demands placed on patients and doctors has resulted in piecemeal changes being made to BMs and the ways in which they are processed.

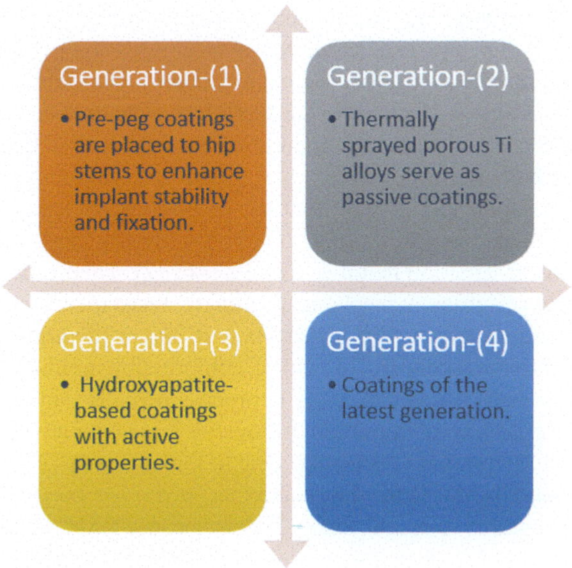

Fig. 12.1 The many different coating eras that have been utilised for the surface modification of prosthetics

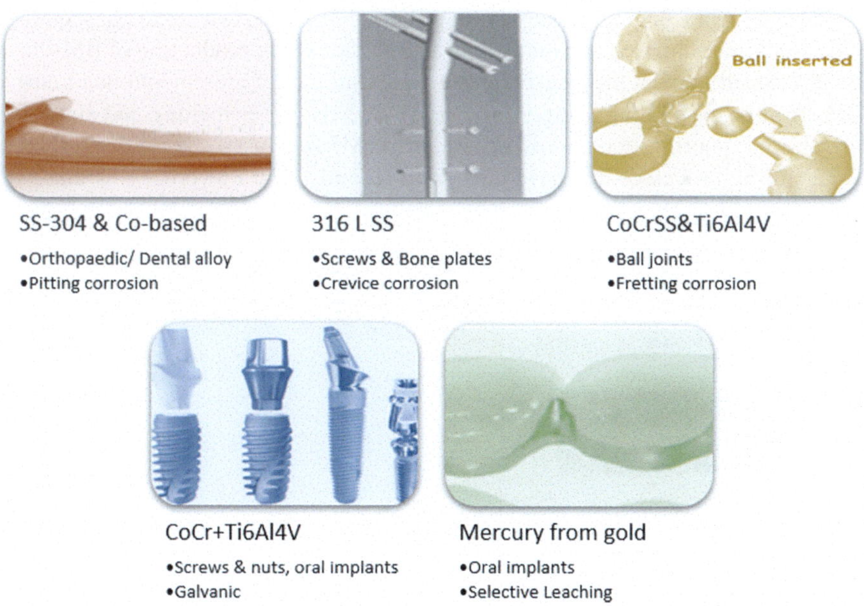

Fig. 12.2 Categorization of corrosion in conventional techniques that are utilized in clinical settings for the purpose of making biomedical implants

The synthetic BMs need to be able to live with the host tissues as well as the biological fluids that are responsible for ensuring that the pH level in the body is kept at the appropriate level [48]. In a manner that is akin to this, different tissues have different stress values. depending on the activities that are conducted daily by the individual. When conditions are normal, the hip joint is subjected to three times the weight that the body does, however, when situations involve hopping, the hip joint is subjected to ten times the weight that the body does. The way in which stress goes through cycles and shifts as a function of the activities that are being carried out is one of the most fascinating features of stress [46, 47]. Therefore, it is essential for biomaterials to have sufficient mechanical strength, resistance to corrosion, and biocompatibility.

As shown in (Fig. 12.3) Many different methods, including material synthesis, slip casting, centrifugal casting, compression molding, and powder metallurgy (PM), have been used historically to process BMs. Because of the granular nature of these criteria, several important discoveries and major advancements in the field of biological sciences are necessary. Due to the necessity of qualities and properties of such materials, the field of biomaterials (BMs) has become one of the most active fields of research in recent times. This is because BMs are used in a wide variety of applications [27]. As a direct result of this, the applications of BMs are fast expanding to encompass the manufacturing of surgical instruments and other sorts of equipment, as well as the replacement of many types of human tissues. As a direct consequence of this, both time-honored and innovative approaches to the production of BMs have been developed to meet the ever-increasing demand for the aforementioned items.

Electron-beam melting, stereolithography, selective laser melting and sintering, spark plasma sintering, and microwave sintering (MS) are just some of the alternative manufacturing processes that have been tried recently. The PM process is made up

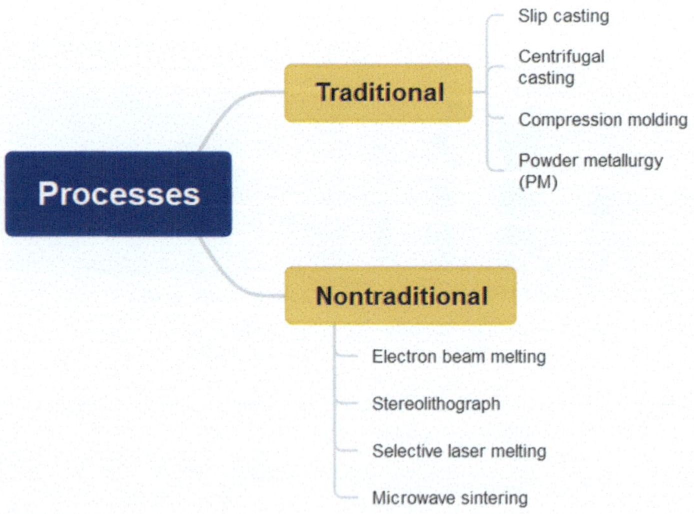

Fig. 12.3 Material processes techniques

of many kinds of fabrication methods that are all performed in sequence. The initial phase of this process is referred to as "sintering," and it includes compressing a green substance while exposing it to high temperatures (Rabiee et al. 2021). This step is followed by further phases of the process. This heating can be achieved through the application of either tried-and-true or innovative sintering processes, depending on one's preference.

According to the findings, additive manufacturing can be used to create porous structures out of polymers as well as composites of polymers and ceramics. In the field of tissue engineering, porous structures are utilized to a significant degree. In addition, it was discovered that the spark plasma sintering technique is best suited for the manufacturing of goods with a limited surface area. This was discovered because of constraints placed on the size of the die, which led to these findings. On the other hand, the field of biomedicine has increased the amount of research that it does on MS and has carried out extensive studies on the disease (Bin et al. 2020). Because of its one-of-a-kind heating mechanism and minimal influence on the environment around it, the MS is a cutting-edge method of sintering that is at the forefront of the industry. The production of high-temperature ceramics was what first led to the development of this technique. It was also noted that while treating metals with a microwave, each variety of metal forms plasma on its surface and reflects microwave radiation. This fact was mentioned in the previous point. Plasma stops metals from heating up because microwaves operating at ambient temperature are unable to permeate the material to sufficient depths. However, more recent research has shown that microwave energy may be utilized in the treatment of metal particles (Pal et al. 2022). This opens new possibilities for the field. It was reported at the turn of the microwave sintering (MS) resulted in a higher density in addition to improved mechanical and metallurgical properties when compared to conventional sintering. This contrasted with the results obtained by conventional sintering (CS). Because of this, it is possible that MS will turn out to be the superior option when it comes to the manufacturing of a variety of BMs. On the other hand, information regarding this topic has not been reported on a very consistent basis. BMs need to possess a certain set of qualities, such as low levels of friction and a prolonged lifespan that may be anticipated for the implants. The unique demands of a patient will determine which traits are most important [25]. These characteristics are influenced by a wide range of factors, some of which are the processing method and the parameters of the processing, among other things. In this overview article, we take a cursory glance at the many different types of artificial BMs that are often utilized in hard tissue implants, as well as the historical evolution of these types of implants up until this point in time (Rabiee et al. 2022). Several different theories and equations are utilized in the process of explaining the fundamentals of microwave material processing. Each of these theories and equations explains the way microwaves interact with a particular kind of material. In addition, research has been done to investigate how the characteristics of biomaterials created expressly for use as orthopedic implants are altered using microwave energy.

2 Biomaterials

Biomaterials are substances, either natural or artificial, that have been produced to interact with the tissues of the host without releasing any harmful or allergenic compounds. Biomaterials can be either natural or synthetic. It is common practice to employ them in order to replenish and rebuild human body tissues that have been damaged as a result of a variety of diseases [12].

Because of the nature of the BMs sector, there have been several advancements in the material realm, as well as the invention of novel graded materials. Metal implants came into being as a direct and immediate consequence of considerable technological developments made in BMs. The selection of a suitable material that is also capable of providing biological functions is one of the most critical variables that goes into determining how long implants will continue to work normally. Metals, ceramics, polymers, and the composites of these three types of materials are all examples of biomaterials that can be utilized in arthroplasty; each category of material has its own unique set of properties [33]. These biomaterials behave in a manner that is analogous to that of engineered materials, except for the qualities that contribute to their biocompatibility and render them appropriate for use in biomedical settings. As shown in (Fig. 12.4) different kinds of organic biomaterials were found in nature. Because of this, the processing that they go through is comparable to the processing that is done with engineering materials, both of which need additional characterization in order to ensure that they are compatible with the tissues of the human body. One of the most important aims of research and development efforts relative to advancements in material processing technologies is the creation of better commodities at reduced processing costs [37]. This goal is one of the most important goals of research and development initiatives. Most of the time, manufacturers hunt for cutting-edge technology that will enable them to produce the goods that customers want in a manner that is faster, better, and more reasonably priced. Because of this, they take into consideration contemporary processing methods, which can increase product quality in terms of both efficiency and cost. The processing of BMs in today's world makes use of a diverse array of different methods [49].

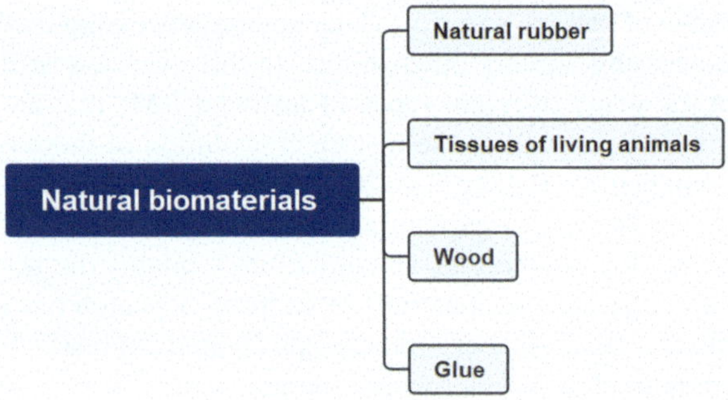

Fig. 12.4 Varieties of natural biomaterials

Processing advanced materials can be done in several different ways, each of which has a unique combination of advantages and disadvantages. Because biomaterials are required to have attributes, such as acceptable porosity, sufficient strength, and an elastic modulus, the processing procedures that are employed on them need to be specialized. These features include suitable porosity, sufficient strength, and an elastic modulus. It is important for the biomaterials that are utilized in orthopedic implants to have porous characteristics on their exteriors. As a result, this will make it possible to create a strong relationship between the orthopedic implant and the host tissues, in addition to promoting the synthesis of bone and tissue around the implant [10]. These kinds of biomaterials need to furthermore have a functionally graded elastic modulus to be able to withstand the stress shielding effects, which are a significant problem in arthroplasty due to the fact that they increase the risk of failure and surgery. The transplant is going to end up being more expensive because of all these problems, which is going to make it challenging to find the money to pay for it. Even though powder metallurgy is one of the more time-honored manufacturing processes, adding the more modern sintering approach can assist you in accomplishing your goals and achieving these characteristics. The properties of the various biomaterials that were considered for use in orthopedic implants as Shown in (Fig. 12.5). This is because sintering is a process in which powder particles are heated to a point where they fuse together [16].

When utilized with the optimum density, this technology has the potential to manufacture products that are essentially net-shaped, practically free of waste, and possess remarkable mechanical and metallurgical qualities.

Adjusting the processing variables that govern the manufacturing process allows for this method to also be used to fabricate components made of metals, ceramics, polymers, and their composites with a quality that is tailored to the specifications of

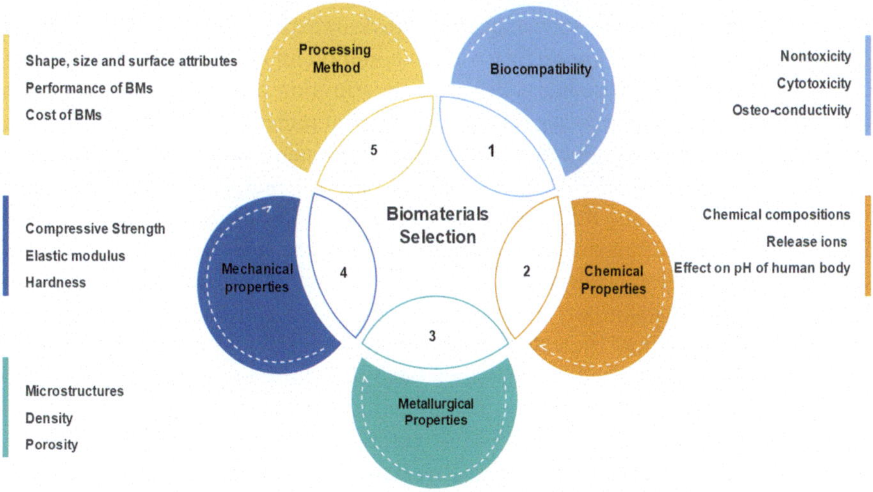

Fig. 12.5 Characteristics of the biomaterials chosen for orthopedic implants

the user. This is accomplished by changing the values of the processing variables. These variables include the composition of the material, the pressure applied during compacting, and the temperature during sintering. The energy required to carry out the sintering process can come from a wide number of places, such as spark plasma, electric induction heating, resistance furnace heating, or microwave radiation [8]. During the sintering process, the temperature of the material will be increased either directly or indirectly in one of these two ways. The first approach involves moving heat from the energy source and the material being processed to the surface of the material being targeted for processing, and then from there to the interior core of the material being targeted for processing. This method is used to heat the material. On the other hand, the mechanism for indirect heating takes advantage of induction as well as the warmth produced by microwave radiation. This type of heating is referred to as "radiant heating." By converting the energy into heat energy and using electromagnetic waves as a catalyst, the substance is heated up. This is performed with the help of electromagnetic waves. "Volumetric heating" refers to the type of heating that occurs at the atomic level because of microwave heating. This type of heating is caused by microwaves [3].

3 Microwave Heating Mechanisms

When electric and magnetic fields that are moving in the opposite direction from one another combine, the result is a phenomenon known as microwave radiation. The preparation of food and the transmission of information are the two main areas in which microwaves have been discovered to be useful. After that, people started applying heat to materials like ceramics and cement, which needed a very high temperature to melt. Microwaves are a form of electromagnetic radiation that, when they come into touch with the substances they are designed to heat, transform into heat [26]. Microwave ovens are commonly used in commercial kitchens. Because of this, the characteristics of a substance will be the determining factor in how it responds when it is subjected to microwaves. Different types of materials will have distinctive responses when exposed to microwaves. Major components of the microwave material interaction include the dielectric loss in non-magnetic materials, eddy current and hysteresis loss in magnetic materials, and hysteresis loss in magnetic materials. It is more common in materials that are not magnetic to experience dielectric loss [50]. A reflector is defined as an electrically conducting material that can reflect all incident microwaves without absorbing any of the energy from those microwaves. Because these materials have small dielectric losses, microwave radiation can travel through them without being absorbed because they have the same properties as transparent materials. Microwave absorbers can take up incident microwaves and remove them from the system, even though they have a significant dielectric loss. In microwaves, on the other hand, the manipulation of materials that are at room temperature and transparent or reflective has led to the creation of several distinct strategies. Direct heating is accomplished by having microwave radiation interact directly with the material that is being heated. This kind of heating is known as convection heating [31]. The mechanism by which heat passes through a substance when it is irradiated by microwaves as well as by conventional light is referred to as thermal conduction, as shown in (Fig. 12.6). As a result of direct interaction, the temperature rises, which in turn causes the material to overheat, which in turn causes changes in temperature, uneven heating, and the formation of cracks. Direct interaction also causes the material to overheat. Because of this, in addition to the processing of metallic powders that have a high loss factor, this method is also effective when used to treat ceramics and food. Because of these constraints, the employment of various methods of heating, such as selective heating and hybrid heating, is required [29]. Because the other sections of the material are masked, the only part of the material that gets heated by this procedure is the area that is supposed to be heated. Other parts of the material remain at their original temperature. The issues that are brought on by direct heating inspired the development of a method of heating known as hybrid heating. When carrying out this method of heating, susceptors, materials with a high dielectric loss factor, and masking materials are all utilized [21].

The effectiveness of these various heating processes is significantly impacted by the characteristics of the substance being heated. The mechanism that causes biomaterials, such as metals, ceramics, polymers, and their composites, to heat up is determined by the dielectric properties of the biomaterials as well as the type of crystal structure they have.

The exchange of energy that takes place between microwaves and the substances that they travel through can be explained using a variety of different fundamental equations and theories. These ideas explain how electromagnetic fields are distributed throughout the process, as well as how electric and magnetic fields interact with the material that is being processed. The values of complex permittivity and complex permeability in magnetic and non-magnetic materials, respectively, have an effect on the degree of material interaction that takes place during the process of microwave material processing [17]. After that, many thermoset and thermoplastic-based polymer composites were cured in a microwave, and it was discovered that the mechanical properties of the cured material were greatly improved as compared to the usual technique. Most polymers are poor microwave absorbers, hence, selective

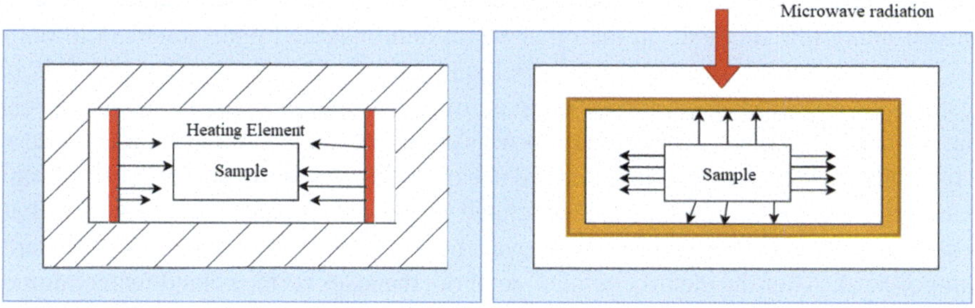

Fig. 12.6 Heat transmission mechanism in a sample during conventional heating and microwave heating

heating with microwaves was utilized to cure the polymers. The characteristics of polymers made it possible to achieve this result. In the past, microwave heating was also utilized on occasion for the treatment of more complicated materials. Particularly in orthopedic treatments, there was an inordinate degree of trust placed on the use of metallic implants for joint replacement [24]. This was especially true in younger patients. Steel is a material that is widely used for implants because it possesses a range of desirable qualities, including sufficient strength. One of the other reasons that steel is regularly used for implants is because it is inexpensive. Because of this, steel is a material that is frequently used for implants. The scientific community is currently conducting research on stainless steel in the hope that it will one day be suitable for use as a material for orthopedic implants. This is because stainless steel contains a high percentage of nitrogen and it does not have any negative side effects. Because of the one-of-a-kind characteristics it possesses, such as its chemical similarity to real bone, hydroxyapatite (HA), which is found in bioactive material, has garnered a lot of attention in recent years. One of the reasons why is because of this similarity. Because of the mechanical similarities that exist between HA and the substance that makes up human bone, coatings made of HA are frequently used on orthopedic implants. This is due to the fact that HA is a biocompatible material [11]. In addition to this, it has been demonstrated that HA coatings are biocompatible with the bone tissues of humans. Coatings are an essential component of bioactive behavior because they encourage the formation of new tissue and increase the material's resilience to corrosion. Coatings also play a role in bioactive behavior. Because of this, the quantity of metal ions that are released into the biological systems that are in the immediate region is decreased. Steel possesses a high mechanical strength, while HA has the potential to provide an adequate level of biocompatibility. This is why the combination of the two materials is advantageous. The creation of HA-based coatings on implant devices is currently accomplished using plasma spraying, which is a process that is both widely utilized and commercially accessible. This is as a result of the fact that plasma spraying is capable of producing coatings that are uniform and have a relatively high density when compared to other ways of coating, such as those that were discussed earlier in this paragraph [44]. In this method, the creation of bonds between the powder particles took place because of the impact of high-velocity particles, as well as the rapid cooling of the partially melted and melted powder particles that took place during the process. The coating that was formed using this method, on the other hand, demonstrated poor adhesion to both the substrate and the coatings, and it is likely to rip and wear away soon. This is because the coating was produced using an improper method. The undesired debris that is formed is a result of the natural process of wear and tear, however this process is typically unnecessary for the extended use of these coatings in a biological setting. In addition, the coatings that are generated using this method have a shape that is not constant and have poor control over the thickness of the layers that they are composed of. After the electrochemical method, the laser surface cladding technique is currently in the position of being in second place when it comes to the practice of placing bioactive coatings on metallic bioimplant material. Utilizing this technology in the metalworking sector carries with it several benefits, some of which include adaptability in the process, little surface preparation, efficient diffusion bonding with the substrate, and precision in the deposition of coatings. Nevertheless, this approach does have a few important drawbacks, the most significant of which is that it results in extremely high operational costs [15]. The extremely brittle character of pure HA coatings is to blame for the poor mechanical performance of the as-sprayed HA coating over surgical grade stainless steel as the substrate material. This poor performance was caused by the coating's interaction with the substrate. However, the findings of studies carried out by a wide number of researchers suggest that the incorporation of ceramic particle reinforcement may have the ability to improve the mechanical stability of HA matrix. To determine whether it is possible to coat the surface layer of steel with HA that is reinforced using the hybrid MH method, the purpose of the work that is being done right now is to determine whether it is possible to do so in order to determine whether or not this can be done in a practical manner. It is anticipated that the incorporation of HA into Al_2O_3 will result in an enhancement in the bioactivity of the composite modified layer that was created by hybrid MH. This is the outcome that is being looked forward to. The matrix of the molten substrate was used as a container for the molten Al_2O_3 and HA that was encapsulated using this approach [7]. During this procedure, melting was performed on both of these different kinds of material. Because of this, the metallurgical bonding that needs to take place between the reinforced Al_2O_3 particles, the molten substrate material, and the HA can take place using this approach.

4 Effect of Post-Heat Treatment

With the assistance of the post-heat treatment, the crystallinity of the HA coating can be improved. The influence that post-heat treatment has on the characteristics of high-performance oxyfuel (HVOF) coatings that have been sprayed via high-velocity spraying. Crystallization took place in the amorphous phase, as the research has shown. Additionally, preheating the substrate before depositing the HA coating on it is an efficient method for increasing the crystallinity of the coating that was formed by plasma spraying. This is done before the coating is deposited on the substrate. According to studies that were carried out in the past, preheating the substrate led to an increase in the crystallinity of the HA coatings that were applied. A lot of study has been done on how the post-heat treatment influences the phase structure of MPS coatings, particularly how it influences the preferred orientation [14]. The MPS procedure was used to apply three distinct coating types, each of which had a different crystallinity, to a substrate. This was carried out to determine the effect that the post-heat treatment had on the varying degrees of crystallinity as well as the strength of the changes in preferred orientation. Tetra calcium phosphate (TCP) and octa-calcium phosphate (OCP), both of which are undesirable phases, have a possibility of forming in the high temperature-induced layer of steel that has been microwave-modified with HA powder in the hybrid MH process. Both phases have a chance of forming in the high temperature-induced layer of steel. The terms "TCP" and "OCP" are used to refer to these phases, respectively. This is because these phases break down at an accelerated rate. The outcomes of this research suggest that the impact of these phases can be greatly reduced by making use of an adequate post-heat treatment.

The usual XRD spectra of samples with microwave-modified surfaces and starting HA powder in a range of various settings each with their own set of parameters as shown in (Fig. 12.7) [40]. The XRD spectra of samples with microwave surface modifications revealed many additional peaks corresponding to unique reaction-induced phases [40]. These peaks were found in addition to the primary HA structure and the Ni–Fe-based (fcc) structure in the spectra. These peaks all corresponded to various stages that were caused by the reaction. These one-of-a-kind phases may have been the result of a reaction that took place at a high temperature between molten HA and steel. The reaction may have taken place as a byproduct. During the MH step, both the HA powder and the surface layer of the substrate based on Fe were totally melted. Because of this, after the changed surface layer has been rapidly quenched, an interaction takes place between the various components present in the surface layer of the melted substrate and the dissolved HA, which results in the creation of new phases. This interaction is what leads to the formation of the new phases [23]. During microwave exposures, it was found that both the steel surface layer and the HA powder had entirely melted. This was a surprising discovery. [Further citation is required] As a result, two distinct liquids are produced: a liquid metal and a liquid mineral. There is no way that these two liquids could ever be mutually soluble with one another due to the fundamental disparities that exist between their bonding properties. The molten material will develop dendrites as it cools, which will result in the formation of an austenite matrix with a foundation composed of Fe and Ni. Calcium and phosphorus, two of the most important elements found in the HA. As a result, the liquid that was left over became increasingly rich in Ca and P as more austenite developed. Calcium and phosphorus are two of the most important elements found in the HA (Muthaiah, Indrakumar et al. 2022).

Because of this, when the operation for solidification was ultimately accomplished, the interdendritic zones were the key areas where HA and a great deal of other reaction-induced chemicals appeared. In addition, when comparing samples that were treated to high-temperature heat treatment with samples that had microwave surface modifications applied while they were being deposited, similar phases are still present. This can be seen when contrasting the two types of samples. Because of the microwave modification to the surface, the presence of these phases was significantly reduced.

As-sprayed HA coatings' crystalline and textural intensities are influenced by subsequent heat treatment, as demonstrated in (Fig. 12.8).

It showed that after heat treatment, the texture intensity of coatings applied via spray decreased. Therefore, it may be inferred that during the recrystallization of ACP and the conversion of impurity phases into HA, the development of grains with a preferred orientation did not take place.

Before and after the Simulated Body Fluid (SBF) immersion test, the weights of the as-deposited, heat-treated, and unaltered SS-316L were measured and compared

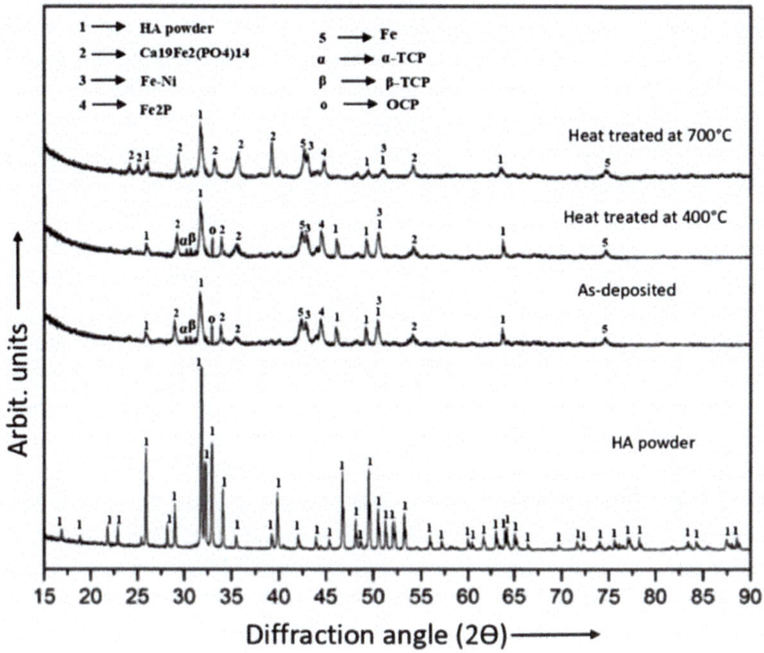

Fig. 12.7 X-ray diffraction patterns of **a** HA powder, **b** an as-deposited modified substrate, **c** a modified substrate that has been heated to 400 °C, and **d** a modified substrate that has been heated to 700 °C [40]

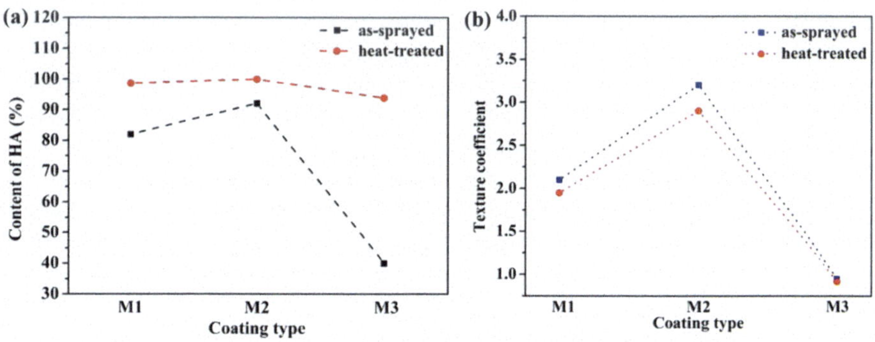

Fig. 12.8 The effect of **a** acrystallinity and **b** preferred orientation intensity changes after heat treatment [19]

to determine the growth of the secondary apatite layer. This was done so that the growth of the secondary apatite layer could be computed. This was carried out so that the rate of expansion of the secondary apatite layer could be calculated. As Shown in (Fig. 12.9) a micrograph is utilized to demonstrate the existence of secondary apatite on the surfaces of the tested, modified, and unmodified substrates that SEM analyzed.

It was found that increasing the temperature of the heat treatment led to a reduction in the ability to create apatite. This was a surprising finding. During this inquiry, the internal section of the steel served the function of an anchor for the implant, while the surface layer of the steel was modified to serve the purpose of a biocompatible material. Both functions were carried out simultaneously. Microstructures of the sample, as seen through a cross-section with an optical microscope, after the surface of the sample was altered by microwaves. Both the modified substrate layer and the unmodified substrate have been demonstrated to be capable of forming a robust metallurgical bond with one another [5, 23]. The interface that is produced by the microwave cladding approach is one that is both transparent and diffused, and it is produced at the point where the surface-modified layer meets the substrate that has not been altered in any way. On the other hand, conventional thermal spray methods leave behind HA coatings with a sharp interface, which is often inappropriate for applications that are designed to continue functioning for a significant amount of time. In either the layer of composite material that was generated or the contact between the solid and the liquid, there are no visible cracks or other types of mechanical discontinuities. In addition, the entire changed solidified layer has a microstructure that is primarily cellular. This is because the hybrid MH process is related to a phenomenon of heating that is both uniform and volumetric. After the radiation was switched off, the molten composite layer stopped converting microwave energy into heat energy and instead began to lose heat. This occurred because the layer was no longer able to convert microwave energy into heat energy. This took place because of the layer losing its ability to convert microwave energy into heat energy. Because the molten composite layer communicated a large quantity of heat to the unaltered substrate, the substrate quickly cooled after the transfer took place. Because of its higher thermal conductivity, the unaltered substrate acts as a heat sink for the molten composite layer as it cools and develops on top of it. This occurs as the layer is being formed. The path that heat will travel is the most important aspect that will determine the way in which cellular development will proceed. [Cellular Development] Due to the fact that the temperature difference was greatest at the boundary between the solid and the liquid in this scenario, cell development will take place in this particular direction. The composite layer starts to adopt the form of an extended cell and a regular hexagonal array as it moves further and further away from its point of origin. This occurs when the layer develops further and further away from its starting point [40].

When compared to the microstructure of the as deposited and heat-treated layer that was created at a lower temperature, the microstructure of the composite layer that was manufactured at a higher temperature exhibited a more uniform microstructure. This occurred because, after being subjected to heat treatment, the composite layer that was created recrystallized. The high temperatures aided in both the diffusion and the recrystallization processes, with the diffusion process being responsible for the creation of HA phases at the grain boundaries [42]. The recrystallization process was enhanced by the high temperatures.

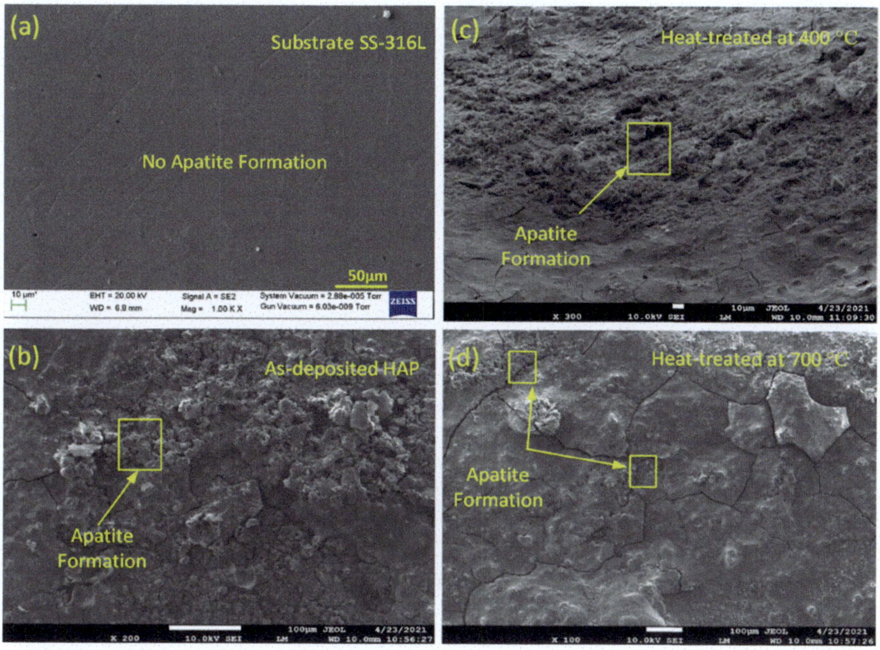

Fig. 12.9 Scanning electron micrographs taken after being submerged in SBF: **a** unmodified SS-316L, **b** as-deposited surface modified substrates, **c** heat treated modified substrates at 400 C, and **d** heat-treated modified substrates at 700 C [40]

5 Function of Microhardness in Microwave Surface-Modified

Porosity and microhardness are two additional key features that have a significant impact on the qualities of the changed layer. It is conceivable for a material to have higher rates of wear and corrosion because of having a bigger porosity, while at the same time experiencing a decrease in its microhardness. When this happens, the porosity of the coatings increases, which in turn causes a proportional decrease in the amount of hardness possessed by the coatings. The porosity, on the other hand, has a detrimental effect on the improved composite layer's mechanical capabilities. In order to enable the growth of bone into the surface-modified layer, a specific level of porosity is required [1]. This, in turn, speeds the patient's healing process and shortens the amount of time that is required for recovery. Even though a higher porosity percentage assists in the initial attachment of the bone, it is unknown whether a high porosity content will have an effect on the long-term mechanical durability of HA coatings. This can be accomplished by making certain that the material's porosity content is distributed uniformly across its whole surface area [45].

As shown in (Table. 12.1) prior to the presentation of this research, analyses on the porosity and microhardness of samples that had microwave surface modifications

Table 12.1 Mechanical properties of Reinforcement micro composites

Reinforcement elements	Particle size (μm)	Elastic modulus (GPa)	Hardness (HV)	Ref
Al_2O_3 + HAP (15 vol.%)	<50–60	44.7	255	[20]
Mg + Al2O3 (10.9 vol.%)	<35	50.8	51.7	[9]
(Mg, Ti 10 wt.% and Al 1 wt.%)	<42	12	55	[51]
Mg + ZnO	<10	–	65	[4]

and that had been heat-treated and deposited were carried out. These analyses were performed on the samples. In addition, it has been found that the quantity of porosity decreases in a proportional manner as the temperature of the heat treatment increases. This was discovered in the study that was mentioned before.

6 Biomaterials Microwave Processing Challenges

Because of the specific properties of the materials that are being processed by microwaves, such as their thermal conductivity, electric permittivity, magnetic permeability, the amount of dielectric loss, of the material that is being processed, etc., microwave processing of materials is an individualised process that is also a challenging phenomenon. These properties include: thermal conductivity; electric permittivity; magnetic permeability; heat capacity; size and shape of the material being processed; etc. Because of their high electrical conductivity and reflective nature, bulk metallic materials bring their own set of issues to the table when it comes to microwave treatment [22]. In addition to this, fire can easily start from these materials when they encounter the sharp corners of the target block.

Processing metal powder, on the other hand, is an uncomplicated process since the size of the powder particles is quite close to being the same size as the size of the penetration depth. According to the findings, the nanopowder possesses remarkable mechanical and metallurgical properties, in addition to a dense microstructure. As shown in (Fig. 12.10) these qualities can be found in the nanopowder. In addition to this, it is believed that the nanopowder can reach the desired temperature in a speedy and efficient manner. Because of the large amount of surface area that micro- and nano-sized particles have, they respond to microwaves in a manner that is similar to that of susceptors when they are exposed to the energy from the microwaves. These particles are utilized in the process of reinforcing composites. As a result, the manufacturing process must contend with a number of obstacles brought on by these two characteristics [6]. Similar to how this works, the vast majority of biopolymers do not generate heat when they are exposed to microwave radiation. This is because of the structure of the polymer, in addition to the exceptionally poor electrical or heat conductivity of the biopolymers. Technicians and researchers are confronted with a wide variety of difficulties whenever BMs are being processed as a result of the cumulative effects of all of these factors. The construction of the equipment, the characteristics of the biomaterials, and the processing carried out with microwaves all contribute to these difficulties. These difficulties can be separated into two distinct categories, and we will examine each of those categories in turn in the following sections [18]. The application of a processing method that makes use of heating via microwaves is one that is both highly quick and very effective. On the other hand, this method has a number of significant drawbacks, such as the development of internal tensions and the uneven heating of materials, both of which lead to the deterioration of the material's thermal properties. There hasn't been a lot of research done on the microwave processing of BMs because there are a lot of obvious issues involved, such as the diverse composition of the materials, the retention of some specific properties, changes in processing temperature, etc. There hasn't been a lot of research done on the microwave processing of BMs because there are a lot of evident issues involved. It is sometimes necessary for BMs to have a higher density while simultaneously retaining a sufficient level of strength and, on the other hand, a porous structure, which is extremely challenging to produce [13]. This creates a considerable obstacle for the production process. Because there is a paucity of experimental data, it is difficult to mimic the process. This is one of the problems that limits the application of microwave processing in the biomedical sector. Another factor is the high cost of using microwaves. The construction of biocomposites involves the utilization of a wide range of materials, each of which brings with it a distinct collection of qualities. It can be challenging to process materials that have non-uniform temperature profiles, non-uniform field intensities, and non-uniform energy consumption [36]. This is because most of these materials do not have temperature profiles that are uniform. Because of this, there is a high possibility that microwaves are the source of non-uniformity and heterogeneity in the final items. This is because microwaves heat food at different rates. The tool design for many microwave operations, such as sintering, differs depending on the target material. This is due to changes in the material's characteristics as well as a lack of data that is adequate for the technique. This is the case despite the fact that there is a paucity of data. Because of the one-of-a-kind qualities of the commodities that are used to replace human body parts, the processing of BMs requires the creation of specialized equipment [2]. This is because BMs are used to replace human body parts.

7 Conclusion and Future Scope

In this review study, the most current discoveries pertaining to HAp-based biocomposites and bioalloys were discussed. Despite the importance of alloys, HAp composites have the capacity to maintain their mechanical integrity despite the rapid rate of deterioration, which enables them to circumvent the problem. To counteract these changes, it is possible to select the appropriate metallic and ceramic reinforcements,

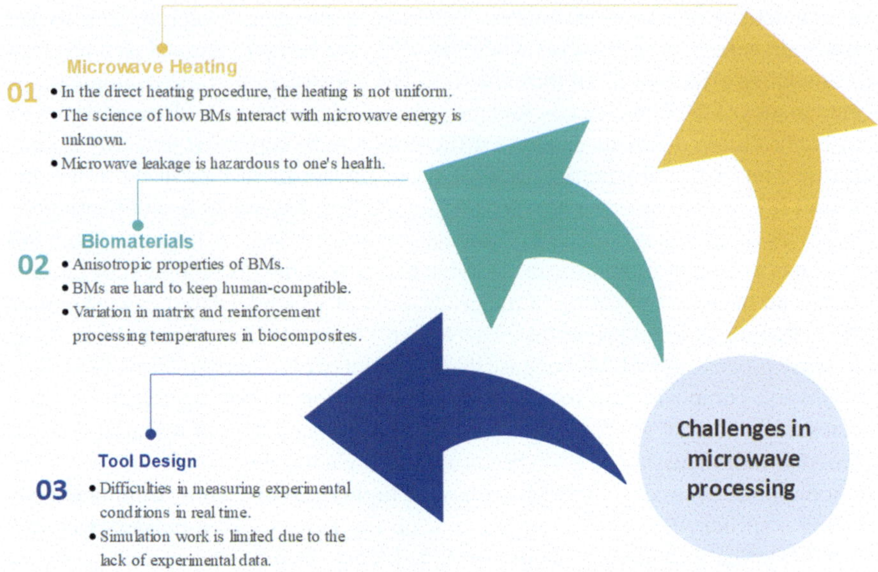

Fig. 12.10 Challenges that come up when using microwaves to process biomaterials

considering the grain size of each material as well as its inherent biocompatibility. Numerous studies have shown that the use of metallic or ceramic reinforcements can have a substantial impact on the mechanical properties and corrosion resistance of HAp-based biocomposites. Microwaves are electromagnetic radiation that, when they come into touch with substances they are designed to heat, transform into heat. Different types of materials will have distinctive responses when exposed to microwaves. The manipulation of materials that are at room temperature and transparent or reflective has led to a number of distinct strategies. In the past, microwave heating was also utilized on occasion for the treatment of more complicated materials. Most polymers are poor microwave absorbers.

Preheating the substrate before depositing the HA coating on it is an efficient method for increasing the crystallinity of the coating. The MPS procedure was used to apply three distinct coating types, each of which had a different crystallinity. Post-heat treatment had an impact on the varying degrees of crystallinity as well as the strength of changes in preferred orientation. Porosity and microhardness are two key features that have a significant impact on qualities of the changed layer. A study has shown that porosity decreases in a proportional manner as the temperature of the heat treatment increases. It is unknown whether this will influence the long-term mechanical durability of HA coatings. This can be accomplished by making certain that the material's porosity content is distributed uniformly across its surface area. Biomaterials are difficult to process because of several factors, including the structure and electrical or heat conductivity of the biopolymers. Processing biomaterials via microwaves is one that is both highly quick and very effective. But it has significant drawbacks, such as internal tensions and uneven heating, which lead to the deterioration of the material's thermal properties. The construction of biocomposites involves the utilization of a wide range of materials, each of which brings with it a distinct collection of qualities. There is a high possibility that microwaves are the source of non-uniformity and heterogeneity in the final items. This is because microwaves heat at different rates and increase the ability of HAp-based composite materials to withstand corrosion.

References

1. Arul KT et al. (2021) Enhanced anticorrosion properties of nitrogen ions modified polyvinyl alcohol/Mg-Ag ions co-incorporated calcium phosphate coatings. 261:124182
2. Atuonwu J, et al. (2019) Energy issues in microwave food processing: a review of developments and the enabling potentials of solid-state power delivery. 59(9):1392–1407
3. Awasthi S et al. (2021) A review on hydroxyapatite coatings for the biomedical applications: Experimental and theoretical perspectives. 9(2):228–249
4. Bahadur NM et al. (2010) Rapid synthesis, characterization and optical properties of TiO_2 coated ZnO nanocomposite particles by a novel microwave irradiation method. **45**(10):1383–1388
5. Basirun WJ et al. (2018) Overview of hydroxyapatite–graphene nanoplatelets composite as bone graft substitute: mechanical behavior and in-vitro biofunctionality. 43(3):177–212
6. Chadha RK et al. (2019) Influence of microwave processing and sintering temperature on the structure and properties of Si/Zr doped hydroxyapatite. 223:319–324
7. García JA et al. (2018). Advanced surface treatments for improving the biocompatibility of prosthesis and medical implants. 199–219
8. George SM et al. (2022). Multifunctional hydroxyapatite composites for orthopedic applications: a review. 8(8):3162–3186
9. Ghasali E et al. (2019) Corrosion behavior and in-vitro bioactivity of porous Mg/Al_2O_3 and Mg/Si_3N_4 metal matrix composites fabricated using microwave sintering process. 225:331–339
10. Ghosh S et al. (2018) A succinct overview of hydroxyapatite based nanocomposite biomaterials: Fabrications, physicochemical properties and some relevant biomedical applications. 12(2):143–158
11. Gorejová R et al. (2019) Recent advancements in Fe-based biodegradable materials for bone repair. 54(3):1913–1947
12. Gupta S, et al. Processing biomaterials using microwave energy and its futuristic scopes. Adv Microw Process Eng Mater crc press:191–211
13. Gupta S et al. (2019) Sintering of biomaterials for Arthroplasty: a comparative study of microwave and conventional sintering techniques. 895:83–89
14. Harun W et al. (2018) A comprehensive review of hydroxyapatite-based coatings adhesion on metallic biomaterials. 44(2):1250–1268
15. Hasan, A., et al. (2018). "Nano-biocomposite scaffolds of chitosan, carboxymethyl cellulose and silver nanoparticle modified cellulose nanowhiskers for bone tissue engineering applications." **111**: 923–934.
16. Jayasathyakawin S et al. (2020) Magnesium matrix composite for biomedical applications through powder metallurgy–Review. 27:736–741
17. Jyoti J et al. (2021) Improved nanomechanical and in-vitro biocompatibility of graphene oxide-carbon nanotube hydroxyapatite hybrid composites by synergistic effect. 117:104376
18. Lee H-J, Yook J-GJM (2019) Graphene nanomaterials-based radio-frequency/microwave biosensors for biomaterials detection. 12(6):952

19. Liu X et al. (2019) Effect of post-heat treatment on the microstructure of micro-plasma sprayed hydroxyapatite coatings. 367:225–230
20. Liyana M et al (2014) Mechanical properties of microwave sintered 60YSZ-Al_2O_3/10HAP Bioceramics Composites. Trans Tech Publ, Applied Mechanics and Materials
21. Majhooll AA et al. (2019). A brief review on biomedical applications of hydroxyapatite use as fillers in polymer. 13:112–119
22. Metwally S et al. (2019). Surface potential and charges impact on cell responses on biomaterials interfaces for medical applications. 104:109883
23. Muthaiah VS et al. (2022) Surface engineering of additively manufactured titanium alloys for enhanced clinical performance of biomedical implants: a review of recent developments. 25:e00180
24. Naqvi RB et al. (2018) Next-generation biomaterials for bone-tissue regeneration: Mg-Alloys on the move. 778:306–315
25. Panda S et al. (2021) A comprehensive review on the preparation and application of calcium hydroxyapatite: a special focus on atomic doping methods for bone tissue engineering. 47(20):28122–28144
26. Pathak DK et al. (2021) Effect of 2 wt.% addition of fe on the compressive properties and corrosion behavior of Zn-HAp-Fe Material. 30(5):3510–3523
27. Pazarlioglu SS, Salman SJJotACS (2020) The effect of alumina additive and sintering temperature on the microstructural, physical, mechanical, and bioactivity properties of hydroxyapatite–alumina composites. 56(2):413–431
28. Peng Q et al. (2020) "Facile fabrication of boronized Ti6Al4V/HA composites for load-bearing applications. 825:153102
29. Pietrzykowska E et al. (2020) Preparation of a ceramic matrix composite made of hydroxyapatite nanoparticles and polylactic acid by consolidation of composite granules. 10(6):1060
30. Piri F et al. (2019) Enhanced adsorption of dyes on microwave-assisted synthesized magnetic zeolite-hydroxyapatite nanocomposite. **7**(5):103338
31. Pupo YM et al. (2021) Effect of Hydroxyapatite Microspheres, Amoxicillin–Hydroxyapatite and Collagen–Hydroxyapatite Composites on Human Dental Pulp-Derived Mesenchymal Stem Cells. 14(24):7515
32. Sarkar R (2017) Synthesis of hydroxyapatite (HAp) and microwave assisted TiO_2-HAp composite coating on commercially pure (CP)-titanium for biomedical applications
33. Sezer N et al. (2018) Review of magnesium-based biomaterials and their applications. 6(1):23–43
34. Shamami DZ et al. (2021) Use of rapid microwave sintering technique for the processing of magnesium-hydroxyapatite composites. 47(9):13023–13034
35. Shamami DZ et al. (2022) Characterization of magnesium-hydroxyapatite functionally graded composites prepared by rapid microwave sintering technique. 48(9):12641–12653
36. Sharma AK, Gupta SJJ (2020) Microwave processing of biomaterials for orthopedic implants: Challenges and possibilities. **72**(3):1211–1228
37. Siddiqui HA et al. (2018) A review on the use of hydroxyapatite-carbonaceous structure composites in bone replacement materials for strengthening purposes. **11**(10): 1813
38. Singh G et al. (2020) Customized hydroxyapatites for bone-tissue engineering and drug delivery applications: a review. **94**(3):505–530
39. Singh P et al. (2022) Hydroxyapatite reinforced surface modification of SS-316L by microwave processing. 28:101701
40. Singh P et al. (2022) Effect of post-heat treatment on the microstructural, mechanical, and bioactivity behavior of the microwave-assisted alumina-reinforced hydroxyapatite cladding. 09544089221116168
41. Singhal C, Murtaza QJMTP (2018) Microwave sintering of advanced composites materials: a review. 5(11):24287–24298
42. Su Y et al. (2019) Biofunctionalization of metallic implants by calcium phosphate coatings. **4**:196–206

43. Su Y et al. (2017) Calcium phosphorus bio-coating on carbon/carbon composites: preparation, shear strength and bioactivity. 419:503–511
44. Tao S et al. (2020) Microstructure, mechanical properties and antibacterial properties of the microwave sintered porous Ti–3Cu alloys. 812:152142
45. Thanigai Arul K et al. (2020) Enhanced Anticorrosion of Nitrogen ions modified polyvinyl alcohol/Mg-Ag ions co-incorporated calcium phosphate coatings. 261
46. Verma N, et al (2022) Multi-scale modelling for adhesive wear in tribo-pairs of nano-hydroxyapatite reinforced UHMWPE bio-composite and Co-Cr alloy. 236(3):567–582
47. Verma V et al. (2022) Experimental and numerical investigation on ZnP coated Mg-Mn based composite materials for internal fixation of tibia midshaft fracture: a comprehensive study. 447:128785
48. Yang H et al. (2017) Solvothermal synthesis of hydroxyapatite nanorods with assistance of green polymer. 79: 9–14
49. Yang J et al. (2018) Material processing and design of biodegradable metal matrix composites for biomedical applications. 46(9):1229–1240
50. Zakaria SM et al. (2013) Nanophase hydroxyapatite as a biomaterial in advanced hard tissue engineering: a review. 19(5):431–441
51. Zhou H et al. (2013) Microwave assisted apatite coating deposition on Ti6Al4V implants. 33(7):4435–4443

Future Trends of Microwave Processing in Composite Industries

Manjeet Rani and Sunny Zafar

Abstract In recent times, there has been a growing search for novel manufacturing techniques that are energy-efficient, eco-friendly, and time-saving, particularly for high-performance fiber-reinforced polymer (FRP) composites. While FRP composites have found increased utilization in various industrial applications, their efficient processing continues to pose challenges. Microwave curing has emerged as an energy-efficient alternative to the traditional autoclave curing process, and it has been employed for processing thermoset-based composites since the 1980s. This chapter provides an overview of the state-of-the-art and future trends in microwave processing of FRP composites. It explores the developments in this field and highlights the challenges encountered during the microwave processing of composites.

Keywords Microwave curing · Fiber-reinforced polymer composites · Mechanical properties · Energy-efficient · Rapid curing time

1 Introduction

1.1 Background Information on Composite Materials and Their Manufacturing

Composite materials, characterized by their lightweight nature, high strength, and high stiffness, have been recognized as a crucial crosscutting technology. The utilization of these materials holds significant potential in enhancing the efficiency of the transportation sector, enabling efficient power generation, improving the storage and transportation of reduced-carbon fuels, and increasing the production of renewable power [1–3]. The fibers in the composite material carry the load, while the

M. Rani (✉) · S. Zafar
Composite Design and Manufacturing Research Group, School of Mechanical and Materials Engineering, Indian Institute of Technology Mandi, Mandi, Himachal Pradesh 175005, India
e-mail: d20063@students.iitmandi.ac.in

M. K. Singh et al. (eds.), *Composite Materials Processing Using Microwave Heating Technology*, Composites Science and Technology,
https://doi.org/10.1007/978-981-97-2772-8_13

matrix material transfers the load between the fibers [4–6]. The final properties of the composite material are superior to those of the constituent materials alone. Fiber-reinforced composite materials offer several advantages over other materials. The fiber-reinforced polymer (FRP) composites are lightweight and have a high strength-to-weight ratio and high stiffness, making them ideal for use in aerospace, automobiles, wind energy, and marine structures. The FRP composites are highly resistant to corrosion and chemical damage, making them well-suited for harsh environments [7–9].

Composite materials can be manufactured using various techniques, which depend on their size and structure. Conventional manufacturing processes include filament winding [10, 11], pultrusion [12–14], injection molding [15, 16], resin transfer molding [17–20], and prepreg layup processes (spray, wet, and hand layup) [21, 22]. Conventional heating routes and room temperature curing methods are typically used to cure composite materials [23, 24]. Curing is crucial in composites manufacturing, ensuring the material develops its desired properties. Different curing methods can be used depending on the type of composite material and the manufacturing process used.

1.2 Importance of Processing Methods in Composite Manufacturing

The selection of manufacturing methods is crucial for manufacturing the composites. Mechanical performance and properties depend on the selected manufacturing methods. Other factors include the type of fibers, matrix material, composite structure application, the complexity of the composite structure, and the mechanical and physical properties of the composite structure [25]. The manufacturing methods generally used to fabricate the composites are open and closed molding, including hand layup, wet layup, resin transfer molding, vacuum infusion process, vacuum assisted resin transfer, or infusion process. The heating routes can be conventional or microwave.

1.3 Introduction of Microwave Processing as an Emerging Technology

Microwave heating has established itself as a well-established technology in the food and communication industries, and its applications have now extended to the field of composite processing [26]. The utilization of microwaves for processing composites offers numerous advantages, including accelerated processing speeds, improved energy efficiency, and a positive environmental impact. Moreover, microwave processing yields composite materials of comparable quality to those manufactured using conventional methods. In response to the manufacturing sector's pursuit of cost and time reduction, microwave energy has emerged as a promising solution while maintaining high-quality composite structures. With its superior energy efficiency, reduced CO_2 emissions, shorter processing time, and enhanced uniform heating capabilities, microwave energy holds the potential to revolutionize the manufacturing industry. The application of microwave energy in composite processing is an area of research that is rapidly developing and has the potential to bring significant and revolutionary changes to the field. It suggests that the advancements and innovations in using microwave energy for composite processing can revolutionize the way composites are manufactured and processed, leading to transformative improvements in the industry.

1.4 Purpose of the Chapter and Its Significance

The book chapter provides a comprehensive overview of the state-of-the-art microwave processing of FRP composites. The authors discuss the different types of microwave processing, the advantages and disadvantages of microwave processing, and the recent developments and challenges in microwave processing technology.

The significance of this chapter is that it provides a clear overview of the potential of microwave processing for the composite manufacturing industry. The authors highlight the advantages of microwave processing, such as faster processing times, improved quality, and reduced energy consumption.

The research also identifies the challenges and limitations of microwave processing, such as the need for specialized equipment and the potential for thermal runaway. However, the authors conclude that the advantages of microwave processing outweigh the challenges and that microwave processing is a promising technology for the future of composite manufacturing.

2 Microwave Processing: An Overview

2.1 Definition of Microwave Processing and Its Principles

Over the past two decades, microwave ovens have established their indispensable presence in domestic kitchens, offering more rapid cooking and energy savings than conventional methods. Expanding beyond culinary applications, microwave energy utilization in material processing has emerged as a promising field. This article explores the potential benefits of microwave-assisted processing, including reduced processing times and enhanced energy efficiency, compared to traditional thermal processing methods.

Thermal heating transfers heat from the surface to the core by conduction, convection, and radiation. In contrast, electromagnetic waves interact with the material at the molecular level and transfer heat from core to surface more uniformly [27]. This distinctive characteristic offers numerous advantages for material processing. Microwave energy can penetrate materials and uniformly deposit heat throughout the volume, eliminating reliance on surface heat diffusion and enabling rapid and homogeneous heating, even in thick materials. Notably, the extended cycle times resulting from slow heating rates, primarily adopted to mitigate process-induced stresses in traditional methods, can be significantly reduced for materials with low thermal conductivities, such as polymers and ceramics, through microwave-assisted processing. Consequently, an inherent trade-off between cure time and mechanical performance in conventional methods can be overcome by leveraging microwaves' volumetric energy transfer capability.

The unique nature of microwave energy delivery revolutionizes material processing by facilitating rapid and uniform heating without compromising product integrity. Microwave-assisted processing capitalizes on the volumetric energy deposition throughout the material, enabling faster and more efficient heating than traditional methods. By bypassing the constraints of surface heat diffusion, microwave energy allows for rapid temperature elevation, leading to reduced processing times. Moreover, the uniform heating achieved through microwave energy distribution minimizes thermal gradients, enhancing overall product quality. This research explores the potential of microwave-assisted processing to streamline manufacturing processes, optimize energy consumption, and improve overall productivity. Further studies are warranted to investigate the precise mechanisms underlying the interaction between microwave energy and different materials to unlock this emerging technology's full potential.

Microwave-assisted processing represents a promising avenue for efficient and time-saving material processing. By capitalizing on the volumetric energy deposition and bypassing the limitations of surface heat diffusion, microwave energy offers significant advantages in reducing processing times and enhancing product quality. The application of this technology holds great potential for a wide range of materials, opening new doors for enhanced efficiency and productivity in various industries.

2.2 *Types of Microwave Processing and Its Applications*

- Curing: Microwave curing has become important in the composite industry due to rapid cure time and less energy consumption. This method enables effective heat transfer across the entire thickness of the laminate, leading to quicker and more consistent temperature distribution. Moreover, microwave curing has the ability to improve the mechanical performance of the FRP composites by facilitating stronger bonding at the interface between the fibers and matrix material. Furthermore, microwave curing is more efficient than traditional oven heating techniques, as it selectively heats only the composite material. Li et al. have proposed an innovative approach for 3D printing utilizing microwaves to heat continuous carbon fiber-reinforced polymer (CFRP) filaments [28]. A temperature control method was developed to analyze the variations in the temperature during the different printing speeds. They achieved a remarkable printing speed of 20 mm/s using microwave heating, resulting in specimens exhibiting an average tensile strength of approximately 358 MPa. In contrast, the conventional heating method achieved a lower printing speed of 5 mm/s, yielding an average tensile strength of approximately 303 MPa. Li et al. developed a novel technique for curing multidirectional CFRP composites using microwave irradiation at 2.45 GHz and 3 kW power [29]. Metal strips with an insulation film were applied to create a coplanar waveguide (CPW) structure, resulting in an electric field perpendicular to the composite surface. Thermal imaging and optical fiber sensors were used to monitor the curing of a 2.3 mm thick composite sample.

- Joining: Microwave joining provides energy-efficient and time-saving advantages compared to adhesive bonding and mechanical fastening methods. It enables the direct joining of composite parts or the incorporation of microwave susceptible implants. Bajpai et al. explored the microwave joining of natural fiber-reinforced polymer composites [30]. They used natural fibers, such as nettle and grewia optiva, reinforced with polylactic acid (PLA) and partially biodegradable composites based on polypropylene (PP). Microwave heating was applied to lap joints using a charcoal powder susceptor near the joint area. The results showed that microwave joining achieved higher joint strength compared to adhesive bonding. Factors like frequency, power, and processing time influence the joining process. Variable-frequency processing and avoiding excessive power levels are important for uniform heating and preventing overheating. Longer exposure times increased joint strength to a threshold beyond which it decreased. Microwave joining offers energy efficiency and time-saving benefits, particularly for thermoplastic composites in in-situ repair applications.
- Recycling: In the manufacturing of FRP composites, a sufficient amount of prepreg material, approximately 30%, is often discarded as waste during the production process, contributing to landfill accumulation. To adhere to stricter environmental regulations, several recycling methods have been developed. These techniques encompass mechanical processes (such as grinding), thermal processes (including fluidized bed and microwave-assisted pyrolysis), and chemical recycling methods [31]. Microwave pyrolysis, an up-and-coming recycling technique, involves heating the resin component of composites in an inert environment, causing it to decompose into smaller molecules. This decomposition enables the separation and reuse of the fibers while the resin transforms into hydrocarbons. Microwave pyrolysis offers advantages such as rapid heat transfer, high decomposition efficiency, and being environmentally friendly. Lester et al. recovered carbon fibers using microwave pyrolysis [32]. Their experimental setup involved suspending composite sheets within a bed of quartz sand, ensuring an inert atmosphere through a continuous flow of nitrogen gas. The composite sheets were then subjected to a brief heating period within a multimode microwave cavity. The recovered carbon fibers exhibited minimal surface alterations, and their strength retention rate reached an impressive 80%, surpassing the performance of fibers recovered through fluidized bed processing. In another study, Rani et al. employed

microwaves for recycling glass fiber-reinforced polymer (GFRP) composites to recover the glass fibers [33]. They utilized a microwave-assisted chemical recycling technique called MACR, where GFRP samples were treated with a solution containing environmentally friendly chemical solvents. Subsequent microwave treatment resulted in the successful recovery of smooth glass fibers while maintaining their mechanical and physical characteristics comparable to those of virgin fibers. In summary, various recycling techniques have been developed to address the waste generated while manufacturing FRP composites. Microwave pyrolysis has shown promise in efficiently separating and recovering fibers while decomposing the resin matrix. Studies have indicated that microwave pyrolysis can yield clean fibers with good strength retention. Additionally, microwave-assisted chemical recycling methods have been successfully employed to recover fibers from GFRP composites while preserving their essential properties. These recycling techniques contribute to the reduction of waste and promote the sustainable use of composite materials.

- Damage inspection: non-contact type damage detection can be enabled using high-power microwaves with thermography techniques. Galietti et al. utilized microwave lock-in thermography to assess impact damage in CFRP composites [34]. They subjected the samples to microwave heating in an oven while an infrared (IR) camera captured the resulting damaged areas. The detected damage areas correlated well with X-ray imaging results. In their work, microwave pulsed thermography was used, and quantitative analysis was done using a data processing algorithm. The consistency in results with X-ray imaging and conventional lock-in thermography was reported. Chady et al. developed a microwave pulsed thermography system to detect delamination in GFRP composite samples [35]. The system consisted of a 2.45 GHz magnetron, an IR camera, an open-ended rectangular waveguide, and a cooling device. They detected delamination by subjecting the composite sample to microwave heating and using the IR camera.

2.3 Advantages and Disadvantages of Microwave Processing

Advantages:

- The volumetric heating characteristics of microwave heating enhanced the diffusion rate that reduces the power consumption (up to 10 to 100 times less) and significantly shorter processing times (up to 10 to 200 times less) compared to conventional methods.
- The high heating rates and rapid diffusion associated with microwave processing have facilitated improvements in the physical and mechanical properties of the treated materials.
- The formation of defects is lower owing to the uniform heating phenomenon.
- Transfer of heat in a short time owing to enhanced heating rate.

Disadvantages:

- Direct heating can pose a fundamental challenge in certain materials due to thermal instabilities, leading to thermal runaway.
- In the case of thick materials, variations in temperature gradients across different areas can result in a poor microstructure within the core.
- Non-uniform temperature distribution can induce internal cracking within the processed material.

3 Microwave Processing in Composite Manufacturing

In past decades, thermoset and thermoplastic composites were cured using microwave energy. To avoid the hot spot formation due to uneven microwave distribution and acing in CFRP was addressed by the [36]. To address both obstacles, variable frequency microwave was used. The experimental work in which fiber composites were cured by microwave routes carried out by various researchers is summarized in Table 13.1.

3.1 Comparison of Microwave Processing with Traditional Methods

Thermal processing of thick composite laminates poses significant challenges due to the difficulty in efficiently transferring heat from the surface to the core, resulting in substantial thermal gradients across the thickness. A slow heating rate typically addresses this issue, although this approach has drawbacks such as prolonged cure times and increased energy costs [38, 40, 41]. The clear picture of the cure time difference between microwave and thermally cured FRP composites is shown in Fig. 13.1.

The high thermal gradients observed during composite processing are primarily attributed to the material's low thermal conductivity and the exothermic chemical reactions during curing. These thermal gradients can lead to undesirable effects such as differential curing, non-uniform solidification, thermal degradation, and the generation of excessive stresses [26, 42, 43]. Conventional methods often involve using slow heating rates and lower temperatures during the cure cycle for thick laminates to overcome these challenges. This approach minimizes thermal gradients and temperature fluctuations throughout the curing process. However, a drawback of this strategy is the extended cure cycles, which can impact production efficiency. In contrast to conventional processing methods, microwave curing offers an alternative approach by utilizing high-energy electromagnetic radiation for energy transfer. Microwave curing is based on the interaction between microwaves and dielectric materials, taking advantage of the alignment of molecular or atomic dipoles within the

Table 13.1 Microwave curing of the composite materials

Author	Material	Parameters	Characterization techniques/testing methods	Findings
Colangelo et al. [23]	Epoxy resin: Mapewrap 31, unidirectional glass fiber: Mapewrap UNI-AX G	Microwave power: 1500 W, Time: 600 s	Flexural and dynamic mechanical tests, DSC, pull-off test	Comparable degree of cure, more stiff composite material
Boey and Lee [37]	Continuous cross-woven E-glass fabric (0/90 pattern)	Microwave power: 400 W, 600 W, 800 W, Time: 8 min to 20 min	Four-point bending test, DMA test	Improved flexural modulus and curing were consistent throughout the 10 mm thick composite laminate
Chen et al. [38]	Glass fiber, Resin: diglycidyl ether of bisphenol A (DGEBA), hardener: 4,40diamino-diphenyl-methane (DDM)	Microwave power: 400 W, 500 W, time: 6 min, 8 min	DSC, Flexural and tensile test, scanning electron microscope (SEM)	Improved T_g and tensile strength, time-saving
Roy et al. [39]	Matrix 1: Epoxy resin LY5052 and hardener HY5052, Matrix 2: Epoxy resin LY556 and hardener HT 972, Bi-directional glass fiber fabric	Continuous and pulsed microwaves	DSC, Tensile strength, and interlaminar shear strength tests	Improved T_g, 20% and 30% improvement in tensile and ILSS, respectively
Papargyris et al. [40]	Matrix and hardener: Araldite LY5052/Aradur HY5052, T300 carbon fiber fabric	–	Flexural test, Interlaminar shear testing, DSC, DMTA, and SEM	Enhancement in ILSS and flexural strength

Fig. 13.1 Comparison of cure time of the fiber-reinforced polymer (FRP) composites. X-axis represents the reference. Data Source: 1 [23], 2 [38], 3,4 [39], 5 [40]

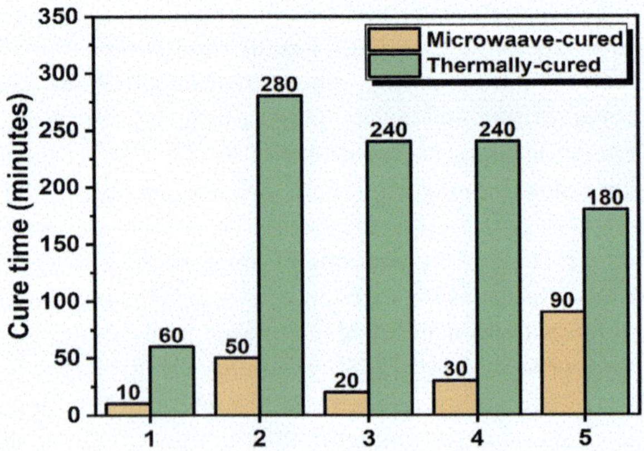

material, resulting in heat dissipation through dipole motion. This unique characteristic enables electrical energy's volumetric and instantaneous transmission through direct interaction with applied electromagnetic fields. The limitations associated with thermal gradients can be mitigated by microwave curing. High-energy electromagnetic radiation facilitates more efficient heat transfer throughout the composite, reducing the need for slow heating rates. This leads to faster and more uniform curing processes, enhancing production efficiency and decreasing energy costs. The application of microwaves in curing processes offers several advantages. Firstly, since energy can be transferred throughout the volume of the laminate, it becomes possible to minimize thermal gradients during the ramp segments of the cure cycle and promote inside-out solidification. Additionally, the instantaneous heat generation provided by microwaves allows control over temperature excursions during the cure cycle, particularly when exothermic chemical reactions generate internal heat. Microwave curing combines the principles of susceptor heating and radiation curing, where the material itself acts as a susceptor, enabling deep penetration depths in glass or organic fiber-based composites, reaching over 500 mm. Furthermore, microwave curing does not require specialized resin systems, as it can heat and cure most thermoset resin systems. This makes microwave curing a cost-effective and practical solution for processing thick laminates. However, microwave curing presents new processing challenges despite its potential due to its distinct energy transfer mechanism. A comprehensive understanding of fundamental aspects such as dipole polarization, dielectric relaxation, and microwave processing equipment is essential to harness the benefits of microwave heating. Previous studies have extensively reviewed these technical challenges, microwave fundamentals, and recent advances in microwave technology, providing valuable insights for the successful implementation of microwave curing in composite manufacturing.

Papargyris et al. fabricated CFRP composites using thermal and microwave curing [40]. The results showed an improvement in the interlaminar shear strength (ILSS) of CFRP that cured with the microwave. This could be correlated to the reduction in epoxy viscosity during the initial stage of microwave curing before a rapid increase in cross-linking density makes the resin highly viscous. The decrease in viscosity facilitates better adhesion between the resin and fibers and improved wetting of the fibers by the resin.

Previous research has also indicated that lowering resin viscosity can enhance resin-fiber adhesion and improve fiber wet-out [44–46]. The interfacial properties of composites are linked proportionally with the adhesion and wetting behavior of fiber and matrix. The lower resin viscosity during microwave curing enables better penetration of the resin into the fibers and stronger bonding, ultimately increasing ILSS. Therefore, the findings from Papargyris et al.'s study suggest that microwave curing effectively reduces resin viscosity during the early stages of curing, leading to improvement in the adhesion of fiber-matrix and fiber wet-out, which contributes to the enhancement of interlaminar shear strength in CFRP composites.

3.2 *Effect of Microwave Processing on the Properties of Composite Materials*

Several studies, including those conducted by Boey and Yue [37, 47, 48], have reported improved mechanical performance of the composites cured with microwave energy. Several studies were conducted in the past decade and generated reported tensile and flexural strength data by experimental research. Some data in the form of a graph is shown in Fig. 13.2. Here, only microwave-cured composites' tensile and flexural strength is given in Fig. 13.2 for detailed study; readers can refer to the literature provided in the Fig. 13.2 caption.

Chen et al. determined the glass transition temperature (T_g) of thermal and microwave-cured composite samples [38]. The findings revealed that microwave-cured samples exhibited a higher T_g than thermal-cured. From the results, it can be concluded that microwave curing enhances the heat resistance of samples. Several studies have used differential scanning calorimetry (DSC) analysis to determine the degree of cure of the cured samples [23, 38]. Different heating rates and temperatures have been used to investigate the degree of cure, with a standard heating ramp rate ranging from 8 °C/min to 10 °C/min [38]. Furthermore, some studies have implemented a multistep heating approach when using microwaves for composite curing [51, 52]. These investigations have demonstrated that multistep heating improves the mechanical performance of composites compared to single-step or thermal heating methods.

In a study conducted by Tanrattanakul and Jaroendee, experiments were carried out to evaluate the storage modulus of composites using both one-step and three-step microwave heating processes [52]. The results indicated that the three-step heating process resulted in a higher storage modulus than the single-step heating process. DSC analysis showed different T_g for the one-step, two-step, and three-step heating processes, measured at 78 °C, 106 °C, and 115 °C, respectively. While higher T_g

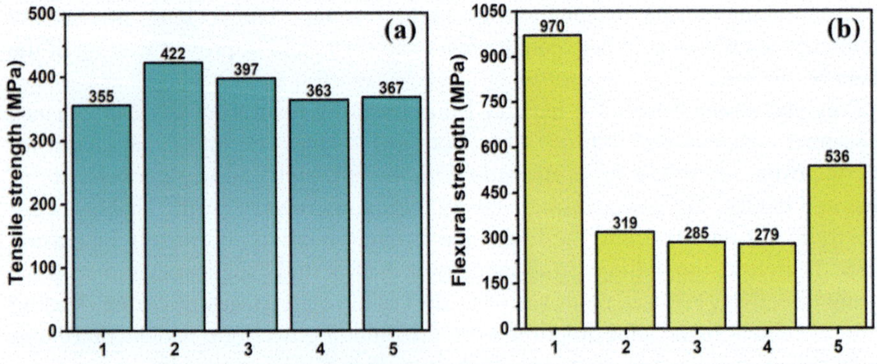

Fig. 13.2 Tensile and flexural strength data of microwave-cured fiber-reinforced polymer (FRP) composites recreated from the previous studies. X-axis represents the reference. Data source: Fig. 13.2(a), 1[38], 2 & 3 [39], 4 [49], 5 [50], Fig. 13.2(b), 1 [23], 2, 3 & 4 [38], 5 [40]

values generally correlate with increased strength and modulus in composites, the mechanical properties are influenced by T_g only up to a certain extent. In the case of CFRP composites, thermally cured composites exhibited lower T_g but showed better mechanical performance than microwave-cured composites. These findings could be correlated to the fiber and matrix material adhesion strength [40, 52].

X-ray micro-computed tomography (micro-CT), an advanced technology, was used to investigate the void characteristics of Fiber-Reinforced Polymer (FRP) composites [53]. It analyzes various void parameters, including void content, location, distribution frequency, shape (morphology), and size. Studies focusing on void characteristics in FRPs have examined the micro-void percentage for both transverse and axial flow at the same capillary number, revealing a higher presence of micro-voids in the transverse flow of the resin. Furthermore, micro-CT was used to analyze the voids content % and their shape and size in the composites. Fewer voids were found in the fibers' direction than in the tangential direction. In the case of carbon-fiber-reinforced Polyether Ether Ketone (PEEK) composites, Comer et al. [54] utilized the micro-CT technique to observe elongated voids aligned in the fiber direction. Chen et al. compared thermal and microwave-cured composites [38]. The microwave-cured specimens exhibited a flat fracture surface and demonstrated significantly improved adhesion between the fiber-matrix interface. This suggests that the epoxy (matrix) had better wetting of the fibers in the microwave-cured composites compared to the thermally cured composites. This can be corelated to the volumetric heating characteristics.

The analysis of the fractured composite surface revealed that under tensile loading, the reinforced fibers initially took the majority of the load and were then transferred to the matrix resin through the interface. Subsequently, the tensile load was redistributed within the fiber and the matrix, leading to a cyclic load redistribution until the load reached its peak value. Eventually, both the fiber and the matrix resin fractured simultaneously. The microwave-cured specimens exhibited higher tensile strength and modulus. The enhancement in the properties was due to the good interfacial bonding strength. Similar improvements in interfacial strength have been observed in the curing of GFRP composites [38]. In contrast, composites produced using the conventional thermally cured method showed poorer interfacial bonding and more extensive fiber pull-out [40]. These composites had a significant amount of clean fibers, indicating weaker adhesion between the fibers and the matrix. Both types of composites exhibited regions of brittle matrix failure.

4 Recent Developments in Microwave Processing of Composites

In this section, recent developments in microwave processing have been discussed. This section provides an overview of the previous studies in which new methods have been used to encounter the existing problems.

4.1 Current State of the Art in Microwave Processing of Composites

Microwave dielectric heating has various applications, including rubber vulcanization, food heating, and composite bonding. In the case of epoxy curing, electromagnetic waves interact with the material at the molecular level. That induces vibrations and rotations of the molecules due to alternating electromagnetic fields. These vibrations increase the temperature of the material. The primary mechanism for dielectric heating is dipolar loss or reorientation loss. When a dielectric material with permanent dipoles is exposed to an alternating electromagnetic field, the dipoles vibrate intensely but struggle to keep up with the high-speed reversals in the field. This lag in molecular response causes a phase lag with the field, leading to power dissipation and heat generation in the dielectric material. However, this phenomenon also introduces challenges in achieving uniform curing across the material, as the heating process becomes highly dependent on the material properties and geometry.

Non-uniform heating can occur in thick or complex structures due to the delay and lag in heating at different locations within the material. Microwaves interact with dielectric materials, such as thermosetting polymers, at a molecular level and transmit energy through polarization. Different types of polarization contribute to energy transfer, including electronic, atomic, dipole, and interfacial polarization. Among these, dipole polarization and interfacial polarization play significant roles in energy transfer in composite materials. The dielectric properties, like the dielectric constant and the dielectric loss factor, govern the level of polarization. Materials with high dielectric loss factors, such as carbon fibers, have low microwave penetration depths and act as reflectors. In contrast, materials with low dielectric loss factors have high penetration depths and absorb less energy from microwaves. Thermosetting polymers come in the medium conductivity range and are heated by microwave energy transfer. In conventional heating, high-conductivity materials are most efficiently heated. Crosslinking occurs during the curing process of thermosetting resins, causing changes in the dielectric properties at a molecular level. When the resin viscosity is low in the early curing stages, the dielectric properties allow efficient coupling with microwaves as the molecular dipoles can freely move and align with the electromagnetic field. The dielectric loss factor decreases as crosslinking progresses, and the resin becomes more viscous. At this stage, the resin becomes less responsive to microwave heating as the molecular dipoles have limited mobility and struggle to align with the field. The microwave process can be optimized by including the variation of dielectric properties of the resin during the cure. Zhang et al. developed an off-line dielectric measurement technique to facilitate the variation in dielectric properties during microwave curing. While other electromagnetic radiation bands could be used for heating and curing thermosetting polymers, microwaves are the most suitable choice. Microwaves can penetrate polymers (resin, epoxy, etc.) without health hazards associated with other electromagnetic waves like gamma radiation.

4.2 Advances in Microwave Processing Equipment and Technology

The out-of-autoclave method, known as microwave curing, has emerged as an appealing alternative for producing high-performance composites. Various approaches have been employed in electromagnetic cure studies, including using single-mode microwave cavities, multimode microwave curing ovens or applicators [39, 40], and waveguides [55]. In another study, the authors designed an advanced microwave curing system consisting of an octagon chamber, operating at 2.45 GHz with a power of 20 kW, and equipped with 16 microwave sources and waveguide slot antennas [56]. To achieve uniform temperature distribution during the curing process, a Labview-PCI data acquisition system, utilizing two PCI-1727U cards, was developed to control the microwave source separately. The temperature measurement was done using two optical fiber fluorescent sensors placed in the vacuum bag during the microwave curing. For detailed information, please refer to the mentioned study [56]. In various studies, fiber Bragg grating (FBG) sensors have been used to measure temperature and process-induced strains during microwave processing [57, 58, 58–62]. Incorporating such sensors within the composite material enables the mapping of stress and strain states under different loading conditions and provides information about curing and cooling-induced shrinkage through strain detection.

To ensure uniform heating and a stable cure temperature, Papargyris et al. utilized a computer control system [40]. This study used parameters like microwave power, frequency, and temperature as process control parameters. The General Purpose Interface Bus (GPIB) connection between the computer and the network analyzer (NA) was used to adjust the frequency and power levels. The computer control system was designed to continuously change the excitation frequencies of the NA while simultaneously regulating the microwave power to maintain the desired cure temperature. The computer continuously monitored the sample temperature through a serial connection (RS-232) with the fluorotic thermometer. The computer control system implemented a quadratic algorithm to regulate the microwave power:

$$Power = [\frac{T_{cure} - T}{T_{cure} - T_{min}}]^2 \quad (1)$$

The amplifier power output used in the microwave curing process was determined based on the temperature conditions. It was governed by an equation (Eq. 1) where T denotes the maximum or actual temperature of the sample, T_{cure} is the desired cure temperature, and T_{min} is pre-set temperature limit.

During the heating phase, thermal expansion stresses occurred in both the composite and the tool, leading to interactions at their interface. In microwave curing, the composite absorbs microwave energy and heats up while the surrounding air and tool material remains at a lower temperature. There is strong heat convection and radiation at the edge where the composite and tool meet, resulting in the tool having a lower temperature than the composite. In contrast, during autoclave curing, the surrounding air and the tool have higher temperatures than the composite (before the resin's exothermic reaction). Therefore, in microwave curing, the composite releases heat to the surrounding, while in autoclave curing, it absorbs heat from the surrounding. The expansion stress induced by microwave curing is lower than autoclave curing due to the temperature of the tool remaining low. The stress gradients in the microwave-cured composite caused by the coefficient of thermal expansion (CTE) mismatch between the composite and the tool are also low. For a more comprehensive understanding, readers can refer to the mentioned article [61]. The work done by Nanya Li et al. presents an effective measurement approach for assessing the tool-part interaction stress between the tool and the composite part during microwave curing. Additionally, they analyzed the comprehensive friction coefficient, which considers the CTE mismatch between the tool and the composite, as well as the chemical shrinkage of the resin [61]. Nanya Li et al. also proposed a method based on a cyclic heating–cooling process to reduce cure-induced stress during curing. This method involves alternating stress reduction through repeated heating and cooling stages. The stresses accumulated during the heating stage are calculated using semi-empirical equations provided in [56]. Further, cooling stage, the temperature variation can be calculated using the accumulated stresses. It is important to note that the cooling and heating stages have different paths that can be attributed to the change in the degree of cure and viscoelastic properties of the polymer.

Traditional microwave curing methods often experience slight temperature fluctuations during the dwell stage due to heat dissipation from the composite to the surrounding environment. The control of change in temperature was the challenge of the traditional microwave curing process. However, the new microwave curing method regulates temperature fluctuations very well. it can minimize cure-induced stresses. Traditional autoclave curing with rapid heating rates often results in uneven temperature distribution within the composite. In contrast, microwave curing, with its fast heating rate, maintains a more uniform temperature distribution across both the composite and the tool, thereby reducing residual stress.

5 Future Trends in Microwave Processing of Composites

5.1 Predicted Impact of Microwave Processing on the Composite Industry

Microwave processing of FRP composites has the potential to revolutionize the composite industry by uncovering new and unexplored possibilities. One of the key benefits lies in the improved design flexibility it offers for complex parts. With microwave processing, uniform curing throughout the material is more likely, resulting in enhanced structural integrity and reduced defects. Figure 13.3 illustrates the wide array of advantages that can be unlocked by adopting microwave curing as

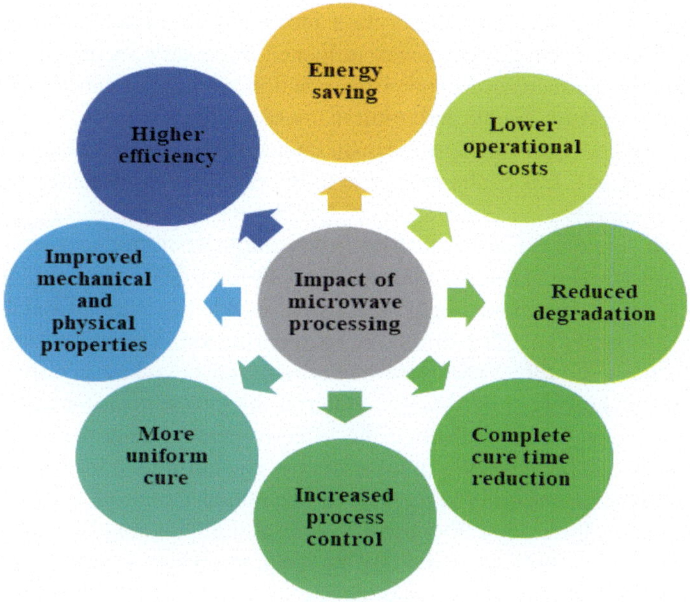

Fig. 13.3 Represents the hidden advantages of microwave processing in the composite industry

the standard method in industries. These hidden advantages include increased production efficiency, reduced manufacturing costs, enhanced mechanical properties, and improved product quality.

5.2 *Challenges and Limitations in the Adoption of Microwave Processing in Composite Manufacturing*

- Thermal conductivity of resin: Thermosets and liquid resins are transparent to microwaves, making them suitable for microwave curing processes. Rapid curing and improvement in resin stability could be due to microwave exposure [3]. However, previous research studies (e.g. [27, 63]) suggested that uneven temperature distribution in the FRP composite sample was due to the uneven microwave absorption capability of the material. These studies were conducted on structural-scale FRP composite samples. This non-uniform heating can lead to temperature variations and the formation of hot spots, primarily caused by the material's low thermal conductivity. Furthermore, a maximum reachable temperature threshold limits the temperature rise in a specimen at a constant microwave power. The hot spots are formed due to heat accumulation in the sample because of the material's low thermal conductivity. The change in dielectric properties of the material during the vitrification stage of a thermoset polymer reduces the dielectric mobility, affecting microwave absorption and heat generation within the material. Additionally, the curing process can influence the resin's thermal conductivity. These factors and increased exothermic reaction restrict the curing rate in some part of the composite sample and result in uneven curing. Various research studies have observed fully curing the entire resin sample without causing harm to specific areas [37, 38, 64]. It is challenging due to the potential for extreme exposure to microwave energy.

- Microwave shielding: Carbon fiber in FRP composites introduces another challenge in microwave curing related explicitly to microwave shielding [65]. This issue affects the penetration depth of microwaves into the material and the dielectric loss parameter (ε″) due to interfacial polarization. The presence of carbon fiber can limit the direct exposure of the material beyond a certain thickness to microwave heating. Additionally, material thickness, fiber content, and fiber morphology can further influence the effectiveness of microwave curing. To address these challenges and optimize microwave curing strategies, it would be beneficial to have the ability to monitor temperatures at multiple locations within the composites. Infrared or fiber optic sensing techniques could provide valuable information for controlling and adjusting the microwave curing process. Monitoring temperature variations during curing makes it possible to ensure uniform heating and optimize the curing strategy based on real-time data.
- Microwave arcing: One significant issue encountered in the microwave curing of thermoset composites is the occurrence of microwave arcing, particularly associated with carbon fibers. Numerous studies have investigated the detrimental effects of carbon fiber arcing during microwave curing, as carbon fibers possess strong microwave absorption properties. To prevent this phenomenon, researchers have focused on controlling microwave power to avoid arcing while maintaining the desired structural integrity of the cured composites, comparable to conventionally heated composites [40, 63, 66]. The absorption of microwave energy occurs due to the molecular mobility in dipolar materials. The difference in energy dissipation between absorption by epoxy and carbon fibers, the microwave exposure results in the partial transmission of energy through the material. In the case of carbon fibers, this transmission can occur between molecules that are relatively close but apart. When microwave exposure is high, the release of energy at this level becomes significant and leads to the phenomenon of arcing. Efforts are made to control microwave power levels and minimize the potential for arcing during curing [67].
- Microwave radiation characteristics: In the microwave cavity, non-uniform distribution of power lead to hot spot formation. It has been considered an important issue of the microwave heating phenomena. This issue is often addressed by considering the geometric shape of the cavity as well as cured structure. One approach to mitigate this problem is rotating the subject on a turntable, ensuring more uniform exposure to microwave radiation. Another alternate method is mode stirrers, which are rotating reflectors. It can be used to alter the distribution of the electromagnetic field within the cavity. Both methods aim to make more unifrom power distribution and reduce the formation of hot spots [26]. In some cases,

multiple magnetrons have been employed to achieve even better power distribution [58]; however, when using a multi-mode microwave system, caution must be exercised in adjusting the size of the cavity, as compared to a single-mode microwave, to ensure effective and uniform heating throughout the material. The configuration and design of the microwave system play a crucial role in achieving a more homogeneous power distribution and minimizing hot spots during the curing process.

6 Conclusion

Employing microwave energy in the composite industry offers energy savings and reduced production costs. The interaction between materials and microwaves is well-documented in the available literature. As researchers explore this field, focusing on inventing new processes that require minimal tooling and equipment costs is essential. By doing so, the composite industry can thrive by adopting microwave energy, taking full advantage of its numerous reported benefits.

While microwave energy has been applied in various areas within the composite industry, there are still unexplored opportunities to enhance its efficiency further. Integrating microwave energy into additive manufacturing or 3D printing processes is one promising area. Composite manufacturing can become more sustainable and environmentally friendly by incorporating microwave energy into additive manufacturing. The inherent ease of the 3D printing process can amplify the benefits of microwave energy in the composite industry.

References

1. Karthikeyan N, Anand RB, Suthakar T, Barhate S (2019) Materials, innovations and future research opportunities on wind turbine blades—insight review. Environ Prog Sustainable Energy 38:1–10. https://doi.org/10.1002/ep.13046
2. Kretsis G (1987) A review of the tensile, compressive, flexural and shear properties of hybrid fibre-reinforced plastics. Composites 18:13–23. https://doi.org/10.1016/0010-4361(87)90003-6
3. Abliz D, Duan Y, Steuernagel L, Xie L, Li D, Ziegmann G (2013) Curing methods for advanced polymer composites—A review. Polym Polym Compos 21:341–348. https://doi.org/10.1177/096739111302100602
4. Jones AR, Blaiszik BJ, White SR, Sottos NR (2013) Full recovery of fiber/matrix interfacial bond strength using a microencapsulated solvent-based healing system. Compos Sci Technol 79:1–7. https://doi.org/10.1016/j.compscitech.2013.02.007
5. Karger-Kocsis J, Mahmood H, Pegoretti A (2015) Recent advances in fiber/matrix interphase engineering for polymer composites. Prog Mater Sci 73:1–43. https://doi.org/10.1016/j.pmatsci.2015.02.003
6. Kazemi ME, Shanmugam L, Dadashi A, Shakouri M, Lu D, Du Z, et al. (2021) Investigating the roles of fiber, resin, and stacking sequence on the low-velocity impact response of novel hybrid thermoplastic composites. Compos B: Eng 2021:207. https://doi.org/10.1016/j.compositesb.2020.108554

7. Naqvi SR, Prabhakara HM, Bramer EA, Dierkes W, Akkerman R, Brem G (2018) A critical review on recycling of end-of-life carbon fibre/glass fibre reinforced composites waste using pyrolysis towards a circular economy. Resour Conserv Recycl 136:118–129. https://doi.org/10.1016/j.resconrec.2018.04.013
8. Lee SM, Kang DS, Roh JS (2015) Bulk graphite: Materials and manufacturing process. Carbon Letters 16:135–146. https://doi.org/10.5714/CL.2015.16.3.135
9. Abbood IS, Odaa SA, Hasan KF, Jasim MA (2021) Properties evaluation of fiber reinforced polymers and their constituent materials used in structures—A review. Mater Today: Proc 43:1003–1008. https://doi.org/10.1016/j.matpr.2020.07.636
10. A. Endruweit, M.S. Johnson ACL (2006) Curing of composite components by ultraviolet radiation: a review. Polymer Compos 27:119–28
11. Paciornik S, D'almeida JRM (2009) Measurement of void content and distribution in composite materials through digital microscopy. J Compos Mater 43:101–112. https://doi.org/10.1177/0021998308098234
12. Alqam M, Bennett RM, Zureick AH (2002) Three-parameter vs. two-parameter Weibull distribution for pultruded composite material properties. Compos Struct 58:497–503. https://doi.org/10.1016/S0263-8223(02)00158-7
13. Asensio M, Esfandiari P, Núñez K, Silva JF, Marques A, Merino JC, et al. (2020) Processing of pre-impregnated thermoplastic towpreg reinforced by continuous glass fibre and recycled PET by pultrusion. Compos B: Eng 2020:200. https://doi.org/10.1016/j.compositesb.2020.108365
14. Ribeiro MCS, Meira-Castro AC, Silva FG, Santos J, Meixedo JP, Fiúza A et al (2015) Re-use assessment of thermoset composite wastes as aggregate and filler replacement for concrete-polymer composite materials: a case study regarding GFRP pultrusion wastes. Resour Conserv Recycl 104:417–426. https://doi.org/10.1016/j.resconrec.2013.10.001
15. Meng F, Cui Y, Pickering S, McKechnie J (2020) From aviation to aviation: environmental and financial viability of closed-loop recycling of carbon fibre composite. Compos B Eng 200:108362. https://doi.org/10.1016/j.compositesb.2020.108362
16. Steenkamer DA, Sullivan JL (1998) On the recyclability of a cyclic thermoplastic composite material. Compos B Eng 29:745–752. https://doi.org/10.1016/S1359-8368(98)00016-X
17. Nicholas J, Mohamed M, Dhaliwal GS, Anandan S, Chandrashekhara K (2016) Effects of accelerated environmental aging on glass fiber reinforced thermoset polyurethane composites. Compos B Eng 94:370–378. https://doi.org/10.1016/j.compositesb.2016.03.059
18. Kuentzer N, Simacek P, Advani SG, Walsh S (2007) Correlation of void distribution to VARTM manufacturing techniques. Compos A Appl Sci Manuf 38:802–813. https://doi.org/10.1016/j.compositesa.2006.08.005
19. González EV, Maimí P, Sainz de Aja JR, Cruz P, Camanho PP (2014) Effects of interply hybridization on the damage resistance and tolerance of composite laminates. Compos Struct 108:319–31. https://doi.org/10.1016/j.compstruct.2013.09.037
20. Sharma S, Wetzel KK (2010) Process development issues of glass-carbon hybrid-reinforced polymer composite wind turbine blades. J Compos Mater 44:437–456. https://doi.org/10.1177/0021998309347569
21. Banakar P, Shivananda HK, Niranjan HB (2012) Influence of fiber orientation and thickness on tensile properties of laminated polymer composites. Int J Pure Appl Sci Technol 9:61–68
22. Elanchezhian C, Ramnath BV, Hemalatha J (2014) Mechanical behaviour of glass and carbon fibre reinforced composites at varying strain rates and temperatures. Procedia Materials Science 6:1405–1418. https://doi.org/10.1016/j.mspro.2014.07.120
23. Colangelo F, Russo P, Cimino F, Cioffi R, Farina I, Fraternali F et al (2017) Epoxy/glass fibres composites for civil applications: Comparison between thermal and microwave crosslinking routes. Compos B Eng 126:100–107. https://doi.org/10.1016/j.compositesb.2017.06.003
24. Rao S Jai (2008) INCAST CURE PROCESS FOR RAPID AN EMERGING , ENERGY-EFFICIENT
25. He X, Shi X, Chung C, Lei Z, Zhang W, Yu K (2021) A sustainable manufacturing method of thermoset composites based on covalent adaptable network polymers. Compos B Eng 221. https://doi.org/10.1016/j.compositesb.2021.109004

26. Thostenson ET, Chou TW (1999) Microwave processing: fundamentals and applications. Compos A Appl Sci Manuf 30:1055–1071. https://doi.org/10.1016/S1359-835X(99)00020-2
27. Mishra RR, Sharma AK (2016) Microwave-material interaction phenomena: Heating mechanisms, challenges and opportunities in material processing. Compos A Appl Sci Manuf 81:78–97. https://doi.org/10.1016/j.compositesa.2015.10.035
28. Li N, Link G, Jelonnek J (2020) Rapid 3D microwave printing of continuous carbon fiber reinforced plastics. CIRP Ann 69:221–224. https://doi.org/10.1016/j.cirp.2020.04.057
29. Li Y, Cheng L, Zhou J (2018) Curing multidirectional carbon fiber reinforced polymer composites with indirect microwave heating. Int J Adv Manuf Technol 97:1137–1147. https://doi.org/10.1007/s00170-018-1974-1
30. Bajpai PK, Singh I, Madaan J (2012) Joining of natural fiber reinforced composites using microwave energy: Experimental and finite element study. Mater Des 35:596–602. https://doi.org/10.1016/j.matdes.2011.10.007
31. Rani M, Choudhary P, Krishnan V, Zafar S (2021) A review on recycling and reuse methods for carbon fiber/glass fiber composites waste from wind turbine blades. Compos B Eng 108768. https://doi.org/10.1016/j.compositesb.2021.108768.
32. Lester E, Kingman S, Wong KH, Rudd C, Pickering S, Hilal N (2004) Microwave heating as a means for carbon fibre recovery from polymer composites: a technical feasibility study. Mater Res Bull 39:1549–1556. https://doi.org/10.1016/j.materresbull.2004.04.031
33. Rani M, Choudhary P, Krishnan V, Zafar S (2022) Development of sustainable microwave-based approach to recover glass fibers for wind turbine blades composite waste. Resour Conserv Recycl 179:106107. https://doi.org/10.1016/j.resconrec.2021.106107
34. Galietti U, Palumbo D, Calia G, Pellegrini M (2012) Non destructive evaluation of composite materials with new thermal methods. In: ECCM 2012—Composites at Venice, Proceedings of the 15th European Conference on Composite Materials, pp 24–8
35. Chady T, Sikora R, Lopato P, Psuj G, Szymanik B, Balasubramaniam K, et al. (2016) Wind turbine blades inspection techniques. Przeglad Elektrotechniczny 92:3–6. https://doi.org/10.15199/48.2016.05.01
36. Kwak M, Robinson P, Bismarck A, Wise R (n.d.) Curing of composite materials using the recently developed hephaistos microwave 1–6
37. Boey FYC, Lee TH (1991) Electromagnetic radiation curing of an epoxy/fibre glass reinforced composite. Int J Radiat Appl Instrum 38:419–423. https://doi.org/10.1016/1359-0197(91)90118-L
38. Chen Y, Li Y, You Y, Xiao J, Song Q (2014) Research on mechanical properties of epoxy/glass fiber composites cured by microwave radiation. J Reinf Plast Compos 33:1441–1451. https://doi.org/10.1177/0731684414524031
39. Rao RMVGK, Rao S, Sridhara BK (2006) Studies on tensile and interlaminar shear strength properties of thermally cured and microwave cured glass-epoxy composites. J Reinforce Plast Compos 25:783–95. https://doi.org/10.1177/0731684406063542
40. Papargyris DA, Day RJ, Nesbitt A, Bakavos D (2008) Comparison of the mechanical and physical properties of a carbon fibre epoxy composite manufactured by resin transfer moulding using conventional and microwave heating. Compos Sci Technol 68:1854–1861. https://doi.org/10.1016/j.compscitech.2008.01.010
41. Joshi SC, Bhudolia SK (2014) Microwave-thermal technique for energy and time efficient curing of carbon fiber reinforced polymer prepreg composites. J Compos Mater 48:3035–3048. https://doi.org/10.1177/0021998313504606
42. Bai SL, Djafari V (1995) Interfacial properties of microwave cured composites. Composites 26:645–651. https://doi.org/10.1016/0010-4361(95)98913-6
43. Thostenson ET, Chou TW (2001) Microwave and conventional curing of thick-section thermoset composite laminates: experiment and simulation. Polym Compos 22:197–212. https://doi.org/10.1002/pc.10531
44. Stringer LG (1989) Optimization of the wet lay-up/vacuum bag process for the fabrication of carbon fibre epoxy composites with high fibre fraction and low void content. Composites 20:441–452. https://doi.org/10.1016/0010-4361(89)90213-9

45. Kedari VR, Farah BI, Hsiao KT (2011) Effects of vacuum pressure, inlet pressure, and mold temperature on the void content, volume fraction of polyester/e-glass fiber composites manufactured with VARTM process. J Compos Mater 45:2727–2742. https://doi.org/10.1177/0021998311415442
46. Il Lee W, Springer GS (1984) Microwave curing of composites. J Compos Mater 18:387–409. https://doi.org/10.1177/002199838401800405
47. Boey F, Gosling I, Lye SW (1992) High-pressure microwave curing process for an epoxy-matrix/glass-fibre composite. J Mater Process Tech 29:311–319. https://doi.org/10.1016/0924-0136(92)90445-X
48. Lye SW, Boey FY (1994) PC-Based Monitoring and Control System for Microwave Curing of Polymer Composites, Materials and Manufacturing Processes. Mater Manuf Process 9:851–868. https://doi.org/10.1080/10426919408934957
49. Maenz S, Mühlstädt M, Jandt KD, Bossert J (2015) Mechanical properties of microwave cured glass fibre epoxy composites prepared by resin transfer moulding. J Compos Mater 49:2839–2847. https://doi.org/10.1177/0021998314557295
50. Mooteri PS, Sridhara BK, Rao S, Prakash MR, Rao RMVGK (2006) Studies on mechanical behavior of microwave and thermally cured glass fiber reinforced polymer composites. J Reinforce Plast Compos 25:503–12. https://doi.org/10.1177/0731684406058290
51. Yang J, Yang W, Chen W, Tao X (2020) An elegant coupling: Freeze-casting and versatile polymer composites. Prog Polym Sci 109. https://doi.org/10.1016/j.progpolymsci.2020.101289
52. Tanrattanakul V, Jaroendee D (2006) Comparison between microwave and thermal curing of glass fiber-epoxy composites: Effect of microwave-heating cycle on mechanical properties. J Appl Polym Sci 102:1059–1070. https://doi.org/10.1002/app.24245
53. Mehdikhani M, Gorbatikh L, Verpoest I, Lomov SV (2019) Voids in fiber-reinforced polymer composites: A review on their formation, characteristics, and effects on mechanical performance. J Compos Mater 53:1579–1669. https://doi.org/10.1177/0021998318772152
54. Comer AJ, Ray D, Obande WO, Jones D, Lyons J, Rosca I et al (2015) Mechanical characterisation of carbon fibre-PEEK manufactured by laser-assisted automated-tape-placement and autoclave. Compos A Appl Sci Manuf 69:10–20. https://doi.org/10.1016/j.compositesa.2014.10.003
55. Mijović J, Wijaya J (1990) Review of cure of polymers and composites by microwave energy. Polym Compos 11:184–191. https://doi.org/10.1002/pc.750110307
56. Li N, Li Y, Jelonnek J, Link G, Gao J (2017) A new process control method for microwave curing of carbon fibre reinforced composites in aerospace applications. Compos B Eng 122:61–70. https://doi.org/10.1016/j.compositesb.2017.04.009
57. Muflikhun MA, Higuchi R, Yokozeki T, Aoki T (2020) The evaluation of failure mode behavior of CFRP/Adhesive/SPCC hybrid thin laminates under axial and flexural loading for structural applications. Compos B Eng 185:107747. https://doi.org/10.1016/j.compositesb.2020.107747
58. Mgbemena CO, Li D, Lin MF, Liddel PD, Katnam KB, Kumar VT et al (2018) Accelerated microwave curing of fibre-reinforced thermoset polymer composites for structural applications: a review of scientific challenges. Compos A Appl Sci Manuf 115:88–103. https://doi.org/10.1016/j.compositesa.2018.09.012
59. Zhao X, Huang Z, Song P, Chen H, Yang H, Xie L et al (2020) Curing kinetics and mechanical properties of epoxy resin/1-benzyl-2-methylimidazole/isophorone diamine system. Thermochim Acta 690:178657. https://doi.org/10.1016/j.tca.2020.178657
60. Barkoula NM, Alcock B, Cabrera NO, Peijs T (2008) Flame-retardancy properties of intumescent ammonium poly(phosphate) and mineral filler magnesium hydroxide in combination with graphene. Polym Polym Compos 16:101–113
61. Li N, Li Y, Wu X, Hao X (2017) Tool-part interaction in composites microwave curing: experimental investigation and analysis. J Compos Mater 51:3719–3730. https://doi.org/10.1177/0021998317693674
62. Zhou J, Li Y, Li N, Hao X, Liu C (2016) Interfacial shear strength of microwave processed carbon fiber/epoxy composites characterized by an improved fiber-bundle pull-out test. Compos Sci Technol 133:173–183. https://doi.org/10.1016/j.compscitech.2016.07.033

63. Kwak M, Robinson P, Bismarck A, Wise R (2015) Microwave curing of carbon-epoxy composites: penetration depth and material characterisation. Compos A Appl Sci Manuf 75:18–27. https://doi.org/10.1016/j.compositesa.2015.04.007
64. Teufl D, Zaremba S (2018) 2.45 GHz microwave processing and its influence on glass fiber reinforced plastics. Materials 11. https://doi.org/10.3390/ma11050838
65. Li Z, Haigh A, Soutis C, Gibson A (2018) Principles and applications of microwave testing for woven and non-woven carbon fibre-reinforced polymer composites: a topical review. Appl Compos Mater 25:965–982. https://doi.org/10.1007/s10443-018-9733-x
66. Nightingale C, Day RJ (2002) Flexural and interlaminar shear strength properties of carbon fibre/epoxy composites cured thermally and with microwave radiation. Compos A Appl Sci Manuf 33:1021–1030. https://doi.org/10.1016/S1359-835X(02)00031-3
67. Kwak M (2016) Microwave curing of carbon-epoxy composites: process development and material evaluation 1–175

The manufacturer's authorised representative in the EU is Springer Nature Customer Service Centre GmbH, Europaplatz 3, 69115 Heidelberg, Germany. If you have any concerns regarding our products, please contact ProductSafety@springernature.com

Printed and bound by CPI Group (UK) Ltd, Croydon, CR0 4YY
30/09/2026
02225253-0008